QUANTUM GROUP SYMMETRY AND q-TENSOR ALGEBRAS

QUANTUM GROUP SYMMETRY AND q-TENSOR ALGEBRAS

L. C. Biedenharn
Univ. Texas, Austin

M. A. Lohe
Northern Territory Univ., Australia

Published by

World Scientific Publishing Co. Pte. Ltd.

P O Box 128, Farrer Road, Singapore 912805

USA office: Suite 1B, 1060 Main Street, River Edge, NJ 07661

UK office: 57 Shelton Street, Covent Garden, London WC2H 9HE

British Library Cataloguing-in-Publication Data
A catalogue record for this book is available from the British Library.

First published 1995
Reprinted 1999

QUANTUM GROUP SYMMETRY AND q-TENSOR ALGEBRAS

ISBN 981-02-2331-5

Printed in Singapore.

Preface

The impact of quantum groups has been immediate and long-lasting, but although the initial impetus to the subject came from physics in areas such as the quantum inverse scattering method and statistical mechanical models, the more vigorous development has more recently been in mathematics. Yet there is the promise of a much more extensive application of quantum groups to physics in a way that mimics the many successful applications of group theory in the past, in which the symmetry of a physical system is extended to a quantum symmetry. For example, there is the possibility that any Hamiltonian which is invariant under a Lie group may be generalized to a quantum group invariant, and that there is an extension of the algebraic methods which lead to a solution of the physical system with a dependence on an arbitrary deformation parameter q.

The aim of this monograph is to develop and extend to quantum groups the symmetry techniques familiar from the application of classical groups to models in physics. Our exposition is intended to be accessible to graduate physics students and to physicists wishing to gain an introduction to quantum groups. However, we hope that experts in quantum groups will also find some topics of interest, or perhaps a different viewpoint that offers some insight into the properties of quantum groups.

We have taken a uniform approach to quantum groups based on the fundamental concept of a tensor operator. Properties of both the quantum algebra and co-algebra are developed from a single point of view using tensor operators, which is especially helpful for an understanding of the noncommuting coordinates of the quantum plane interpreted as elementary tensor operators. Representations are constructed using a generalization of the boson calculus in which q-boson operators, later to be interpreted as tensor operators, play a central role, including the case when q is a root of unity.

After some introductory remarks and definitions in Chapter 1 we investigate representations of the quantum unitary groups in Chapter 2, beginning with the q-analog of the angular momentum group. In Chapter 3 we introduce the concept of a tensor operator and systematically develop the fundamental properties such as the 'multiplication' of two tensor operators using the q-analog of the Clebsch-Gordan coefficients to produce a third tensor operator. Whereas our initial discussion is based on the definition of a quantum group as the deformation of the classical Lie algebra, in Chapter 4 we determine properties of the q-analog of the Lie group, usually defined in terms of the dual of the universal enveloping algebra, but in our approach constructed using tensor operators. Then in Chapter 5 we further develop properties of the representation matrices and their interpretation as tensor operators. In Chapter 6 we analyse the interesting specialization when q is a root of unity, a case which appears in significant applications to physics. Algebraic induction is a method of constructing representations of the classical groups which expresses group properties in terms of the subgroup and which we generalize to quantum groups in Chapter 7. Finally, we consider several special topics in Chapter 8 which in some way illustrate the ideas, particularly those of tensor operators, developed in previous chapters.

We regret that our exposition is necessarily limited in scope and that we have not been able to include details of the more significant applications to physics, such as to statistical mechanics and conformal field theory. We have, however, provided references that will direct the reader to these applications. Nevertheless, partly because of the rapidly expanding literature there are inevitably many omissions here too and we apologize in advance to those authors whose work has not been accorded due credit. Several conference proceedings, for example [1, 2, 3, 4, 5, 6], provide windows looking onto the wide range of activities relating to quantum groups, and we refer to these for further references. There have also been several recent monographs on quantum groups, including those by Kassel [7], Chari and Pressley [8] and also Shnider and Sternberg [9] (which provides a very extensive bibliography), with a mathematical style and presentation to which we also refer the reader seeking more details than we have provided. By contrast, although our presentation is less precise, we have focussed more directly on the concepts relevant to symmetry techniques in physics.

We gratefully acknowledge and thank our many colleagues for their support, both direct and indirect over many years, who helped us to formulate and develop our investigation, in particular (but not only) R. Askey, V. Dobrev, H.-D. Doebner, D. Flath, B. Gruber, J. Louck, M. Tarlini, J. Towber, P. Truini, and also Professor M. Nomura for his interest in the project. We thank Brenda Gage for her efficient secretarial help, and especially our wives Sarah and Thilagam for their encouragement and support.

L. C. Biedenharn and M. A. Lohe, May 1995.

NOTATION: We use standard notations, but in particular note that $\mathbb{R}^+$ denotes the set of positive real numbers and $\mathbb{C}^\times$ denotes the nonzero complex numbers. The symbol $\stackrel{\text{def}}{=}$ is used in order to emphasize that the equation constitutes a definition. Where necessary we denote q-analog functions and operators with a suffix q (for example, q-integers are denoted $[n]_q$, generators $J^q_\pm$) which we omit for convenience when confusion with the $q=1$ case is unlikely.

Equations are numbered, where necessary, consecutively within each chapter together with items such as theorems, lemmas, remarks and examples; the first digit specifies the chapter in which this item appears.

Contents

Chapter 1

Origins of Quantum Groups

The concept of a quantum group arose from physics as an abstraction from the problem of understanding common features of exactly solvable models in quantum mechanics. The first quantum group written down was the q-analog of $SU(2)$ by Kulish, Reshetikhin [10] and Sklyanin, Takhtajan and Faddeev [11] who used the quantum inverse scattering method to study the behaviour of integrable systems in quantum field theory and statistical mechanics. It was known earlier (by Baxter [12]) that a *sufficient* condition for solvability of two-dimensional Ising models in statistical mechanics was the Yang-Baxter [13] equation, which Faddeev [11] has interpreted conceptually as playing the role of the Jacobi condition for quantum groups. Hence, quantum groups arose from physics (albeit two-dimensional) and the insights into quantized symmetry and noncommuting geometry that quantum groups afford will almost certainly have application to the real world of nature and lead to new physics.

Quantum groups are a generalization of the fundamental symmetry concepts of classical Lie groups and Lie algebras and involve two fundamental and distinct ideas: *deformation* and *noncommutative co-multiplication.*

Consider first the concept of deformation. Let G be a simple Lie group and $\mathcal{U}(\mathfrak{g})$ its associated universal enveloping algebra (the algebra over $\mathbb{C}$ based on all sums and products of generators, a factor algebra of the algebra of tensor operators). Let now the structure constants be continuous functions of a parameter q, such that for $q = 1$ we regain the original Lie algebra. The algebra $\mathcal{U}_q(\mathfrak{g})$ generated over $\mathbb{C}$ by all products and sums of the deformed generators defines the quantum group (which is really an algebra and not a group). Deformation is an old and very useful concept in physics: quantum mechanics is a deformation[1] of classical mechanics, in which $q = e^h \to 1$, that is, $\hbar \to 0$ and Einsteinian relativity is a deformation of Galilean relativity, in which $q = e^{1/c} \to 1$, that is, $c \to \infty$.

Noncommutative co-multiplication is a new idea for physics, and the implications are not fully known. Currents and momenta are observables in quantum physics

[1] This accounts for the name "quantum group" chosen by Drinfeld.

and are additive quantities both classically and quantum mechanically. This addition comprises, in mathematical language, a *commutative co-product.* For a quantum group, this structure becomes noncommutative (but commutative for $q = 1$). It is a consequence of noncommutative co-multiplication that the q-analog of the group manifold has noncommutative coordinates. This could have profound consequences for physics, and it has been suggested [14] that q might function to eliminate singularities in quantum field theory. The existence of a co-multiplication has the important implication that the deformation is smooth in the sense described by Witten [15].

There are two basic approaches to the study of quantum groups, corresponding to these two concepts given above. The first approach emphasizes the Lie algebra aspects, that is, state vectors, tensor operators, commutation relations, and matrix elements of the operators used in quantum physics. The second approach, which is dual to the first, emphasizes functions of the noncommuting coordinates of the quantum group analog to the group manifold, and has as one aim the development of the noncommutative differential geometry required for quantum gauge field theory. We discuss both approaches, but take as a starting point the first, in which we assume the q-deformed commutation relations of the quantum angular momentum group, and then determine irreducible representations (irreps) and properties of tensor operators. The noncommuting properties of tensor operator components lead us then to a discussion of quantum matrices and functions of these noncommuting coordinates.

1.1 Quantum Inverse Scattering Method

The development of the inverse scattering method applied to quantum mechanical systems was fundamental to the formulation of quantum groups, and arose from an investigation of solitary waves, or solitons as they became known when interpreted as particles, and integrable systems which admit soliton solutions. The remarkable properties of solitons, which are localized nondispersive solutions to certain nonlinear evolution equations, were observed experimentally some time ago, but their origin was established only more recently in the 1960's by means of several methods of solution culminating in the inverse scattering method, developed by Gardner, Greene, Kruskal and Miura[2]. This method was first applied to the Korteweg-de Vries (KdV) equation but was soon found to enjoy much wider applicability, and lead directly to properties such as an infinite number of conservation laws and the integrability of the corresponding Hamiltonian systems.

One way of formulating the method of solution by inverse scattering is due to Lax (1968), in which the KdV equation is written in the form of the equations:

$$Lv = \lambda v, \quad v_t = Mv,$$

[2]There are numerous accounts of solitons and the application of the inverse scattering method, and we refer to the monographs by Ablowitz and Clarkson [16] and by Newell [17] for an introduction to this subject, and for further references.

where L, M (the Lax pair) are operators which depend on the unknown function $u(x,t)$, and λ is a time-independent parameter. In order for nontrivial eigenfunctions $v(x,t)$ to exist, the compatibility equation $L_t+[L, M]=0$ must hold and is equivalent to the KdV equation for a suitable choice of L, M. The solution of this equation proceeds by solving the linear equation $Lv=\lambda v$, which is actually the Schrödinger equation for the potential $u(x,t)$ in which t is fixed, at the initial time $t=0$ given an initial value $u(x,0)$, to determine the scattering data. This is the direct scattering problem. As $u(x,t)$ evolves in time according to the KdV equation the scattering data also evolves, but in a way which is simple and linear. Hence, the scattering data is known at any later time and then, by inverse scattering using the Gel'fand-Levitan-Marchenko equation, the potential $u(x,t)$ is constructed from the scattering data at time t to solve the original nonlinear partial differential equation. For a linear problem this method of solution, the inverse scattering transform, reduces to the Fourier transform.

A more general way of writing the Lax pair of equations which applies to a wider class of models is

$$\boldsymbol{v}_x = U(x,t,\lambda)\boldsymbol{v}, \quad \boldsymbol{v}_t = V(x,t,\lambda)\boldsymbol{v},$$

where $\boldsymbol{v}=\boldsymbol{v}(x,t)$ is an n-dimensional vector and U, V are $n\times n$ matrices which depend on x, t implicitly through the unknowns $\boldsymbol{u}(x,t)$ and their derivatives as well as on a parameter λ. The matrix V is the time evolution operator and U is a potential. The integrability condition $\boldsymbol{v}_{xt}=\boldsymbol{v}_{tx}$ requires that $U_t - V_x + [U, V] = 0$, an equation which is equivalent to the original nonlinear equation and which can be viewed as a zero curvature condition (see the 1982 Les Houches lectures of Faddeev [11], also Faddeev and Takhtajan [18]). Many examples of nonlinear evolution equations which can be written in this form are known [16], but let us mention just one, the nonlinear Schrödinger equation $i\Psi_t = -\Psi_{xx} + 2g|\Psi|^2\Psi$ for which

$$U = i\begin{pmatrix} -\frac{\lambda}{2} & -\sqrt{g}\Psi^* \\ \sqrt{g}\Psi & \frac{\lambda}{2} \end{pmatrix}, \quad V = i\begin{pmatrix} \frac{\lambda^2}{2}+g|\Psi|^2 & \sqrt{g}(i\Psi_x^*+\lambda\Psi^*) \\ \sqrt{g}(i\Psi_x-\lambda\Psi) & -\frac{\lambda^2}{2}-g|\Psi|^2 \end{pmatrix},$$

where $\lambda\in\mathbb{C}$ is known as the spectral parameter.

In general, we can translate solutions of $\boldsymbol{v}_x = U\boldsymbol{v}$ according to

$$\boldsymbol{v}(x) = T(x,y,\lambda)\boldsymbol{v}(y) \qquad x \geqslant y,$$

where we suppress the time variable. Here, the transition matrix T is formally defined by

$$T(x,y,\lambda) = \overrightarrow{\exp}\int_y^x U(z,\lambda)\,dz\,,$$

where $\overrightarrow{\exp}$ denotes a path-ordered exponential which can be defined by its series expansion, obtained by iterating the integrated equation

$$\boldsymbol{v}(x) = \boldsymbol{v}(y) + \int_y^x U(z,\lambda)\boldsymbol{v}(z)\,dz,$$

and in which the noncommuting factors are ordered such that matrices $U(z,\lambda)$ with larger z stand to the left. These path-ordered exponentials satisfy several properties, in particular the semigroup property

$$T(x,y,\lambda) = T(x,z,\lambda)T(z,y,\lambda), \quad x \geqslant z \geqslant y.$$

Suppose we wish to solve the original time evolution equation on the interval $[0,L]$ with periodic boundary conditions. The operator $T(L,0,\lambda)$, called the monodromy matrix, has the important property that its trace is independent of time for arbitrary λ, which leads to the existence of an infinite number of conserved quantities [11], and in turn leads to the explicit construction of action-angle variables for completely integrable Hamiltonian systems [18].

Whereas in the classical theory the aim is to solve the nonlinear evolution equation for the unknowns $\boldsymbol{u}(x,t)$ on some domain, given appropriate initial data $\boldsymbol{u}(x,0)$, in the corresponding quantum theory one wishes to diagonalize the quantum Hamiltonian operator and determine the correlation functions. The inverse scattering transform can be extended to the quantum theory to accomplish this, with the help of the Bethe ansatz. Taking the nonlinear Schrödinger equation as a model, the Hamiltonian is

$$H = \int_0^L \left(\Psi_x^* \Psi_x + g|\Psi|^4\right) \, dx,$$

where the integration is over the interval $[0,L]$ (in order to avoid infrared divergences) on which periodic boundary conditions are imposed, and where the quantum fields Ψ and Ψ^* are now operator-valued and satisfy the equal time commutation relations

$$[\Psi(x),\, \Psi^*(y)] = \delta(x-y),$$

with other commutators equal to zero. In order to regularize ultraviolet divergences the interval $[0,L]$ is replaced by a lattice of points, and the transition matrix T, which is now operator-valued, is equal by the semigroup property to a product of elementary transition operators L_n, where L_n is the transition operator from the nth to the $n+1^{\text{th}}$ site. The commutation relations of L_n and hence also T are determined by calculating their Poisson brackets. These may be represented in a form which involves an $n^2 \times n^2$ matrix $r(\lambda,\mu)$, the classical r-matrix, which depends on spectral parameters λ,μ (see Faddeev [11], also the monograph by Korepin et al. [19, Chapter 5] and Takhtajan [20]). As a consequence of the Jacobi identity which the Poisson brackets satisfy, the r-matrix satisfies an identity known as the classical Yang-Baxter identity.

In the quantized model the Poisson bracket relations are replaced by commutation relations, and for the transition matrix $T(n,m,\lambda)$ (defined on the lattice of points (n,m)) these relations have the form [11, 19]

$$R(\lambda,\mu)\Big(T(n,m,\lambda) \otimes T(n,m,\mu)\Big) = \Big(T(n,m,\mu) \otimes T(n,m,\lambda)\Big) R(\lambda,\mu),$$

where $R(\lambda,\mu)$ is an $n^2 \times n^2$ c-number matrix, depending only on λ,μ. We can write these relations in a different notation by defining the permutation operator P according to $P(A\otimes B)P = B\otimes A$, and letting $\widetilde{R}(\lambda,\mu) = PR(\lambda,\mu)$. Then we have

$$\widetilde{R}_{12}(\lambda,\mu)T_1(\lambda)T_2(\mu) = T_2(\mu)T_1(\lambda)\widetilde{R}_{12}(\lambda,\mu), \tag{1.1}$$

where we have suppressed the space coordinates n, m and the subscripts denote the vector space on which the matrix acts. (Hence, $T_1 = T\otimes I$ and $T_2 = I\otimes T$). The matrix R satisfies the Yang-Baxter relation:

$$\big(I\otimes R(\lambda,\mu)\big)\big(R(\lambda,\mu)\otimes I\big)\big(I\otimes R(\lambda,\mu)\big) = \big(R(\lambda,\mu)\otimes I\big)\big(I\otimes R(\lambda,\mu)\big)\big(R(\lambda,\mu)\otimes I\big). \tag{1.2}$$

The restrictions on R implied by these relations are sufficient in order to assure compatibility of the relations (1.1), see [19, §VI.2]. The Yang-Baxter relation guarantees the associativity of the abstract algebra generated by the elements of T, see Jimbo [21, p. 111-134] for a discussion. Together with (1.1) the Yang-Baxter relations form the basic ingredients in the quantum inverse scattering method of solution of two-dimensional models of quantum field theory and statistical mechanics. Indeed, solutions of the Yang-Baxter equation form the starting point for constructing integrable models.

The equations (1.1) and (1.2) can also be regarded as the starting point for the development of quantum matrix groups, which we discuss in Chapter 4. A solution R of the Yang-Baxter equation is said to be quasi-classical (Jimbo [21]) if R depends on a parameter $\hbar$ such that as $\hbar \to 0$ we can recover the classical r-matrix. Beginning with such solutions, constructed using the Lie algebra $\mathfrak{g}$, Drinfeld [22] presented in 1986 a general theory of quantum groups. Deformations of Lie algebras were considered also by Jimbo [23] who introduced the Hopf algebra $\mathcal{U}_q(\mathfrak{g})$ in 1985, expressed in terms of the Chevalley generators. These Hopf algebras appear naturally via a duality condition with the R-matrix, and provides an alternate algebraic approach to the matrix quantum group implied by the relation (1.1).

1.2 Applications of Quantum Groups

In order to demonstrate the far-reaching implications of quantum groups, let us mention briefly some examples of applications. A useful reference is the reprint volume [24] (ed. M. Jimbo) which discusses many developments flowing from quantum groups. As already outlined, quantum groups appeared in the application of inverse scattering to 2-dimensional models in quantum field theory and also statistical mechanical models. Examples of such models are the nonlinear Schrödinger model and the sine-Gordon model, formulated on a lattice but still preserving their integrability and other significant classical properties, such as the structure of the action-angle variables. These lattice models are related to models of statistical mechanics, including the six and eight vertex models of Baxter [12]. Solutions of the Yang-Baxter equation led to the study of IRF (interacting round a face) models and their connection with conformal field theory (see Pasquier [25]) and involve the Wigner calculus of coupling coefficients extended to quantum groups, which we discuss in Chapter 3 (but not this application). Cyclic representations of the quantum group at roots of unity can occur, as has been analyzed in the chiral Potts model and generalizations [26, 27].

Quantum groups have found a significant application in two space-time dimensions to conformal field theory. Conformal transformations can be expressed in terms of complex coordinates $z = x + it$, $\overline{z} = x - it$ as holomorphic and anti-holomorphic coordinate transformations. An example of such a field theory is the Wess-Zumino-Witten (WZW) model which is defined by a chirally invariant Lagrangian constructed from boson fields taking values in a Lie group, and includes a Wess-Zumino term with an integer coefficient, the level k, which is related to the coupling constant when conformal invariance holds. The current algebra contains a Schwinger term (a central extension) which arises from correction terms in the quantum theory, and is an example of an affine algebra. It is convenient to distinguish the primary fields $\varphi_i(z, \overline{z})$, defined to be tensors under conformal transformations, and which form an associative algebra over $\mathbb{C}$ as determined from the operator product expansion. This algebra satisfies rules (known as fusion rules) of the form

$$\varphi_i \times \varphi_j = \sum_k \mathcal{N}_{ij}^k \varphi_k,$$

where $\mathcal{N}_{ij}^k \in \mathbb{N}$. These rules, which determine the coupling of the primary fields, are analogous to Wigner-Clebsch-Gordan (WCG) coupling rules, indeed there is a correspondence between concepts in group theory and conformal field theory which extends to other aspects, as has been discussed at length by Moore and Seiberg [28]. The quantum group emerges when we compute the braiding properties of the vertex fields (which are normal-ordered exponentials of the boson fields) and acts as the centralizer of the braid group action on the conformal blocks. The braiding, monodromy and modular properties of WZW-type models are determined by the representations of a quantum group with a deformation parameter $q = \exp \frac{2i\pi}{k+g}$, where k is the level and g is the dual Coxeter number of the algebra (equal to 2 for $SU(2)$). The appearance of q at roots of unity causes difficulties with unitarity conditions of the field theory and also with indecomposable representations which is typical of the application of quantum groups to conformal field theory. We refer to the monograph by Fuchs [29] and the articles by Alvarez-Gaumé et al. [30] (which also discusses solvable lattice models and conformal field theory), Pasquier and Saleur [31] and also Fröhlich and Kerler [32] for detailed descriptions.

Quantum groups have been proposed as q-analog extensions of the space-time symmetry groups such as the Poincaré group, although there are difficulties in defining inhomogeneous quantum groups. The κ-Poincaré group can be obtained as a contraction of the quantum conformal group, in which the space-time coordinates and four-momenta remain commutative, and constitutes a Hopf algebra (see [4, p. 287-

326] and [33, 34]). Another approach (Podleś and Woronowicz [35]) is to develop the quantum deformation of the Lorentz group, which may be identified with $SL_q(2,\mathbb{C})$, and the q-analog of Minkowski space in which space-time coordinates are noncommuting and can be identified as quadratic combinations of q-spinors [36]. We refer to the articles [2, pages 157,469,477] for further details and references.

Other applications and connections of quantum groups to physics which we merely mention (not for any lack of importance!) are those involving quantum gravity (Gervais [37], Boulatov [38, p. 39], Castellani [39]), Chern-Simons gauge theory (Witten [15], Guadagnini et al. [1, p. 307]), hidden quantum group symmetries in quantum field theory (Reshetikhin and Smirnov [40]), deformed rotational spectra of nuclei and molecules (Bonatsos and Daskaloyannis [2, p. 89]), quantum optics (Solomon [2, p. 705]), and condensed matter physics (Rasetti [41]), amongst others.

Quantum groups have provided significant and at times spectacular connections between areas of mathematics, an example being the link invariants discovered by Jones and Conway (1984) which can be related to properties of quantum groups. Isotopy invariants of knots and links in $\mathbb{R}^3$ can be constructed from certain solutions of the Yang-Baxter equation. The $6j$ symbols of the quantum group can be used to construct invariants of 3-manifolds, as pointed out by Reshetikhin and Turaev [42] but although we discuss q-$6j$ symbols in Chapter 3, we do not touch on this aspect here and refer to [42], the monograph by Turaev [43], and also the paper by Carter et al. [44] for details. A discussion of one-dimensional submanifolds of $\mathbb{R}^3$ such as knots, links and braids, and their connection with quantum groups is given in [7] and [8, Chapter 15]. The braid group B_n is generated by elements $\sigma_1, \ldots \sigma_{n-1}$ satisfying the following relations (for $n \geqslant 3$ and $1 \leqslant i, j \leqslant n-1$):

$$\sigma_i \sigma_{i+1} \sigma_i = \sigma_{i+1} \sigma_i \sigma_{i+1}, \qquad \sigma_i \sigma_j = \sigma_j \sigma_i, \quad |i-j| > 1. \tag{1.3}$$

We can construct braid group representations from any solution of the Yang-Baxter equation [7, Chapter X], [20]. The connection between quantum groups, the theory of knots and braids and the relevance to physics has been explored by many authors, see Kauffman [45] and the collection of articles [21] for a description.

There are other mathematical developments related to quantum groups, such as noncommutative geometry (Manin [46]), and the abstraction of their properties leading to the concept of a tensor category, which we do not discuss but are covered in the references already cited.

1.3 Special Functions and Quantum Groups

Calculations with tensor operators for Lie groups inevitably involve special functions, a situation which extends to quantum groups and which we now discuss in detail. The name "special functions" is possibly something of a misnomer; Askey in his 1975 lectures [47] cites Turan's suggestion that perhaps "useful functions" may better capture the real meaning. Certainly this is how special functions entered theoretical physics, as those very useful functions that arose from separation of variables in the partial differential equations defined by physical problems. Special functions were accordingly both numerous and, as a discipline, somewhat chaotic. This situation — at least for theoretical physics — changed markedly in the 1960's. Although Klein, Lie and Cartan had indicated the symmetry origins of some special functions, it was the 1955 lectures of Eugene Wigner [48] that, at least for physicists, brought order out of chaos. Wigner pointed out that large classes of special functions arise as matrix elements of the representations of the symmetry groups so fundamental to theoretical physics. Thus, for example, the Legendre and Jacobi functions stem from representations of the quantum rotation group, $SU(2)$; the Bessel functions from representations of the Euclidean motion group, $E(2)$; and the Laguerre and Hermite functions from the Heisenberg group.

Not only did the special functions themselves receive order, but so did their relations: addition theorems arise uniformly from the group multiplication rule; differential equations arise as limits of the multiplication by infinitesimal elements (the generators); and integral and completeness relations appear as the orthonormality relations for matrix elements, as typified by the Peter-Weyl theorem. Moreover, one could understand intuitively such results as Mehler's formula, asymptotically relating the Legendre functions to Bessel functions, as the contraction of groups: $SO(3) \to E(2)$, that is, the rotation group $SO(3)$ acts on the sphere S^2 which for an asymptotically large radius is replaced by a tangent plane on which the contracted group $E(2)$ acts.

This point of view for special functions was developed further in the monograph by Talman [49], based on the Wigner lectures, and the more extensive monograph by Vilenkin [50], based partly on Gel'fand's work. A very general setting for the spherical functions (Helgason [51]) considers a group and a two-point homogeneous space acted on by the group. The spherical functions are orthogonal polynomials of ${}_2F_1$ type (Askey [47]) with the variable being the distance function.

One should not be too categoric about this "group representation viewpoint", important as it is, for there are special functions that do not fall into this context. Askey gave different settings for some special functions and noted, for example, that statistical mechanics and critical phenomena involve special functions arising from certain positivity constraints and inequalities. Another example is furnished by the Appell functions F_2 and F_3 that occur in physics in the theory of Coulomb excitation (Biedenharn and Brussaard [52]), concerning energy transfer by the electromagnetic field of accelerated charged particles during scattering, and do not appear in the context of matrix elements of group representations.

A widening of the group representation context appears in the book by Miller [53], which is based on the recognition that the special functions defined by separation of variables in second-order linear partial differential equations can be characterized as eigenfunctions defined by commuting elements in the enveloping algebra of the Lie algebra of the symmetry group of the equation. This structure encompasses naturally the factorization method of Schrödinger (see Kaufman [54]), the approaches of Truesdell [55] and Weisner [56], and leads to a uniform approach to hypergeometric special functions as well as nonhypergeometric functions such as Mathieu and Lamé functions amongst others. One may discern a trend toward algebraic, as opposed to purely group-theoretic, methods in the approach to spherical functions in terms of convolution algebras.

Conspicuous by their absence in this group-theoretical approach are the q-analog functions, and a group-theoretical understanding of their properties, especially the basic hypergeometric functions which were introduced by Heine in 1846, and developed systematically by him and others over a span of many years, and which have appeared in ever more diverse applications (see the review of Andrews [57]). Significant contributions to the development of the theory of hypergeometric functions and their q-analogs are due particularly to Bailey [58] and Slater [59], and more recently to Askey [61], Andrews [57] and Milne [62] amongst others; we refer to Gasper and Rahman [64] for a wide-ranging account. One understands now that *quantum groups* provide the missing piece of the puzzle regarding the absence of q-functions in group theory, by supplying an algebraic explanation of the rich and varied properties of basic hypergeometric functions and demonstrating why these properties are natural generalizations of the better-known special functions. The appearance of basic hypergeometric functions was noted at an early stage in the development of quantum groups when the q-analogs of Clebsch-Gordan and Racah coefficients were calculated [65], and the coupling coefficient ${}_3\phi_2$ and ${}_4\phi_3$ functions were seen to appear naturally. We will find many such examples in our development and we discuss them as they arise. We outline the fundamental properties of q-analog functions and basic hypergeometric functions in §2.8,§2.8.3 and properties of such functions are used in §3.5.2, §3.6 (where they are related to $3j$ and $6j$ symbols), in §4.4.1 (as basis states), in §8.3.2 (as components of a tensor operator), and in §7.7.4 in connection with the method of algebraic induction.

Let us turn now to less widely known approaches to the understanding of special functions and their q-analogs. The main theme is that we now go beyond group representations and Lie algebras, and beyond even the enveloping algebra of a Lie algebra. We may gain an intuitive idea of this by considering the prototype for generalizations: Abelian groups G defining the elementary transcendentals. The characters $\{\chi^j\}$, which form representations, obey the product rule given by

$$\chi^j(a)\chi^j(b) = \chi^j(ab), \quad a, b \in G.$$

For compact groups, this multiplication is generalized to the matrix product law for representations $\mathfrak{D}$, which satisfy

$$\mathfrak{D}^j(a)\,\mathfrak{D}^j(b) = \mathfrak{D}^j(ab), \quad a, b \in G,$$

and which characterized the approach ascribed to Wigner and Vilenkin above. Moreover, the characters of an Abelian group obey a *second* multiplication rule expressed by

$$\chi^j(a)\chi^{j'}(a) = \chi^{j+j'}(a), \tag{1.4}$$

where we have indexed the distinct characters by the symbol j. One recognizes this equation as the **product law** of the dual group.

A result for $SU(2)$, very similar in structure to (1.4), is known. This is the **Wigner product law** where, in analogy to (1.4), different representations for the same group element are to be multiplied. If we express this result in matrix form we have the following relation:

$$\mathfrak{D}^{j_1}_{m_1,m_1'}(g)\,\mathfrak{D}^{j_2}_{m_2,m_2'}(g) = \sum_{j_3,m_3,m_3'} C^{j_1\ j_2\ j_3}_{m_1\ m_2\ m_3}\ \mathfrak{D}^{j_3}_{m_3,m_3'}(g) C^{j_1\ j_2\ j_3}_{m_1'\ m_2'\ m_3'}.$$

Here $\mathfrak{D}^j_{m\,m'}(g)$ is the matrix element of the representation $\mathfrak{D}^j$ for $g \in SU(2)$ and the real, orthonormal coefficients $C^{\cdots}_{\cdots}$ are the WCG coefficients discussed in detail in, for example, the monograph by Biedenharn and Louck [66] and reviewed in §3.2. We can achieve a form analogous to (1.4) if, using orthonormality, we put the WCG coefficients on the left hand side, thus obtaining:

$$\sum_{m_1,m_1',m_2,m_2'} C^{j_1\ j_2\ j_3}_{m_1\ m_2\ m_3} C^{j_1\ j_2\ j_3}_{m_1'\ m_2'\ m_3'} \mathfrak{D}^{j_1}_{m_1,m_1'}(g)\ \mathfrak{D}^{j_2}_{m_2,m_2'}(g) = \mathfrak{D}^{j_3}_{m_3,m_3'}(g).$$

Now let us regard the two coefficients $C^{j_1\ j_2\ j_3}_{m_1\ m_2\ m_3}$ each as effecting a product, that is, we regard the left hand side above as the formal, symbolic product to give

$$\left[\mathfrak{D}^{j_1} \underset{\textcircled{w}}{\textcircled{w}} \mathfrak{D}^{j_2}\right]^{j_3} = \mathfrak{D}^{j_3}. \tag{1.5}$$

Here, each of the symbols ⓦ denotes a Wigner product implemented by the $C^{\cdots}_{\cdots}$ coefficient, with a sum over the relevant indices. In this form, we can discern that (1.5) is a generalization of (1.4), and constitutes a valid identity that extends the Abelian group model to the non-Abelian group $SU(2)$. (Since $SU(2)$ is compact, the dual space, indexed by j, m, m', is discrete. Thus, instead of a simple monomial product as in (1.4) we now have a complicated sum of products effected on the dual space.)

Our reason for writing the Wigner product law in this curious way is to motivate two valid identities — each a formal analog of (1.5) — that involve algebraic elements outside the enveloping algebra of $SU(2)$ whose matrix elements are special functions of generalized hypergeometric type.

Let us define a Hilbert space $\mathfrak{M}$ as a direct sum of irrep spaces of $SU(2)$, each irrep space occurring once and only once (we call $\mathfrak{M}$ a model space, discussed in §3.3). Thus the irrep j has, as a basis, the $(2j+1)$ orthonormal vectors $|jm\rangle$, $m =$

$j, j-1, \ldots, -j$ for each $j = 0, \frac{1}{2}, 1, \ldots$. We also define the (J, M, Δ) **Wigner operator** to be the operator in $\mathfrak{M}$ given by

$$\left\langle {}_{2J} \begin{smallmatrix} J+\Delta \\ \\ J+M \end{smallmatrix} {}_0 \right\rangle = \sum_{j,m} |j+\Delta, m+M\rangle \, C^{j\ J\ j+\Delta}_{m\ M\ m+M} \, \langle jm|.$$

(This notation is explained in §3.2). The matrix elements of this operator are the WCG coefficients $C^{\cdots}_{\cdots}$. The product of two Wigner operators is

$$\left\langle {}_{2J'} \begin{smallmatrix} J'+\Delta' \\ \\ J'+M' \end{smallmatrix} {}_0 \right\rangle \left\langle {}_{2J} \begin{smallmatrix} J+\Delta \\ \\ J+M \end{smallmatrix} {}_0 \right\rangle = \sum_{J'', M'', \Delta''} W^{J\ J'\ J''}_{\Delta\ \Delta'\ \Delta''} \, C^{J\ J'\ J''}_{M\ M'\ M''} \left\langle {}_{2J''} \begin{smallmatrix} J''+\Delta'' \\ \\ J''+M'' \end{smallmatrix} {}_0 \right\rangle, \qquad (1.6)$$

where $W^{\cdots}_{\cdots}$ denotes the $SU(2)$ invariant operator whose matrix elements are the Racah, or $6j$ coefficients.

We can invert this last equation to define the symbolic **Wigner operator product law**

$$\left[\left\langle {}_{2J'} \begin{smallmatrix} \bullet \\ \bullet \end{smallmatrix} {}_0 \right\rangle \underset{\text{Ⓦ}}{\overset{\text{Ⓡ}}{}} \left\langle {}_{2J} \begin{smallmatrix} \bullet \\ \bullet \end{smallmatrix} {}_0 \right\rangle \right]^{J''} = \left\langle {}_{2J''} \begin{smallmatrix} \bullet \\ \bullet \end{smallmatrix} {}_0 \right\rangle, \qquad (1.7)$$

where Ⓦ denotes the Wigner product and Ⓡ denotes the Racah product using, in effect, the $6j$ coefficients. There is a similar product law involving two Racah operators and two Racah products.

Thus we have three product laws in $SU(2)$, namely (1.5, 1.7) and the product law involving Racah operators, that are formal extensions of the Abelian product law (1.4) for the dual group. Each of these three product laws is an identity in special functions involving, respectively, the ${}_2F_1, {}_3F_2$ and ${}_4F_3$ special functions, as is discussed at length in [66]. Moreover, there is a remarkable series of asymptotic relations $6j \sim 3j \sim \mathfrak{D}^j$ that carry these three special functions, and hence these three product laws, into each other. Furthermore, these properties are also valid for quantum groups! Specifically, the three product laws generalize directly to the quantum algebra $\mathcal{U}_q(\mathfrak{su}(2))$ and are identities for the basic hypergeometric functions ${}_2\phi_1, {}_3\phi_2$ and ${}_4\phi_3$; the asymptotic relations also remain valid and relate these three basic hypergeometric functions and their identities via limits.

It is clear even from this brief sketch that the extension of these ideas to quantum groups holds forth the promise of yielding families of new q-analog special functions and relations [62, 63].

1.4 Definition of Quantum Group

The precise formulation of quantum groups has been given by Drinfeld [22] and Manin [46], and although we will not explicitly require a formal presentation it is useful to outline these definitions. A quantum group is defined to be a (not necessarily commutative) Hopf algebra. Since a Hopf algebra is essentially a bi-algebra with an antipode, (defined below, see (1.8)) we must first define a bi-algebra.

Let A be an associative algebra with unity $\mathbb{1}$, over a field $\mathbb{k}$ which we may take to be the set of complex numbers $\mathbb{C}$. Then a bi-algebra on A is defined by four morphisms:

$$A \otimes A \xrightarrow{m} A \xrightarrow{\Delta} A \otimes A,$$
$$\mathbb{k} \xrightarrow{\eta} A \xrightarrow{\varepsilon} \mathbb{k}$$

satisfying the following axioms, written as commutative diagrams:

Associativity :

$$\begin{array}{ccccc} & & A \otimes A & & \\ & {\scriptstyle m\otimes \mathrm{id}}\nearrow & & \searrow^{m} & \\ A \otimes A \otimes A & & & & A \\ & {\scriptstyle \mathrm{id}\otimes m}\searrow & & \nearrow_{m} & \\ & & A \otimes A & & \end{array}$$

that is, $m(m \otimes \mathrm{id}) = m(\mathrm{id} \otimes m)$. The operation m is the usual product in A: $m(a \otimes b) = ab$ for $a, b \in A$.

Coassociativity :

$$\begin{array}{ccccc} & & A \otimes A & & \\ & {\scriptstyle \Delta}\nearrow & & \searrow^{\Delta\otimes \mathrm{id}} & \\ A & & & & A \otimes A \otimes A \\ & {\scriptstyle \Delta}\searrow & & \nearrow_{\mathrm{id}\otimes\Delta} & \\ & & A \otimes A & & \end{array}$$

that is, $(\Delta \otimes \mathrm{id})\Delta = (\mathrm{id} \otimes \Delta)\Delta$. Co-multiplication Δ is a homomorphism of A.

Unit :

$$\begin{array}{ccc} & A \otimes A & \\ {\scriptstyle \mathrm{id}\otimes\eta,\, \eta\otimes \mathrm{id}}\nearrow & & \searrow^{m} \\ A = A \otimes \mathbb{k} = \mathbb{k} \otimes A & \xrightarrow[\mathrm{id}]{} & A \end{array}$$

that is, $m(a \otimes \mathbb{1}) = m(\mathbb{1} \otimes a) = a$ for all $a \in A$. The operation η is defined by $\eta(c) = c\mathbb{1}$

for all $c \in \Bbbk$.

Counit :

$$\begin{array}{ccccc} & & A \otimes A & & \\ & {}^{\Delta}\nearrow & & \searrow_{\varepsilon\otimes \mathrm{id},\, \mathrm{id}\otimes\varepsilon} & \\ A & & \underset{\mathrm{id}}{\longrightarrow} & & A = \Bbbk \otimes A = A \otimes \Bbbk \end{array}$$

that is, $(\varepsilon \otimes \mathrm{id})\Delta = (\mathrm{id} \otimes \varepsilon)\Delta = \mathrm{id}$, and ε is also a homomorphism: $\varepsilon(ab) = \varepsilon(a)\varepsilon(b)$ for all $a, b \in A$.

Connecting axiom:

$$\begin{array}{ccccc} & & A & & \\ & {}^{m}\nearrow & & \searrow^{\Delta} & \\ A \otimes A & & & & A \otimes A \\ \Big\downarrow {\scriptstyle \Delta\otimes\Delta} & & & & \Big\uparrow {\scriptstyle m\otimes m} \\ A \otimes A \otimes A \otimes A & & \underset{S_{(23)}}{\longrightarrow} & & A \otimes A \otimes A \otimes A \end{array}$$

where $S_{(23)}$ is the morphism exchanging the second and third places in the tensor product.

An antipode of a bi-algebra (A, m, Δ) is a linear map $\gamma : A \to A$ such that the following diagram is commutative:

$$\begin{array}{ccccccc} & & A \otimes A & \overset{\gamma\otimes\mathrm{id}}{\longrightarrow} & A \otimes A & & \\ & {}^{\Delta}\nearrow & & & & \searrow^{m} & \\ A & \overset{\varepsilon}{\longrightarrow} & & \Bbbk & & \overset{\eta}{\longrightarrow} & A \\ & {}^{\Delta}\searrow & & & & \nearrow_{m} & \\ & & A \otimes A & \underset{\mathrm{id}\otimes\gamma}{\longrightarrow} & A \otimes A & & \end{array} \tag{1.8}$$

that is, $m(\mathrm{id} \otimes \gamma)\Delta(a) = m(\gamma \otimes \mathrm{id})\Delta(a) = \varepsilon(a)\mathbb{1}$, where $a \in A$. The antipode is an anti-homomorphism: $\gamma(ab) = \gamma(b)\gamma(a)$. The antipode γ reverses multiplication and co-multiplication, that is, it defines a bi-algebra morphism on the algebra (A, m, Δ).

Now we may define a Hopf algebra as a ring of polynomial functions on an affine group G, which is a bi-algebra with an antipode, but without the condition of commutativity. Hopf algebras were introduced by H. Hopf in 1941 and were studied in several works (Sweedler [67], Abe [68]) but a rich supply of Hopf algebras which are neither commutative nor co-commutative came only with the introduction of quantum groups.

We may define a quasi-triangular Hopf algebra by defining another "opposite" co-multiplication, which will arise in our applications by putting $q \to q^{-1}$. Let $\sigma : A \otimes A \to A \otimes A$ be the permutation operator defined by $\sigma(a \otimes b) = b \otimes a$, then $\overline{\Delta} = \sigma \circ \Delta$ is another co-multiplication in A with antipode $\overline{\gamma} = \gamma^{-1}$. A Hopf algebra is **quasi-triangular** if $\Delta, \overline{\Delta}$ are related by

$$\overline{\Delta}(a) = \mathcal{R}\Delta(a)\mathcal{R}^{-1}, \quad a \in A,$$

where $\mathcal{R}$ (the universal R-matrix for A) is an invertible element in $A \otimes A$, and if the following conditions are satisfied:

$$\begin{aligned} (\Delta \otimes \mathrm{id})\mathcal{R} &= \mathcal{R}_{13}\mathcal{R}_{23}, \qquad & (\mathrm{id} \otimes \Delta)\mathcal{R} &= \mathcal{R}_{13}\mathcal{R}_{12} \\ (\gamma \otimes \mathrm{id})\mathcal{R} &= \mathcal{R}^{-1}, \qquad & (\mathrm{id} \otimes \gamma)\mathcal{R}^{-1} &= \mathcal{R}. \end{aligned} \tag{1.9}$$

Here, $\mathcal{R}_{13} \in A \otimes A \otimes A$ and acts as the identity in the second factor, and as $\mathcal{R}$ in the first and third factors, and similarly for $\mathcal{R}_{12}, \mathcal{R}_{23}$.

It follows (Drinfeld [22]) from these axioms that $\mathcal{R}$ satisfies the Yang-Baxter equation. Writing $\mathcal{R} = \sum_i a_i \otimes b_i$ we have

$$\mathcal{R}_{13}\mathcal{R}_{23} = \sum_{i,j} a_i \otimes a_j \otimes b_i b_j$$

and hence:

$$(\sigma \circ \Delta \otimes \mathrm{id})(\mathcal{R}) = \sigma_{12}((\Delta \otimes \mathrm{id})(\mathcal{R})) = \sigma_{12}(\mathcal{R}_{13}\mathcal{R}_{23}) = \mathcal{R}_{23}\mathcal{R}_{13},$$

where σ_{12} permutes the first two positions. Also,

$$\begin{aligned} (\sigma \circ \Delta \otimes \mathrm{id})(\mathcal{R}) &= \sum_i \overline{\Delta}(a_i) \otimes b_i = \sum_i \mathcal{R}_{12}\Delta(a_i)\mathcal{R}_{12}^{-1} \otimes b_i \\ &= \mathcal{R}_{12}\sum_i \Delta(a_i) \otimes b_i \mathcal{R}_{12}^{-1} = \mathcal{R}_{12}(\Delta \otimes \mathrm{id}(\mathcal{R}))\mathcal{R}_{12}^{-1} \\ &= \mathcal{R}_{12}\mathcal{R}_{13}\mathcal{R}_{23}\mathcal{R}_{12}^{-1}. \end{aligned}$$

By comparing results we determine that the Yang-Baxter equations are satisfied:

$$\mathcal{R}_{12}\mathcal{R}_{13}\mathcal{R}_{23} = \mathcal{R}_{23}\mathcal{R}_{13}\mathcal{R}_{12}. \tag{1.10}$$

The quantum groups we shall encounter are quasi-triangular Hopf algebras and consist of either the q-deformed universal enveloping algebra of the classical Lie algebra or its dual, the matrix quantum group, which may be viewed as the q-analog of the classical Lie group. For further details regarding the axioms and properties of Hopf algebras we refer to the monographs: Chari and Pressley [8, Chapter 4], Kassel [7, Chapter 3] and the review articles: Majid [14], Takhtajan [20, 1], Doebner et al. [1, p. 29].

Chapter 2

Representations of Unitary Quantum Groups

Our investigation into quantum groups begins with the simplest non-Abelian quantum group, the q-analog of the angular momentum group $SU(2)$. Because angular momentum is such a central concept in quantum physics we can expect its q-analog to eventually assume a similarly significant role in applications of quantum groups, and so we now develop the q-generalization of the concepts familiar to the quantum theory of angular momentum. This includes in particular the matrix representations, through which the quantum group may be expected to manifest itself in physical applications.

We introduce $\mathcal{U}_q(\mathfrak{su}(2))$ as the q-deformation of the universal enveloping algebra of $SU(2)$, which is the q-analog of the Lie algebra $\mathfrak{su}(2)$ (we consider the q-analog of the group $SU(2)$ in Chapter 4). Following a description of fundamental properties of $\mathcal{U}_q(\mathfrak{su}(2))$ we determine all its irreducible unitary representations using the techniques of raising and lowering operators familiar from quantum mechanics. We verify for $\mathcal{U}_q(\mathfrak{su}(2))$ the remarkable result (proved for $\mathcal{U}_q(\mathfrak{u}(n))$ by Lusztig and Rosso [69, 70]) that all irreducible representations (irreps) of $\mathcal{U}_q(\mathfrak{su}(2))$ are continuous deformations of those of $SU(2)$, to which they reduce for $q = 1$.

We construct the irreps of $\mathcal{U}_q(\mathfrak{su}(2))$ using operator methods familiar from quantum field theory, by introducing boson operators and their q-analogs which we discuss in detail. We find that the techniques of the boson calculus, well-known from the work of Schwinger [71], generalize to $\mathcal{U}_q(\mathfrak{su}(2))$ and also to all the unitary quantum groups. In an equivalent description we may construct irreps in vector spaces consisting of homogeneous polynomials of complex variables.

This chapter concludes with a collection of useful results and facts, firstly a review of the Gel'fand-Weyl notation for vectors in the irrep spaces, then a discussion of properties of q-numbers which appear in nearly all aspects of quantum groups, and finally a review of the q-analogs of special functions, mainly the basic hypergeometric functions.

2.1 The Prototype for Quantum Groups: $\mathcal{U}_q(\mathfrak{su}(2))$

Angular momentum theory is the prototype for applications of symmetry in quantum physics and it is hardly a surprise that the quantum deformation of this theory, the deformed quantal angular momentum algebra $\mathcal{U}_q(\mathfrak{su}(2))$, is similarly the prototype for general quantum groups. $\mathcal{U}_q(\mathfrak{su}(2))$ is generated by three operators $J^q_\pm, J^q_z$ satisfying the commutation relations

$$[J^q_z, J^q_\pm] = \pm J^q_\pm \tag{2.1a}$$

$$\left[J^q_+, J^q_-\right] = \frac{q^{J^q_z} - q^{-J^q_z}}{q^{\frac{1}{2}} - q^{-\frac{1}{2}}}, \quad q \in \mathbb{R}^+. \tag{2.1b}$$

The algebra, which is over $\mathbb{C}$, has a unit denoted by $\mathbb{I}$. Let us note several significant features:

Remark **2.2** 1. The commutator in (2.1b) is not $2J_z$ as usual, but an infinite series (for generic $q \in \mathbb{R}^+$) involving *all odd powers*: $(J^q_z)^1, (J^q_z)^3, \ldots$. Each such power is a linearly independent operator in the enveloping algebra; accordingly, the algebra $\mathcal{U}_q(\mathfrak{su}(2))$ is not of finite dimension.

2. We may write (2.1a) in the form

$$q^{J^q_z} J^q_\pm q^{-J^q_z} = q^{\pm 1} J^q_\pm$$

showing that the $\mathcal{U}_q(\mathfrak{su}(2))$ algebra may be expressed entirely in terms of $q^{J^q_z}$ and $J^q_\pm$.

3. For $q \to 1$, the right hand side of (2.1b) approaches $2J_z$. Thus, we recover in the limit the usual Lie algebra of $SU(2)$, and we say that the quantum group $\mathcal{U}_q(\mathfrak{su}(2))$ is a **deformation** of the universal enveloping algebra of $\mathfrak{su}(2)$. (The universal enveloping algebra is spanned by polynomials in the group generators). The concept of deformation has been discussed by Witten [15], in particular smooth deformations which preserve the dimension of the algebra.

4. When J^q_z has an eigenvalue m, where m is a half-integer, the right hand side of (2.1b) depends on the parameter q in a characteristic way. Let us define the **q-integer** $[n]_q$ for each $n \in \mathbb{Z}$ by

$$[n]_q \stackrel{\text{def}}{=} \frac{q^{\frac{n}{2}} - q^{-\frac{n}{2}}}{q^{\frac{1}{2}} - q^{-\frac{1}{2}}} = q^{\frac{1}{2}(n-1)} + q^{\frac{1}{2}(n-3)} + \ldots q^{-\frac{1}{2}(n-1)}. \tag{2.3}$$

These q-integers obey the rule $[-n]_q = (-1)[n]_q$, with $[0]_q = 0$ and $[1]_q = 1$ and, since $[n]_q = [n]_{q^{-1}}$, the defining relations (2.1a,2.1b) are invariant to the interchange $q \leftrightarrow q^{-1}$. The use of steps of unity for powers of q accounts for the convention using $q^{\frac{1}{2}}$, although the usage of q instead of $q^{\frac{1}{2}}$ is equally common in the literature. We note that with our convention the basic hypergeometric functions, which frequently arise in quantum groups, correspond to those given by standard definitions with the same q. The designation of $[n]_q$ as a "q-integer" is justified by the fact that q-integers form an additive group isomorphic to $\mathbb{Z}$ (see Lemma 2.90, p. 56).

By using the q-integer notation we can express the commutation relation (2.1b) in a concise notation:

$$\left[J_+^q,\ J_-^q\right] = [2J_z^q]_q,$$

with (2.1a) as before. Further properties of $[n]_q$, which we will use extensively, and the extension to the real numbers $[x]_q$ for $x \in \mathbb{R}$ are discussed in §2.7.

5. We have limited q to be positive and real, for which the generators $\mathbf{J} = (J_\pm^q,\ J_z^q)$ can be chosen to be Hermitean, as is required for physics. More precisely, Eqns. (2.1a) and (2.1b) admit an **anti-automorphism** ω defined by

$$\omega(J_z^q) = J_z^q, \quad \omega(J_+^q) = J_-^q, \quad \omega(J_-^q) = J_+^q$$

(that is, the mapping ω preserves the relations (2.1a,2.1b) provided $\omega(JK) = \omega(K)\omega(J)$ where $J, K \in \mathcal{U}_q(\mathfrak{su}(2))$). Furthermore, ω is **involutive**, that is, $\omega^2 = 1$. Hence, we may impose the Hermiticity conditions

$$(J_+^q)^\dagger = J_-^q, \quad (J_-^q)^\dagger = J_+^q, \quad (J_z^q)^\dagger = J_z^q, \tag{2.4}$$

for which the representations of the quantum group are unitary. If $q \in \mathbb{C}$ the commutation relations are no longer invariant under ω (under which $q \to \overline{q}$) unless either $q = \overline{q}$ or $q^{-1} = \overline{q}$. This latter case arises when q is a root of unity, discussed in Chapter 6, when these same Hermiticity conditions may therefore be imposed.

6. Despite appearances, the inherent symmetry of three-space is not broken. To see this we recognize that three-space would correspond to the three-dimensional irrep of $\mathcal{U}_q(\mathfrak{su}(2))$, and the three states of this irrep are degenerate, as is shown in Lemma 2.16 below, and are thus abstractly equivalent. Also there exists an operator C, defined in (2.11), which commutes with each generator and so the fundamental symmetry of the classical $q = 1$ case is retained for arbitrary real, positive q.

2.1.1 Co-Algebra Structure

Let us now complete the definition of the quantum group $\mathcal{U}_q(\mathfrak{su}(2))$ by giving the explicit co-algebra structure. For $\mathcal{U}_q(\mathfrak{su}(2))$ the co-product is defined by[1]:

$$\begin{aligned} \Delta(J_\pm) &\stackrel{\text{def}}{=} q^{-\frac{J_z}{2}} \otimes J_\pm + J_\pm \otimes q^{\frac{J_z}{2}}, \\ \Delta(J_z) &\stackrel{\text{def}}{=} \mathbb{I} \otimes J_z + J_z \otimes \mathbb{I}. \end{aligned} \tag{2.5}$$

[1] We now omit the superscript q for the generators J.

As indicated in §1.4, the co-product maps from $\mathcal{U}_q(\mathfrak{su}(2))$ to $\mathcal{U}_q(\mathfrak{su}(2)) \otimes \mathcal{U}_q(\mathfrak{su}(2))$, and this may be explicitly verified for the definitions (2.5). This is straightforward for the diagonal generator J_z and we find $[\Delta(J_z),\, \Delta(J_\pm)] = \pm\Delta(J_\pm)$. For the remaining commutator we obtain

$$\begin{aligned}[\Delta(J_+),\, \Delta(J_-)] &= q^{-J_z} \otimes [J_+,\, J_-] + [J_+,\, J_-] \otimes q^{J_z} \\ &= [2(\mathbb{I} \otimes J_z + J_z \otimes \mathbb{I})] = [2\Delta(J_z)],\end{aligned}$$

as required, where we cancelled several terms to get the first equation using (2.1a), and obtained the final result using the identity

$$[n + n'] = q^{-\frac{n}{2}}[n'] + q^{\frac{n'}{2}}[n]$$

with operator values for n, n'.

We note that the $q \leftrightarrow q^{-1}$ symmetry of the commutation relations (2.1a,2.1b) is broken by the co-multiplication, and also that Δ is non-commutative. The summands defining co-multiplication q-commute, that is, if we put

$$a = q^{-\frac{J_z}{2}} \otimes J_\pm, \quad b = J_\pm \otimes q^{\frac{J_z}{2}},$$

then $ab = q^{-1}ba$. For $q \in \mathbb{R}^+$ co-multiplication preserves the Hermiticity properties of the generators, however this is no longer true for q a root of unity.

The remaining Hopf algebra operations, formal definitions and properties of which are given in §1.4, are

$$\begin{aligned}&\varepsilon(\mathbb{I}) = 1, \quad \varepsilon(J_z) = 0, \quad \varepsilon(J_\pm) = 0, \\ &\gamma(J_\pm) = -q^{\mp\frac{1}{2}}J_\pm, \quad \gamma(J_z) = -J_z.\end{aligned}$$

For quantum physics the bi-algebra structure, involving co-multiplication, is new.

What is co-multiplication in physical terms? To answer this, consider the angular momentum operator $\mathbf{J}$ in quantum mechanics. Both classically and quantum mechanically one can add angular momenta, that is,

$$\mathbf{J}_{\text{total}} = \mathbf{J}^{(1)} + \mathbf{J}^{(2)}. \tag{2.6}$$

More precisely, when we add angular momenta we use an action on product kets,

$$|\psi\rangle_{\text{total}} = |\psi\rangle_{(1)} \otimes |\psi\rangle_{(2)}. \tag{2.7}$$

Thus the action implied by (2.6) and (2.7) written in a more explicit form is

$$\mathbf{J}_{\text{total}} = \mathbf{J}^{(1)} \otimes \mathbb{I} + \mathbb{I} \otimes \mathbf{J}^{(2)}.$$

This action actually defines a co-multiplication Δ:

$$\Delta(\mathbf{J}) = \mathbf{J} \otimes \mathbb{I} + \mathbb{I} \otimes \mathbf{J}.$$

Hence the vector addition of angular momentum in quantum physics defines a commutative co-multiplication in a bi-algebra.

The significant new feature of quantum group symmetry is that co-multiplication is not commutative; the addition of q-angular momenta depends on the order. This in turn allows braiding to be possible, that is, paths in two dimensions can cross either over or under. The braid group is defined by the relations (1.3) and representations can be constructed from solutions of the Yang-Baxter equation (see the articles in [21]).

We could equally well define a co-multiplication $\overline{\Delta}$ with q replaced by q^{-1} in (2.5):

$$\begin{aligned} \overline{\Delta}(J_\pm) &\stackrel{\text{def}}{=} q^{\frac{J_z}{2}} \otimes J_\pm + J_\pm \otimes q^{-\frac{J_z}{2}} \\ \overline{\Delta}(J_z) &\stackrel{\text{def}}{=} \mathbb{I} \otimes J_z + J_z \otimes \mathbb{I} \,. \end{aligned}$$

If we denote by σ the permutation operator in $\mathcal{U}_q(\mathfrak{su}(2)) \otimes \mathcal{U}_q(\mathfrak{su}(2))$, defined by $\sigma(u \otimes v) = v \otimes u$ (where $u, v \in \mathcal{U}_q(\mathfrak{su}(2))$), then $\overline{\Delta} = \sigma \circ \Delta$. The co-multiplication $\overline{\Delta}$ and antipode $\overline{\gamma} = \gamma^{-1}$, together with the previous Hopf algebra operations, may be used to define the quantum group $\mathcal{U}_{q^{-1}}(\mathfrak{su}(2))$.

The co-multiplications Δ and $\overline{\Delta}$ are connected by

$$\overline{\Delta}(u) = \mathcal{R}\Delta(u)\mathcal{R}^{-1}, \quad u \in \mathcal{U}_q(\mathfrak{su}(2)), \tag{2.8}$$

where $\mathcal{R}$ is the **universal $\boldsymbol{R}$-matrix** which is an element of $\mathcal{U}_q(\mathfrak{su}(2)) \otimes \mathcal{U}_q(\mathfrak{su}(2))$. If we define $\mathcal{R}^\sigma = \sigma \circ \mathcal{R}$ then from (2.8) it follows that:

$$\overline{\Delta}(u) = (\mathcal{R}^\sigma)^{-1}\Delta(u)\mathcal{R}^\sigma, \quad u \in \mathcal{U}_q(\mathfrak{su}(2)), \tag{2.9}$$

and $\mathcal{R}^\sigma\mathcal{R}$ commutes with Δ, which leads to a general method of constructing invariants of the quantum group, see Reshetikhin [72], Zhang et al. [73], Gould and Biedenharn [74].

For $\mathcal{U}_q(\mathfrak{su}(2))$ we have the following expression for $\mathcal{R}$ (Drinfeld [22]):

$$\mathcal{R} = q^{J_z \otimes J_z} \sum_{n \geqslant 0} \frac{(1-q^{-1})^n}{[n]_q!} E^n \otimes F^n, \tag{2.10}$$

where $[n]_q! \stackrel{\text{def}}{=} [n]_q[n-1]_q \ldots [1]_q$, and $E = q^{\frac{1}{2}J_z} J_+$, $F = q^{-\frac{1}{2}J_z} J_-$.

2.2 Irreducible Unitary Representations of $\mathcal{U}_q(\mathfrak{su}(2))$

As already indicated in the introductory remarks, the irreps of $\mathcal{U}_q(\mathfrak{su}(2))$ are in one-to-one correspondence with those of the Lie algebra $\mathfrak{su}(2)$; let us demonstrate precisely why this is so by determining explicitly all unitary irreps of $\mathcal{U}_q(\mathfrak{su}(2))$ and the matrix elements of the generators. We may adapt the standard derivation, well-known from quantum mechanics, of the irreps of $\mathfrak{su}(2)$.

We begin by defining the Casimir invariant[2] C according to

$$C \stackrel{\text{def}}{=} J_-J_+ + [J_z]_q[J_z+1]_q = J_+J_- + [J_z]_q[J_z-1]_q, \tag{2.11}$$

where the second form of C follows from the commutation relations and the identity

$$[a]_q[b-c]_q + [b]_q[c-a]_q + [c]_q[a-b]_q = 0 \tag{2.12}$$

with $a = 1, b = J_z, c = -J_z$.

LEMMA **2.13** *C is an invariant of $\mathcal{U}_q(\mathfrak{su}(2))$, that is, it commutes with each of the generators $J_\pm, J_z$.*

PROOF: This follows easily for J_z by using the commutators (2.1a), and also for J_+ by using the identity (2.12) again with $a = 1, b = J_z, c = -J_z$. □

A related form of the Casimir invariant is

$$C' \stackrel{\text{def}}{=} J_-J_+ + \left[J_z + \tfrac{1}{2}\right]_q^2, \tag{2.14}$$

which has also been used (see for example [75, 65]) and which differs from C by a constant: $C' - C = \left[\frac{1}{2}\right]_q^2$.

We may simultaneously diagonalize the commuting operators C and J_z, which act in an irreducible vector space $\mathfrak{V}$ with an inner product (,) with respect to which the Hermiticity conditions (2.4) are satisfied. Since $C^\dagger = C$, $J_z^\dagger = J_z$, the eigenvalues of C and J_z are real. Let $v_m \in \mathfrak{V}$ be an eigenvector of C and J_z with eigenvalues λ and m (the weight) respectively:

$$Cv_m = \lambda v_m, \quad J_z v_m = m v_m.$$

From the commutation relations $[J_z, J_\pm] = J_\pm$ we see that $(J_+)^n v_m$ is also an eigenvector of J_z, with an eigenvalue $m+n$ for each $n \in \mathbb{N}$, and $(J_-)^n v_m$ is an eigenvector with eigenvalue $m-n$, for $n \in \mathbb{N}$. From the expression (2.11) for C it follows that J_-J_+ is a diagonal operator, and in particular

$$\|J_+v_m\|^2 = (v_m, J_-J_+v_m) = (\lambda - [m]_q[m+1]_q)\|v_m\|^2. \tag{2.15}$$

Since $\|J_+v_m\|^2 \geqslant 0$ we must have $[m]_q[m+1]_q \leqslant \lambda$ for all possible eigenvalues m. $[m]_q$ increases without bound as m increases (for $q \in \mathbb{R}^+$) and so we find that the set of eigenvalues $\{m\}$ is bounded above and, by a similar argument, also below, and furthermore the maximum and minimum values are each attained. Hence, there exist states of highest and lowest weights. Let j be the largest value of m, then we

[2] This operator is not in the form of the invariant scalar product of a q-tensor operator with itself, as one might expect from the $q = 1$ case. This is discussed in §3.3, p. 92, see (3.48).

must have $J_+ v_j = 0$, where v_j is the state of highest weight. We also deduce that $\lambda = [j]_q[j+1]_q$.

We may apply the lowering generator J_- n times to generate the vectors v_{j-n} in $\mathfrak{V}$, with weight $j-n$, and there exists a value of n such that $J_- v_{j-n} = 0$ (again, in order that (2.15) not be violated). By using the second form of C in (2.11), and allowing C to act on v_{j-n}, we find that the eigenvalue of C is $[j-n]_q[j-n-1]_q$, which must equal $[j]_q[j+1]_q$. By putting $a = j, b = n+1, c = -j+n$ into (2.12), we find

$$[j]_q[j+1]_q - [j-n]_q[j-n-1]_q = [n+1]_q[n-2j]_q,$$

which must equal zero. Hence, we have $n = 2j$ for some $n \in \mathbb{N}$, showing that j is a non-negative half-integer (that is, an integer or a half odd integer). The weight m takes all half-integer values such that $-j \leqslant m \leqslant j$, with integer steps, showing that the dimension of the irrep is $2j+1$.

The matrix elements of $J_\pm$ are readily determined; we have $J_+ v_m = \mathcal{N}_m v_{m+1}$ for some normalization $\mathcal{N}_m$ which, as (2.15) shows, is given by

$$|\mathcal{N}_m|^2 = \lambda - [m]_q[m+1]_q = [j]_q[j+1]_q - [m]_q[m+1]_q = [j-m]_q[j+m+1]_q,$$

where we again used (2.12) with $a = j, b = j+m+1, c = m$. With the usual choice of phase we obtain the matrix elements of J_+ and hence the matrix elements of J_-. We may summarize the results as follows:

Lemma 2.16 *The irreps of $\mathcal{U}_q(\mathfrak{su}(2))$ are labelled by a non-negative half integer j and are of dimension $2j+1$. A vector in an orthonormal basis may be denoted $|jm\rangle$, where $m = j, j-1, \ldots, -j$ and matrix elements of the generators are given by*

$$\begin{aligned} J_\pm |jm\rangle &= \left([j \mp m]_q[j \pm m + 1]_q\right)^{\frac{1}{2}} |j, m \pm 1\rangle, \\ J_z |jm\rangle &= m|jm\rangle. \end{aligned}$$

The invariant C has the eigenvalue $[j]_q[j+1]_q$.

These matrix elements are simple q-integer extensions of those for $\mathfrak{su}(2)$. Although perhaps not surprising, this is actually a remarkable result: for all values of $q \in \mathbb{R}^+$ the unitary irreps of $\mathcal{U}_q(\mathfrak{su}(2))$ are in one-to-one correspondence with the unitary irreps of $\mathfrak{su}(2)$ and have precisely the same dimension. We will see later that this result, which is a special case of a theorem of Rosso and Lusztig [70, 69], extends to the unitary irreps of the q-generalization of all compact simple Lie groups. This remarkable result has its origin in a simple fact: linear integral factors in matrix elements of raising/lowering operators in ordinary Lie algebras extend in quantum groups to linear q-integral factors thereby preserving all structural zeroes, and hence in particular the dimensionality of the irrep.

EXAMPLE **2.17** 1. Let us write out explicitly the matrix elements for the generators for the three simplest nontrivial cases. For $j = \frac{1}{2}$ we find

$$J_z = \begin{pmatrix} \frac{1}{2} & 0 \\ 0 & -\frac{1}{2} \end{pmatrix} \quad J_+ = \begin{pmatrix} 0 & 1 \\ 0 & 0 \end{pmatrix} \quad J_- = \begin{pmatrix} 0 & 0 \\ 1 & 0 \end{pmatrix},$$

and for $j = 1$ we obtain the three-dimensional representation:

$$J_z = \begin{pmatrix} 1 & 0 & 0 \\ 0 & 0 & 0 \\ 0 & 0 & -1 \end{pmatrix} \quad J_+ = \begin{pmatrix} 0 & \sqrt{[2]_q} & 0 \\ 0 & 0 & \sqrt{[2]_q} \\ 0 & 0 & 0 \end{pmatrix} \quad J_- = \begin{pmatrix} 0 & 0 & 0 \\ \sqrt{[2]_q} & 0 & 0 \\ 0 & \sqrt{[2]_q} & 0 \end{pmatrix}.$$

For $j = \frac{3}{2}$ we obtain the following matrices for J_z, J_+, and J_- is represented by the transpose of J_+:

$$J_z = \begin{pmatrix} \frac{3}{2} & 0 & 0 & 0 \\ 0 & \frac{1}{2} & 0 & 0 \\ 0 & 0 & -\frac{1}{2} & 0 \\ 0 & 0 & 0 & -\frac{3}{2} \end{pmatrix}, \quad J_+ = \begin{pmatrix} 0 & \sqrt{[3]_q} & 0 & 0 \\ 0 & 0 & [2]_q & 0 \\ 0 & 0 & 0 & \sqrt{[3]_q} \\ 0 & 0 & 0 & 0 \end{pmatrix}.$$

We observe that the two-dimensional representation does not depend on q, that is, it is not sensitive to the quantum group; one consequence of this is that the Jordan map, with which we construct a realization of $\mathcal{U}_q(\mathfrak{su}(2))$, does not depend on q and so requires modification when applied to $\mathcal{U}_q(\mathfrak{su}(2))$.

2. We may form further representations of $\mathcal{U}_q(\mathfrak{su}(2))$ using co-multiplication (2.5). For example, if we co-multiply the two-dimensional irreps we obtain the following representation of $\mathcal{U}_q(\mathfrak{su}(2))$:

$$\Delta(J_z) = \begin{pmatrix} 1 & 0 & 0 & 0 \\ 0 & 0 & 0 & 0 \\ 0 & 0 & 0 & 0 \\ 0 & 0 & 0 & -1 \end{pmatrix}, \quad \Delta(J_+) = \begin{pmatrix} 0 & q^{-\frac{1}{4}} & q^{\frac{1}{4}} & 0 \\ 0 & 0 & 0 & q^{-\frac{1}{4}} \\ 0 & 0 & 0 & q^{\frac{1}{4}} \\ 0 & 0 & 0 & 0 \end{pmatrix}, \tag{2.18}$$

with $\Delta(J_-)$ represented by the transpose of $\Delta(J_+)$. This four-dimensional representation is completely reducible, and is equivalent to a direct sum of the one-dimensional and three-dimensional irreps, as may be verified directly with the help of the following real orthogonal matrix P:

$$P = \begin{pmatrix} 0 & \frac{q^{\frac{1}{4}}}{\sqrt{[2]_q}} & -\frac{q^{-\frac{1}{4}}}{\sqrt{[2]_q}} & 0 \\ 1 & 0 & 0 & 0 \\ 0 & \frac{q^{-\frac{1}{4}}}{\sqrt{[2]_q}} & \frac{q^{\frac{1}{4}}}{\sqrt{[2]_q}} & 0 \\ 0 & 0 & 0 & 1 \end{pmatrix}.$$

We find that

$$P\Delta(J_z)P^{-1} = \begin{pmatrix} 0 & 0 & 0 & 0 \\ 0 & 1 & 0 & 0 \\ 0 & 0 & 0 & 0 \\ 0 & 0 & 0 & -1 \end{pmatrix}, \quad P\Delta(J_+)P^{-1} = \begin{pmatrix} 0 & 0 & 0 & 0 \\ 0 & 0 & \sqrt{[2]_q} & 0 \\ 0 & 0 & 0 & \sqrt{[2]_q} \\ 0 & 0 & 0 & 0 \end{pmatrix},$$

in which the decomposition into the one- and three-dimensional irreps is evident. The entries of P are examples of q-Wigner-Clebsch-Gordan (q-WCG) coefficients which we investigate in detail in Chapter 3.

3. The R-matrix is given by (2.10) and for the two-dimensional irrep we find

$$E = \begin{pmatrix} 0 & q^{\frac{1}{4}} \\ 0 & 0 \end{pmatrix}, \quad F = \begin{pmatrix} 0 & 0 \\ q^{\frac{1}{4}} & 0 \end{pmatrix}, \quad q^{J_z \otimes J_z} = \begin{pmatrix} q^{\frac{1}{4}} & 0 & 0 & 0 \\ 0 & q^{-\frac{1}{4}} & 0 & 0 \\ 0 & 0 & q^{-\frac{1}{4}} & 0 \\ 0 & 0 & 0 & q^{\frac{1}{4}} \end{pmatrix}.$$

Hence, using $E^2 = 0 = F^2$, we find the following expression for $\mathcal{R}$:

$$\mathcal{R} = \begin{pmatrix} q^{\frac{1}{4}} & 0 & 0 & 0 \\ 0 & q^{-\frac{1}{4}} & q^{-\frac{1}{4}}(q^{\frac{1}{2}} - q^{-\frac{1}{2}}) & 0 \\ 0 & 0 & q^{-\frac{1}{4}} & 0 \\ 0 & 0 & 0 & q^{\frac{1}{4}} \end{pmatrix}. \tag{2.19}$$

If we denote this matrix by $\mathcal{R}_q$ we may verify directly that $(\mathcal{R}_q)^{-1} = \mathcal{R}_{q^{-1}}$, and also that the Yang-Baxter relations (1.10) are satisfied.

4. As mentioned above, we could equally well perform co-multiplication of the generators by replacing q by q^{-1} (to obtain the quantum group $\mathcal{U}_{q^{-1}}(\mathfrak{su}(2))$), and we may explicitly verify, for the two-dimensional irreps, the relations (2.8) which connect these two forms of co-multiplication. We have found the expressions for Δ and $\overline{\Delta}$ in (2.18), (replacing q by q^{-1} to obtain $\overline{\Delta}$), and by direct calculation we obtain:

$$\overline{\Delta}(J_z) = \mathcal{R}\Delta(J_z)\mathcal{R}^{-1}, \quad \overline{\Delta}(J_+) = \mathcal{R}\Delta(J_+)\mathcal{R}^{-1}, \quad \overline{\Delta}(J_-) = \mathcal{R}\Delta(J_-)\mathcal{R}^{-1}.$$

Similarly, we may verify the relations (2.9), where $\mathcal{R}^\sigma = \widetilde{\mathcal{R}}$ (the transpose of $\mathcal{R}$), and also that $\widetilde{\mathcal{R}}\mathcal{R}$ commutes with Δ.

From the matrix elements of the generators we may determine the matrix elements of the R-matrix in any irrep of $\mathcal{U}_q(\mathfrak{su}(2)) \otimes \mathcal{U}_q(\mathfrak{su}(2))$. We have

$$\begin{aligned} E^n|j_1 m_1\rangle &= q^{\frac{1}{4}n(2m_1+n+1)} \left(\frac{[j_1+m_1+n]_q![j_1-m_1]_q!}{[j_1-m_1-n]_q![j_1+m_1]_q!} \right)^{\frac{1}{2}} |j_1, m_1+n\rangle \\ F^n|j_2 m_2\rangle &= q^{-\frac{1}{4}n(2m_2-n-1)} \left(\frac{[j_2-m_2+n]_q![j_2+m_2]_q!}{[j_2+m_2-n]_q![j_2-m_2]_q!} \right)^{\frac{1}{2}} |j_2, m_2-n\rangle, \end{aligned}$$

and hence we obtain the $(2j_1+1)(2j_2+1)$-dimensional R-matrix given by the nonzero elements:

$$\left(\mathcal{R}^{j_1 j_2}\right)^{m_1,m_2}_{m_1+n,m_2-n} = \left(\frac{[j_1+m_1+n]_q![j_1-m_1]_q![j_2-m_2+n]_q![j_2+m_2]_q!}{[j_1-m_1-n]_q![j_1+m_1]_q![j_2+m_2-n]_q![j_2-m_2]_q!}\right)^{\frac{1}{2}} \times \frac{\left(q^{\frac{1}{2}}-q^{-\frac{1}{2}}\right)^n}{[n]_q!} q^{m_1 m_2} q^{-\frac{1}{2}n(n+m_1-m_2)}, \tag{2.20}$$

where $0 \leqslant n \leqslant \min(j_1 - m_1, j_2 + m_2)$, and

$$\left(\mathcal{R}^{j_1 j_2}\right)^{m_1,m_2}_{m_1+n,m_2-n} \stackrel{\text{def}}{=} \langle j_1, m_1+n| \otimes \langle j_2, m_2-n|\, \mathcal{R}\, |j_1 m_1\rangle \otimes |j_2 m_2\rangle.$$

This matrix appears in symmetry relations of the q-WCG coefficients, considered in §3.5. It is expressible as a sum over q-WCG coefficients (Nomura [76]), a fact which we will use in §5.4.

2.3 The Jordan Map and Unitary Symmetry

In 1935 Pascual Jordan [77], one of the pioneers of quantum field theory, realized that boson and fermion operator techniques, particularly the creation-destruction operators devised for field theory, provided an elegant approach to Lie algebras and Lie groups. Jordan's method, later rediscovered for the special case of $SU(2)$ by J. Schwinger [71, p. 229] in 1953, may be summarized in the following way.

Consider the Lie algebra $\mathfrak{g}$ of the Lie group G. Let the generators E_α be realized by the $n \times n$ matrices $(E_\alpha)_{ij}$ of, say, the fundamental irrep. Introduce now the boson operators a_i (creation) and $\overline{a}_i$ (destruction), for $i = 1, \ldots, n$, obeying the commutation relations:

$$[\overline{a}_i,\ a_j] = \delta_{ij}, \quad (i, j = 1, 2, \ldots, n)$$

with all other commutators zero. Define a new realization of the Lie algebra generators to be the operator-valued matrices

$$X_\alpha \stackrel{\text{def}}{=} \sum_{i,j=1}^{n} (E_\alpha)_{ij} a_i \overline{a}_j. \tag{2.21}$$

LEMMA **2.22** *The map (the "Jordan map") $\mathcal{J} : E_\alpha \to X_\alpha$ preserves all commutators:*

$$\mathcal{J}([E_\alpha,\ E_\beta]) = [\mathcal{J}(E_\alpha),\ \mathcal{J}(E_\beta)],$$

that is, the Jordan map is a Lie algebra homomorphism.

This construction, transforming the fundamental, defining, matrix representation of the generators of a Lie group into a generic operator realization, essentially solves, at one stroke, the problem of constructing all unitary irreps and (with a bit more work) all tensor operator matrix elements of all compact Lie groups! In order to accomplish this, one defines a Hilbert space of states by defining a vacuum vector $|0\rangle$, for which

$$\overline{a}_i|0\rangle = 0, \quad i = 1, \ldots, n,$$

and then constructs irreps in the Fock space $\mathfrak{F}$ of quantum field theory. The states in $\mathfrak{F}$ consist of polynomials $\mathcal{P}(a_i)$ in the boson creation operators acting on the vacuum, with an inner product defined by

$$(\mathcal{P}', \mathcal{P}) = \langle 0|\, \mathcal{P}'(\overline{a})\, \mathcal{P}(a)|0\rangle. \tag{2.23}$$

Since $\langle 0|(\overline{a}_i)^{m'}(a_j)^m|0\rangle = \delta_{ij}\delta_{mm'}m!$ we determine that an orthonormal set of basis states in $\mathfrak{F}$ is

$$|m_1, m_2, \ldots, m_n\rangle = \frac{(a_1)^{m_1}(a_2)^{m_2}\ldots(a_n)^{m_n}}{\sqrt{m_1!m_2!\ldots m_n!}}|0\rangle.$$

These states are eigenvectors of the **number operators** $N_i \stackrel{\text{def}}{=} a_i\overline{a}_i$, and the eigenvalue m_i is the **number of quanta** for the mode i in the state $|m_1, m_2, \ldots m_n\rangle$. Matrix elements of the boson operators are given by

$$a_i|m_1, \ldots, m_n\rangle = \sqrt{m_i + 1}\; |m_1, \ldots, m_n\rangle, \quad \overline{a}|m_1, \ldots, m_n\rangle = \sqrt{m_i}\; |m_1, \ldots, m_n\rangle,$$

for each $i = 1, \ldots, n$.

The four elements $a, \overline{a}, N, e$, in which e is central and which we have set equal to the identity, comprise the Heisenberg-Weyl algebra, denoted $\mathfrak{h}_4$, which is a Lie algebra; as noted out by Petersen [78, Chapter 5], the universal enveloping algebra $\mathcal{U}(\mathfrak{h}_4)$ is a Hopf algebra with the maps

$$\Delta(x) = x \otimes \mathbb{I} + \mathbb{I} \otimes x, \quad \varepsilon(x) = 0, \quad \gamma(x) = -x.$$

We will discuss the boson calculus, as the techniques of the Jordan map may be termed, further in §2.5.1 where we consider the explicit construction of all irreps of all unitary groups, but for a detailed description we refer to the monograph by Biedenharn and Louck [66]. We turn now to its generalization to quantum groups.

Remark **2.24** For noncompact symplectic groups Dirac (1944) defined implicitly a mapping similar to the Jordan map, in terms of 'expansors'. This is very useful for physics, see for example [71], since the de Sitter group has the symplectic Lie algebra $C_2 \cong B_2$ and contracts to the Poincaré group[3].

---

[3]**The Firenze group [33] has used this technique to develop, along with others [34], the deformed κ-Poincaré group.**

2.4 The q-Generalization of the Boson Calculus

There is a q-generalization of the Jordan map, with the same elegant constructive properties, which we will now develop for the prototype quantum group $\mathcal{U}_q(\mathfrak{su}(2))$. But first we must define q-boson operators (following Biedenharn, Macfarlane [79, 80], see also Sun and Fu [81], and Hayashi [82]):

DEFINITION **2.25** *The q-Heisenberg-Weyl algebra $\mathcal{U}_q(\mathfrak{h}_4)$ (or q-boson algebra) is a set of elements called q-boson operators comprising a creation operator a^q, a destruction operator $\overline{a}^q$, and a number operator N^q, obeying the commutation rules*

$$\overline{a}^q a^q - q^{\frac{1}{2}} a^q \overline{a}^q = q^{-\frac{N^q}{2}}, \tag{2.26}$$

and

$$[N^q,\ a^q] = a^q, \quad [N^q, \overline{a}^q] = -\overline{a}^q. \tag{2.27}$$

Although this definition is valid for any $q \in \mathbb{C}$ we consider only $q \in \mathbb{R}^+$ and, in Chapter 6, q a root of unity. The importance of q-boson operators lies in the simplifications which they permit in constructing representations of quantum groups, and in the understanding of tensor operators, themselves constructed from q-boson operators, which act in spaces formed using q-boson operators. Indeed, the successful program of the boson calculus, applied to the unitary groups, generalizes fully to the unitary quantum groups although possibly, as we shall see in connection with tensor operators, in ways that requires consideration of subtle points.

In applications we usually extend the q-boson algebra by considering the n-fold tensor product, that is by introducing n commuting sets of q-boson operators, denoted a_i^q and $\overline{a}_i^q$ for $i = 1, \ldots, n$ with number operators N_i^q, which for each i satisfy the relations of Definition 2.25. We may note several properties for $n = 1$ immediately:

REMARK **2.28** 1. For $q = 1$, we see easily that (2.26) becomes exactly the algebra of boson operators, the Heisenberg-Weyl algebra $\mathfrak{h}_4$.

2. We may assume that $\overline{a}^q$ is the Hermitian conjugate of a^q and that N^q is Hermitian, that is, the q-boson algebra admits an involutive anti-automorphism, so that we may impose

$$(a^q)^\dagger = \overline{a}^q, \quad (\overline{a}^q)^\dagger = a^q, \quad (N^q)^\dagger = N^q.$$

This leads to a suitable inner product, considered below, but only for $q \in \mathbb{R}^+$; at roots of unity we impose more general Hermiticity relations.

3. The q-boson algebra is invariant under the $U(1)$ transformation

$$a^q \to e^{i\theta} a^q,\ \overline{a}^q \to e^{-i\theta} \overline{a}^q,$$

that is, under a change of phase.

4. The q-boson algebra is invariant under the dilatation transformation

$$a^q \to q^{\frac{\alpha}{4}} a^q,\ \overline{a}^q \to q^{\frac{\alpha}{4}} \overline{a}^q,\ N^q \to N^q - \alpha, \tag{2.29}$$

where $\alpha \in \mathbb{R}$.

5. The operator Ω defined by

$$\Omega = q^{-\frac{N^q}{2}}([N^q]_q - a^q \overline{a}^q) \tag{2.30}$$

commutes with each of a^q, $\overline{a}^q$, N^q, that is, it belongs to the center of the q-boson algebra. We will generally consider the q-boson algebra factored by the relation $\Omega = 0$, in order to perform the Jordan map. Under the dilatation transformation (2.29) we find Ω is transformed to $q^\alpha \Omega - q^{\frac{\alpha}{2}}[\alpha]_q$ which, of course, also commutes with the q-boson operators.

6. There are many equivalent forms for the relation (2.26). For example we could define operators $A^q \stackrel{\text{def}}{=} a^q q^{\frac{N^q}{4}}$, $\overline{A}^q \stackrel{\text{def}}{=} q^{\frac{N^q}{4}} \overline{a}^q$, which satisfy

$$\overline{A}^q A^q - q A^q \overline{A}^q = I, \tag{2.31}$$

with $[N^q,\ A^q] = A^q$, $[N^q, \overline{A}^q] = -\overline{A}^q$. (The left hand side of (2.31) is often called a **q-commutator**.) The homogeneous form of (2.31), that is with the right hand side equal to zero, is the usual form for non-commuting variables.

7. The q-boson form given in the previous remark is complicated and becomes even more so when generalized to many non-commuting copies. The basic reason for using this form, as we will develop in §4.6, is that it defines a q-boson as a tensor operator with well defined transformation properties (see also §8.3.2 and Example 8.35 (p. 270), where these operators are examples of q-symplecton polynomials). Such q-bosons are often called covariant or contravariant q-bosons, but we wish to emphasize that they are simply special cases of more general tensor operator concepts.

8. As before, one introduces a linear space of states by defining the vacuum ket $|0\rangle$ with the property $\overline{a}^q|0\rangle = 0$, and then constructing vectors in the Fock space $\mathfrak{F}$ by allowing polynomials in the creation operator to act on the vacuum.

9. In contrast to the $q = 1$ case, $\mathcal{U}_q(\mathfrak{h}_4)$ is not a Hopf algebra (Petersen [78, §5.3.6]), however Celeghini et al. [83] have introduced a q-analog of the Heisenberg algebra which does have a nontrivial Hopf algebra structure.

There is one major difference between q-bosons and ordinary bosons: the operator $a^q\overline{a}^q$ is not the number operator but rather is equal, in Fock space, to $[N^q]_q$ (where $[\cdot]_q$ is defined in (2.3)). We may state the following result:

LEMMA **2.32** *In the Fock space $\mathfrak{F}$, and when (2.27) is satisfied, the relation (2.26) is equivalent to*

$$a^q\overline{a}^q = [N^q]_q, \quad \overline{a}^q a^q = [N^q + 1]_q. \tag{2.33}$$

PROOF: Beginning firstly with (2.26), we note that we can choose the vacuum state $|0\rangle$ to carry zero quanta, that is, we may choose $N^q|0\rangle = 0$. This follows from the dilatation invariance (2.29); if $N^q|0\rangle = \alpha|0\rangle$ for some $\alpha \in \mathbb{R}$ then we can transform α to zero. Next, from (2.26) we find by induction that

$$\overline{a}^q(a^q)^n - q^{\frac{n}{2}}(a^q)^n\overline{a}^q = [n]_q(a^q)^{n-1}q^{-\frac{N^q}{2}},$$

for all $n \in \mathbb{N}$ (using the identity $[n+1]_q = q^{-\frac{n}{2}} + q^{-\frac{1}{2}}[n]_q$), from which follows

$$\overline{a}^q(a^q)^n|0\rangle = [n]_q(a^q)^{n-1}|0\rangle. \tag{2.34}$$

Hence $(a^q\overline{a}^q)(a^q)^n|0\rangle = [n]_q(a^q)^n|0\rangle$, that is, $a^q\overline{a}^q|n\rangle = [N^q]_q|n\rangle$ for all states $|n\rangle \in \mathfrak{F}$, and so we may write $a^q\overline{a}^q = [N^q]_q$. The second relation in (2.33) now follows. Conversely, the same identity $[n+1]_q = q^{-\frac{n}{2}} + q^{\frac{1}{2}}[n]_q$ shows that (2.33) implies (2.26). □

REMARK **2.35** 1. The relations (2.33) are simpler to use than (2.26), in particular they have the advantage that q-bosons defined by them are manifestly invariant under $q \leftrightarrow q^{-1}$, unlike (2.26), showing that we can identify q-boson operators with q^{-1}-boson operators.

2. In order to realize the algebra of $\mathcal{U}_q(\mathfrak{su}(2))$ we use the relations (2.33), not (2.26), as the calculation below in the proof of Lemma 2.46 indicates. A generalization of (2.33) is

$$a^q\overline{a}^q = q^{-\frac{\alpha}{2}}[N^q - \alpha]_q, \quad \overline{a}^q a^q = q^{-\frac{\alpha}{2}}[N^q - \alpha + 1]_q$$

for $\alpha \in \mathbb{R}$, obtained from the scaling transformation (2.29), and which also satisfies (2.26). For general α, however, the commutation relations of $\mathcal{U}_q(\mathfrak{su}(2))$ are not satisfied with these relations and, in any case, we adopt a scheme in which the vacuum state carries zero quanta, as measured by N^q, which requires $\alpha = 0$.

3. An alternative way of restricting the q-boson algebra so that the relations (2.33) are satisfied is to factor out the relation $\Omega = 0$, where Ω is the central element defined in (2.30).

4. The formula $a^q\overline{a}^q = [N]_q$ for the number operator can be inverted to express N^q explicitly in terms of $a^q\overline{a}^q$ by using the inverse sinh function, however we will not find this expression useful.

We can define an inner product in $\mathfrak{F}$ by using $(a^q)^\dagger = \overline{a}^q$ and $(N^q)^\dagger = N^q$, and extending the definition (2.23) to its q-analog. We find that the inner product of monomials in $\mathfrak{F}$ is given, in the bra-ket notation, by

$$\langle 0| (\overline{a}^q)^n (a^q)^m |0\rangle = \delta_{m,n}[n]_q! \tag{2.36}$$

which follows by induction on n using (2.33), or by using the identity

$$(\overline{a}^q)^n (a^q)^n = [N^q+1]_q[N^q+2]_q \dots [N^q+n]_q, \quad n \in \mathbb{N}.$$

From the property $[n]_q \geqslant n$ for all $n \in \mathbb{N}$, proved in Lemma 2.94, we see that (2.36) defines a positive definite inner product for $q \in \mathbb{R}^+$ (this contrasts with the case when q is a root of unity, considered in §6.3).

If we now introduce n commuting copies of the q-boson operators to form the Fock space $\mathfrak{F}^n$, we find that an orthonormal set of states is

$$|m_1, m_2, \dots, m_n\rangle = \frac{(a_1^q)^{m_1}(a_2^q)^{m_2}\dots(a_n^q)^{m_n}}{\sqrt{[m_1]_q![m_2]_q!\dots[m_n]_q!}} |0\rangle,$$

where m_i is the number of quanta in the mode i, that is,

$$N_i^q |m_1, \dots, m_n\rangle = m_i |m_1, \dots, m_n\rangle$$

for $i = 1, \dots, n$. The matrix elements of the q-boson operators, with a suitable choice of phase, are given by

$$a_i^q |m_1, \dots, m_n\rangle = \sqrt{[m_i+1]_q}\, |m_1, \dots, m_n\rangle, \quad \overline{a}_i^q |m_1, \dots, m_n\rangle = \sqrt{[m_i]_q}\, |m_1, \dots, m_n\rangle. \tag{2.37}$$

As is the case with ordinary boson operators, we see that q-bosons are unbounded operators in $\mathfrak{F}^n$. Again, this contrasts with the case when q is a root of unity, when a^q and $\overline{a}^q$ can be represented by finite-dimensional matrices.

2.4.1 Realizations of q-Boson Operators

Although it is convenient to use the language of q-boson creation and destruction operators, borrowed from quantum field theory, in order to construct irreps of quantum groups, we could equivalently describe this construction by realizing these operators in terms of complex variables and differentiable operators. This has some advantages, for example in Chapter 7 we construct representations on sections of a line bundle which consist of holomorphic functions, and so a description in terms of complex variables is appropriate. On the other hand, a general algebraic formulation, including the construction of irreps using q-boson operators, is independent of any realization, and serves to emphasize our algebraic approach.

In the classical case ($q = 1$), we can realize boson operators as differential and multiplicative operators in complex variables z_i, $i = 1, \ldots, n$ acting in the space $\mathfrak{P}$ of polynomials in z_i; specifically, if we define (taking now $n = 1$)

$$af(z) = zf(z), \quad \overline{a}f(z) = \frac{\partial f(z)}{\partial z},$$

where $f \in \mathfrak{P}$, then a and $\overline{a}$ satisfy the commutation relations of boson operators. The Fock space of states, consisting of polynomials in a acting on $|0\rangle$, is equivalent to $\mathfrak{P}$, with the vacuum state equal to the constant 1. A suitable inner product on $\mathfrak{P}$ is given by

$$(f\,,\,g) = f\left(\frac{\partial}{\partial z}\right) g(z)\Big|_{z=0}$$

with respect to which $a^\dagger = \overline{a}$.

Although this definition is the most convenient for our purposes, another definition of inner product which is useful in some contexts because it extends to a class of functions larger than $\mathfrak{P}$, is one determined by Bargmann [84]. For any two holomorphic functions f, g we define the inner product

$$(f, g) = \int \overline{f(z)} g(z) e^{-|z|^2}\, dz \tag{2.38}$$

and then we find $(f, \frac{\partial g}{\partial z}) = (zf, g)$, that is, the operation of differentiation is adjoint to the operation of multiplication by z. The Bargmann space $\mathcal{F}$ defined by this inner product consists of holomorphic functions of finite norm, and one may determine a reproducing kernel which maps from the conventional Hilbert space of square integrable functions onto this space $\mathcal{F}$. This mapping can be carried out with the help of coherent states, defined as the eigenstates of the destruction operator, and discussed more fully in §8.1.1.

There is a similar realization for q-boson operators a and $\overline{a}$, in which $\overline{a}$ now acts as a finite difference operator. Specifically, in the same space $\mathfrak{P}$ of polynomials of a complex variable z define the **finite difference operator** D^q by

$$\begin{aligned} D^q f(z) &\stackrel{\text{def}}{=} \frac{f(zq^{\frac{1}{2}}) - f(zq^{-\frac{1}{2}})}{z(q^{\frac{1}{2}} - q^{-\frac{1}{2}})} \\ &= z^{-1}\left(\frac{q^{\frac{N}{2}} - q^{-\frac{N}{2}}}{q^{\frac{1}{2}} - q^{-\frac{1}{2}}}\right) f(z) = z^{-1}[N]_q f(z), \quad f \in \mathfrak{P}, \end{aligned} \tag{2.39}$$

which becomes differentiation in the limit $q \to 1$. We have written D^q in operator form using the following realization of the number operator N, which does not depend on q:

$$N = z\frac{\partial}{\partial z}, \qquad q^{\frac{N}{2}} f(z) = f(zq^{\frac{1}{2}}), \quad f \in \mathfrak{P}.$$

We also define the creation operator a^q as for $q = 1$, that is,

$$a^q f(z) = z f(z), \quad f \in \mathfrak{P},$$

which again is independent of q. Now it is easy to see that these operators do indeed realize the q-boson algebra. In summary, we have found:

LEMMA 2.40 *The operators $a^q, \overline{a}^q, N^q$ defined by*

$$a^q f(z) = z f(z), \quad \overline{a}^q f(z) = D^q f(z), \quad N^q f(z) = N f(z) = z \frac{\partial f(z)}{\partial z},$$

satisfy the q-boson relations (2.33) and (2.27).

The finite difference operator D^q is not new to quantum groups, having appeared in the work of Jackson [85, 86, 87] in 1909 in connection with studies of q-extensions of the classical functions, and is discussed further in §2.8. The appearance of D^q as a q-boson operator indicates that q-generalizations of the classical functions can be expected to play a role in quantum groups, as mentioned in §1.3.

We may define an inner product on $\mathfrak{P}$ by

$$(f, g) \stackrel{\text{def}}{=} f(D^q) g(z)\Big|_{z=0} \quad f, g \in \mathfrak{P}, \tag{2.41}$$

from which follows the orthogonality property

$$(z^m, z^n) = \delta_{m,n} [n]_q!, \quad n, m \in \mathbb{N}.$$

Hence, the monomials $z^n / \sqrt{[n]_q!}$ form an orthonormal basis in $\mathfrak{P}$. With this inner product we have $(\overline{a}^q)^\dagger = (D^q)^\dagger = a^q$ and $N^\dagger = N$. As for $q = 1$ we may seek a q-analog of Bargmann space and its inner product (2.38), and this has been done by Bracken et al. [88]. The exponential weight factor generalizes to the q-exponential $\exp_q$, although convergence of the improper integral requires special consideration of $\exp_q(x)$ for large negative x (see also Gray and Nelson [89]). For our purposes, however, the definition (2.41) suffices and is closer in spirit to our algebraic approach.

We can express the realization of q-boson operators shown in Lemma 2.40 entirely in terms of ordinary boson operators, using the operator form shown in (2.39), according to

$$a^q = a, \quad \overline{a}^q = \overline{a} \frac{[N]_q}{N}, \quad N^q = N = a\overline{a},$$

and we may verify directly that the relations (2.27,2.33) are satisfied. A more symmetric realization, which retains the boson inner product, is

$$a^q = \sqrt{\frac{[N]_q}{N}}\, a, \quad \overline{a}^q = \overline{a} \sqrt{\frac{[N]_q}{N}}, \quad N^q = N = a\overline{a}. \tag{2.42}$$

Indeed, we may generalize this boson realization by seeking expressions for $a^q, \overline{a}^q$ in terms of boson operators $a, \overline{a}$ in the form

$$a^q = f(N)a, \qquad \overline{a}^q = \overline{a}f(N), \qquad N^q = N - \alpha = a\overline{a} - \alpha,$$

where $\alpha \in \mathbb{R}$ and $f(N)$ is determined by satisfying (2.26). This leads to the following generalization of (2.42):

$$a^q = \sqrt{\frac{[N-\alpha]_q}{N}}\, a, \quad \overline{a}^q = \overline{a}\sqrt{\frac{[N-\alpha]_q}{N}}, \quad N^q = N - \alpha,\ \alpha \in \mathbb{R}, \tag{2.43}$$

which satisfies (2.27) and (2.33). For nonzero α this realization has the surprising property that the boson and q-boson vacua do not coincide, and this possibility will be explored in §6.5, when irreps of $\mathcal{U}_q(\mathfrak{su}(2))$ at roots of unity are investigated. We will find, however, that (2.42) enables us to realize all irreps of $\mathcal{U}_q(\mathfrak{su}(2))$ for $q \in \mathbb{R}^+$.

The fact that a^q and $\overline{a}^q$ can be written in terms of boson operators has been noted also by several authors, see for example Kulish and Damaskinsky [90], also Filippov et al. [91]. It is a consequence of (2.42) that the generators of $\mathcal{U}_q(\mathfrak{su}(2))$, when realized as a Jordan-Schwinger map, can be written in terms of those for $\mathfrak{su}(2)$ by means of similar formulas; we return to this property in §2.4.2.

Apart from the finite difference analog of differentiation, there is also the q-analog of integration, essentially a finite sum approximation to the Riemann integral. This concept of q-integration appears naturally in the investigation of q-analogs of classical functions, see for example the q-integral representation of basic hypergeometric series (Andrews [57]), but also dates back to Thomae [92] and Jackson [87].

The q-boson formalism incorporates the concept of q-integration by means of a realization of q-boson operators which differs from that in Lemma 2.40, and which we may state in the operator form:

$$a^q = ([N]_q)^{-1} z, \quad \overline{a}^q = z^{-1}([N]_q)^2, \quad N^q = N = z\frac{\partial}{\partial z}. \tag{2.44}$$

This realization also satisfies (2.33). In order to determine the meaning of these operators, consider

$$a^q z^n = \frac{z^{n+1}}{[n+1]_q} = z^{n+1} q^{\frac{1}{2}(n+1)} \frac{(q^{\frac{1}{2}} - q^{-\frac{1}{2}})}{(q^{n+1}-1)} = \sum_{r=0}^{\infty} \left(zq^{r+\frac{1}{2}}\right)^n \left(zq^r - zq^{r+1}\right),$$

where we have expanded the denominator, assuming $q < 1$. For a function $f \in \mathfrak{P}$, or one which can be represented by its Taylor series, we have

$$(a^q f)(z) = \sum_{r=0}^{\infty} f\left(zq^{r+\frac{1}{2}}\right)\left(zq^r - zq^{r+1}\right). \tag{2.45}$$

The right hand side is a finite sum approximation to the integral of f, for suppose that $z \in \mathbb{R}$ and $q < 1$, then we can choose a partition of the interval $[0, z]$ with grid points located at $z, zq, zq^2, \ldots$ (in descending order). On each subinterval $[zq^{r+1}, zq^r]$ we approximate f by its value at the geometric midpoint $zq^{r+\frac{1}{2}}$, and so the approximate area under the curve $y = f(z)$ is given by the right hand side of (2.45). Hence

$$\lim_{q\to 1}(a^q f)(z) = \int_0^z f(x)dx,$$

where we assume f is Riemann-integrable. (Our formulas differ slightly from those given by Andrews [57] since a^q as defined in (2.44) is invariant under $q \leftrightarrow q^{-1}$). Hence, for $q = 1$ we have realized the boson creation operator by the operation of integration, and for general q the realization (2.44) is a finite sum representation of the q-boson operator.

2.4.2 The q-Boson Realization of $\mathcal{U}_q(\mathfrak{su}(2))$ Unitary Irreps

As indicated in Example 2.17 the matrices of the generators for the fundamental 2×2 irrep of $\mathcal{U}_q(\mathfrak{su}(2))$ are exactly the same as the Pauli matrices which realize $\mathfrak{su}(2)$ without deformation. It follows from this that the Jordan map using q-bosons directly as in (2.21) must be modified in order to realize the quantum group; this can be done by applying the q-boson Jordan map only to off-diagonal matrix generators, obtaining the diagonal matrix generators by replacing $a_i^q \bar{a}_i^q = [N_i^q]_q$ by the number operator N_i^q itself.

Hence, for the prototype quantum group $\mathcal{U}_q(\mathfrak{su}(2))$ we introduce two commuting sets of q-boson operators $\{a_i^q, \bar{a}_i^q, N_i^q\}$, where $i = 1, 2$, and realize the generators of $\mathcal{U}_q(\mathfrak{su}(2))$, which act in the Fock space $\mathfrak{F}^2$, as bilinears in the q-boson operators:

Lemma 2.46 *The following operators satisfy the commutation relations (2.1a-2.1b) of $\mathcal{U}_q(\mathfrak{su}(2))$:*

$$J_+ = a_1^q \bar{a}_2^q, \quad J_- = a_2^q \bar{a}_1^q, \quad J_z = \tfrac{1}{2}(N_1^q - N_2^q). \tag{2.47}$$

Proof: We have

$$J_+ J_- = a_1^q \bar{a}_2^q a_2^q \bar{a}_1^q = [N_1^q]_q [N_2^q + 1]_q$$

and similarly

$$J_- J_+ = [N_1^q + 1]_q [N_2^q]_q,$$

where we used (2.33). Now with the help of the identity (2.12) with $a = N_1^q, b = N_2^q, c = -1$, we find $[J_+, J_-] = [N_1^q - N_2^q]_q$ as required. The remaining relations follow immediately from the property (2.27) of the number operators. □

This realization is Hermitean, that is, $(J_+)^\dagger = J_-$, $(J_z)^\dagger = J_z$ as follows from $(a_i^q)^\dagger = \bar{a}_i^q$, $(N_i^q)^\dagger = N_i^q$. It is easy now to obtain explicitly all unitary irreps, realized in $\mathfrak{F}^2$. These irreps are labelled by the invariant C defined in (2.11) which, in the realization (2.47), becomes

$$C = \left[\tfrac{1}{2}(N_1^q + N_2^q)\right]_q \left[\tfrac{1}{2}(N_1^q + N_2^q) + 1\right]_q$$

where we again used the identity (2.12) with $a = \frac{1}{2}(N_1^q - N_2^q)$, $b = \frac{1}{2}(N_1^q + N_2^q)$, $c = N_1^q + 1$. Hence, the eigenfunctions of C are also eigenfunctions of the total number operator $N_1^q + N_2^q$ with eigenvalue $2j$, which is therefore an integer. If we define the invariant operator J_{op} implicitly by $C = [J_{\text{op}}]_q[J_{\text{op}} + 1]_q$ then for the q-boson realization we have $J_{\text{op}} = \frac{1}{2}(N_1^q + N_2^q)$.

The basis states are homogeneous polynomials in the q-boson operators acting on the vacuum and, for $\mathcal{U}_q(\mathfrak{su}(2))$, are simply monomials in a_1^q and a_2^q. The generators $J_\pm, J_z$ therefore act in the subspace of $\mathfrak{F}^2$ consisting of polynomials of fixed degree $2j$ spanned by the vectors

$$|jm\rangle = \frac{(a_1^q)^{j+m}(a_2^q)^{j-m}}{\sqrt{[j+m]_q![j-m]_q!}}\,|0\rangle, \tag{2.48}$$

where $2j$ is a non-negative integer and $-j \leqslant m \leqslant j$ (with integer steps). These states are orthogonal and correctly normalized, as follows from (2.36). They are eigenfunctions of J_z and from them we find the matrix elements as given in Lemma 2.16 (p. 21).

The realization (2.47) of $\mathcal{U}_q(\mathfrak{su}(2))$ extends easily to $\mathcal{U}_q(\mathfrak{u}(2))$, by defining the diagonal generators $E_{11} = N_1^q$, $E_{22} = N_2^q$. However, as the proof of Lemma 2.46 indicates, this realization satisfies more than just the defining relations of $\mathcal{U}_q(\mathfrak{u}(2))$; we have in addition $J_+J_- = [E_{11}]_q[E_{22} + 1]_q$ and as a result we obtain only the symmetric irreps, discussed more generally for $\mathcal{U}_q(\mathfrak{u}(n))$ in §2.5.1.

It was noted in §2.4.1 that q-boson operators can be expressed in terms of ordinary boson operators (see Eqn. (2.42)), and this extends to a similar property for the generators of $\mathcal{U}_q(\mathfrak{su}(2))$. This was observed independently of the q-boson formulation by Jimbo [93] and later by Curtright and Zachos [94] and also Fairlie [95]. From (2.42) we find that

$$J_+ = \left(\frac{[N_1^q]_q[N_2^q + 1]_q}{(N_1^q)(N_2^q + 1)}\right)^{\frac{1}{2}} j_+$$

where $j_+ = a_1\bar{a}_2$ denotes the $\mathfrak{su}(2)$ raising generator expressed in terms of boson operators. Now we may substitute $N_1^q = J_{\text{op}} + J_z$, $N_2^q = J_{\text{op}} - J_z$, where J_z is also the undeformed diagonal generator of $\mathfrak{su}(2)$, to obtain

$$J_+ = \left(\frac{[J_{\text{op}} + J_z]_q[J_{\text{op}} - J_z + 1]_q}{(J_{\text{op}} + J_z)(J_{\text{op}} - J_z + 1)}\right)^{\frac{1}{2}} j_+$$

which expresses the $\mathcal{U}_q(\mathfrak{su}(2))$ generator J_+ as a functional of the $\mathfrak{su}(2)$ generators. (This is also evident from the matrix elements given in Lemma 2.16). As pointed out by Zachos [5, p. 351-377], this functional mapping from the undeformed generators to $\mathcal{U}_q(\mathfrak{su}(2))$ is invertible and demonstrates why the irreps of $\mathfrak{su}(2)$ yield corresponding irreps of $\mathcal{U}_q(\mathfrak{su}(2))$ which are, moreover, in the form shown in Lemma 2.16. In our approach this appears as a property of the q-boson operators.

This construction of the irreps of $\mathcal{U}_q(\mathfrak{su}(2))$ in $\mathfrak{F}^2$ may equivalently be expressed in terms of complex variables z_1, z_2 by using the q-boson realization described in §2.4.1, in which the generators are expressed in terms of finite difference operators D_1^q, D_2^q which act in the space $\mathfrak{P}^2$ of polynomials in z_1, z_2. Specifically, we have

$$J_+ = z_1 D_2^q, \quad J_- = z_2 D_1^q, \quad J_z = \tfrac{1}{2}\left(z_1\partial_1 - z_2\partial_2\right), \tag{2.49}$$

where $\partial_i \stackrel{\text{def}}{=} \frac{\partial}{\partial z_i}$, and the basis vectors are

$$|jm\rangle = \frac{z_1^{j+m} z_2^{j-m}}{\sqrt{[j+m]_q![j-m]_q!}}. \tag{2.50}$$

These vectors have also been used by Alvarez-Gaumé et al. [75] to construct irreps of $\mathcal{U}_q(\mathfrak{su}(2))$.

2.4.3 Realization on a Projective Space

There is a well-known construction of irreps of $SU(2)$ in which the generators are expressed as differential operators of a single complex variable z acting in the space $\mathfrak{P}$ of polynomials in z. This realization is closely connected to that described in §2.4.2, for we may transform that description, which uses homogeneous polynomials, to one using a projective space by means of a certain similarity transformation, as described for example by Vilenkin [50, p. 108-114]. This transformation generalizes to the quantum group and enables us to realize all irreps of $\mathcal{U}_q(\mathfrak{su}(2))$ in terms of operators of a single complex variable or, equivalently, by using only a single q-boson operator and its conjugate. We shall encounter this same realization in Chapter 7, in an apparently completely different context, where we use the method of Borel and Weil to construct irreps of the quantum groups; this realization is also known (classically) using the construction of vector coherent states. There is another realization which is well-known in the physics literature and which is also closely related to the construction we are about to describe; this is associated with the work of Holstein and Primakoff [96], in which the generators of $\mathfrak{su}(2)$ are realized in terms of a single boson operator and its conjugate. The generalization of this to $\mathcal{U}_q(\mathfrak{su}(2))$ can be accomplished in a way which preserves the q-boson and ordinary boson inner products, and hence also the Hermiticity properties of the generators.

Consider firstly $SU(2)$ and the construction of the irrep labelled j in terms of two complex variables in the space $\mathfrak{P}^2$. The generators and basis states are given by Eqs. (2.49,2.50) for $q = 1$. In the language of the boson calculus we introduce two commuting sets of boson operators $\{a_i, \bar{a}_i\}$, for $i = 1, 2$, and realize the generators by means of the Jordan map acting in the Fock space $\mathfrak{F}^2$. The homogeneous polynomials in $\mathfrak{P}^2$ are uniquely determined by their values on the line $z_1 = 1$ and so, following Vilenkin [50], we may associate to each homogeneous polynomial $f(z_1, z_2)$ in $\mathfrak{P}^2$ a polynomial φ according to

$$\varphi(z_2) = f(1, z_2).$$

Equivalently, we may consider representations in the space of polynomials $\mathcal{P}_j$ of degree $2j$ on a complex projective space $\mathbb{P}_1(\mathbb{C})$ of dimension one; elements $\varphi(\zeta)$ of $\mathcal{P}_j$ are obtained from the formula

$$f(z_1, z_2) = z_1^{2j}\varphi(\zeta), \qquad \zeta \stackrel{\text{def}}{=} \frac{z_2}{z_1},$$

and ζ is therefore the inhomogeneous coordinate for $\mathbb{P}_1(\mathbb{C})$ and is defined for all $z_1 \neq 0$. Since f is homogeneous of degree $2j$, φ is homogeneous of degree zero, and $\mathcal{P}_j$ is spanned by the polynomials $1, \zeta, \zeta^2, \ldots \zeta^{2j}$ (for further discussion, see for example Hermann [97]).

A representation Θ_g, for $g \in SU(2)$, may be defined in $\mathfrak{P}^2$ by

$$\Theta_g f(\boldsymbol{z}) = f(\boldsymbol{z}g), \quad f \in \mathfrak{P}^2, \tag{2.51}$$

where $\boldsymbol{z} = (z_1, z_2)$, and we have $\Theta_{g_1 g_2} = \Theta_{g_1}\Theta_{g_2}$ for all $g_1, g_2 \in SU(2)$. The $SU(2)$ generators can be determined from Θ_g in the usual way, by consideration of suitable one-parameter subgroups of $SU(2)$, and one obtains the $q = 1$ case of (2.49), that is, the Jordan map in the boson operator language. To be explicit, let us write $g \in SU(2)$ as

$$g = \begin{pmatrix} a & b \\ c & d \end{pmatrix}$$

then (2.51) reads

$$\Theta_g f(z_1, z_2) = f(az_1 + cz_2,\ bz_1 + dz_2).$$

There is a natural action of $SU(2)$ on $\mathbb{P}_1(\mathbb{C})$, which is a homogeneous space of $SU(2)$, by linear fractional transformations which can be determined by writing the right hand side (using homogeneity) as

$$(az_1 + cz_2)^{2j} f\left(\frac{bz_1 + dz_2}{az_1 + cz_2}\right).$$

Hence, we may define a representation Θ'_g acting in $\mathcal{P}_j$ by the following formula, comprising the well-known fractional linear transformation:

$$\Theta'_g\, \varphi(\zeta) = (a + c\zeta)^{2j}\varphi\left(\frac{b + d\zeta}{a + c\zeta}\right). \tag{2.52}$$

The mapping $g \to \Theta'_g$ is a representation of $SU(2)$, or more generally, of $GL(2, \mathbb{C})$. The generators can be determined by consideration of the infinitesimal elements as before, and take the form (see for example Vilenkin [50, p. 111]):

$$J'_+ = \frac{d}{d\zeta}, \quad J'_- = 2j\zeta - \zeta^2\frac{d}{d\zeta}, \quad J'_z = j - \zeta\frac{d}{d\zeta}.$$

Here, we may regard j as an invariant operator which commutes with the generators, indeed, the Casimir invariant C for $q = 1$, defined in (2.11), reduces identically to $C = j(j+1)$ and so we may replace j by J_{op}, where $C = J_{\text{op}}(J_{\text{op}}+1)$. A basis in $\mathcal{P}_j$ is the set $\{1, \zeta, \zeta^2, \dots, \zeta^{2j}\}$ which may be normalized once an inner product in $\mathcal{P}_j$ has been selected.

Now let us derive this transformation, from the homogeneous space $\mathfrak{P}^2$ to the set of functions $\mathcal{P}_j$ on the inhomogeneous or projective space, in a slightly different way by consideration of the Lie algebra rather than the group $SU(2)$. Reformulating the transformation so as to consider only $\mathfrak{su}(2)$ allows us to generalize the procedure directly to $\mathcal{U}_q(\mathfrak{su}(2))$. Nevertheless, it is possible to define a quantum group generalization of the fractional linear transformation (2.52) but we defer discussion of this until Chapter 4 when we have introduced non-commuting coordinates and have defined the matrix quantum group $SU_q(2)$ (see §4.7.1, p. 154).

Let $|jm\rangle$ denote the basis vectors in $\mathfrak{P}^2$ in an irrep of $\mathfrak{su}(2)$, then $|jm\rangle$ is given explicitly by (2.50) for $q = 1$. We perform a similarity transformation on the generators and basis vectors by means of the operator

$$U \stackrel{\text{def}}{=} \frac{z_1^{2j}}{\sqrt{(2j)!}}.$$

That is, we put

$$\begin{aligned} |jm\rangle' = U^{-1}|jm\rangle &= \frac{z_1^{-2j}\sqrt{(2j!)}\, z_1^{j+m} z_2^{j-m}}{\sqrt{(j+m)!(j-m)!}} \\ &= \sqrt{\frac{(2j)!}{(j+m)!}}\, \frac{\zeta^{j-m}}{\sqrt{(j-m)!}}. \end{aligned} \tag{2.53}$$

We have chosen the normalization so as to obtain directly the basis vectors in the form derived by the Borel-Weil method of Chapter 7, and which leads to the standard matrix elements for the generators.. Essentially, we have used U to "divide" the vectors in $\mathfrak{P}^2$ by the state of highest weight with respect to $\mathfrak{su}(2)$. Given the generators $\mathbf{J} = \{J_+, J_-, J_z\}$ we define

$$\mathbf{J}' = U^{-1}\mathbf{J}U,$$

and it follows that $\mathbf{J}'$ also satisfy the commutation relations of $\mathfrak{su}(2)$. In fact, we obtain in this way precisely the generators determined from the fractional linear transformation (2.52):

LEMMA **2.54** *The $\mathfrak{su}(2)$ generators defined by the similarity transformation $\mathbf{J}' = U^{-1}\mathbf{J}U$ are given by*

$$J'_+ = \frac{d}{d\zeta}, \quad J'_- = 2j\zeta - \zeta^2\frac{d}{d\zeta}, \quad J'_z = j - \zeta\frac{d}{d\zeta}.$$

PROOF: Since

$$z_1 \frac{\partial}{\partial z_2} f\left(\frac{z_2}{z_1}\right) = \frac{d}{d\zeta} f(\zeta),$$

we may write

$$z_1^{-2j}\left(z_1 \frac{\partial}{\partial z_2}\right) z_1^{2j} = \frac{d}{d\zeta},$$

where the left hand side acts in $\mathfrak{P}^2$, and the right hand side in $\mathcal{P}_j$. Also,

$$z_1^{2j}\left(z_2 \frac{\partial}{\partial z_1}\right) z_1^{2j} = 2j\frac{z_2}{z_1} + z_2 \frac{\partial}{\partial z_1} = 2j\zeta - \zeta^2 \frac{d}{d\zeta}$$

where we used

$$\frac{\partial}{\partial z_1} f\left(\frac{z_2}{z_1}\right) = \frac{-\zeta^2}{z_2}\frac{d}{d\zeta} f(\zeta).$$

The remaining expression for J_z' also follows from these equations.□

By means of the transformation U, therefore, we can map the space $\mathfrak{P}^2$ into $\mathcal{P}_j$ and accordingly realize $\mathfrak{su}(2)$ in terms of a single complex variable. It is useful to note that this construction generalizes to all the unitary, orthogonal and symplectic groups, as described in detail by Zhelobenko [98]. Representations of the general linear groups are constructed over the homogeneous space $GL(n)/ZD$ where Z is the subgroup of lower triangular matrices with unit diagonal elements, and D is the subgroup of diagonal matrices. If we denote by ζ the upper triangular matrices, with elements ζ_{ij}, then the generalization of (2.52) to all general linear groups is

$$\Theta_g \varphi(\zeta) = \Delta_1^{m_1-m_2} \Delta_2^{m_2-m_3} \ldots \Delta_n^{m_n} \varphi(\zeta_g)$$

where $m_i - m_{i+1}$ are non-negative integers, Δ_i are the principal minors of the matrix ζg, and ζ_g is the right component of ζg in its decomposition into lower and upper triangular matrices (see [98, Chapter VII]). The mapping from the homogeneous space to the polynomials on a projective space, which is accomplished by the variable $\zeta = \frac{z_2}{z_1}$ for $SU(2)$, generalizes to

$$\zeta_{ij} = \frac{z_{12\ldots i-1\,j}}{z_{12\ldots i-1\,i}}, \quad j > i$$

where the antisymmetric combination $z_{12\ldots i-1\,j}$ is the minor of the $n \times n$ matrix (z_{ij}), where z_{ij} are the homogeneous coordinates. This is discussed by Lohe and Hurst [99] to which we refer for further details. However, we develop here the generalization to quantum groups only for $\mathcal{U}_q(\mathfrak{su}(2))$.

The transformation from the homogeneous to the inhomogeneous space, as implemented by U, applies almost unchanged to $\mathcal{U}_q(\mathfrak{su}(2))$. The quantum group generators (denoted now $\mathbf{J}^q$) are given by (2.49), and we define

$$U_q \stackrel{\text{def}}{=} \frac{z_1^{2j}}{\sqrt{[2j]_q!}}.$$

Lemma 2.55 *The $\mathcal{U}_q(\mathfrak{su}(2))$ generators defined by the similarity transformation $\mathbf{J}^{q\prime} = U_q^{-1}\mathbf{J}^q U_q$ are given by*

$$J_+^{q\prime} = D_\zeta^q, \quad J_-^{q\prime} = \zeta\left[2j - \zeta\frac{d}{d\zeta}\right]_q, \quad J_z^{q\prime} = j - \zeta\frac{d}{d\zeta},$$

where D_ζ^q denotes the finite difference operator with respect to ζ, defined by (2.39).

Proof: We have

$$z_1^{-2j}(z_1 D_2^q) z_1^{2j} f\left(\tfrac{z_2}{z_1}\right) = \frac{f\left(\frac{z_2}{z_1}q^{\frac{1}{2}}\right) - f\left(\frac{z_2}{z_1}q^{-\frac{1}{2}}\right)}{\frac{z_2}{z_1}(q^{\frac{1}{2}} - q^{-\frac{1}{2}})} = D_\zeta^q f(\zeta),$$

that is, $U_q^{-1}(z_1 D_2^q)U_q = D_\zeta^q$. Also

$$\begin{aligned} U_q^{-1}(z_2 D_1^q)U_q f\left(\tfrac{z_2}{z_1}\right) &= -\zeta^2\frac{q^{-j}f(\zeta q^{\frac{1}{2}}) - q^j f(\zeta q^{-\frac{1}{2}})}{\zeta(q^{\frac{1}{2}} - q^{-\frac{1}{2}})} \\ &= \zeta\left[2j - \zeta\frac{d}{d\zeta}\right]_q f(\zeta), \end{aligned}$$

and similarly, $U_q^{-1}J_z^q U_q = J_z^{q\prime}$. □

The operator U_q transforms the states (2.50) according to

$$|jm\rangle' = U_q^{-1}|jm\rangle = \sqrt{\frac{[2j]_q!}{[j+m]_q!}}\,\frac{\zeta^{j-m}}{\sqrt{[j-m]_q!}}.$$

The Casimir invariant (2.11) evaluates identically to $C = [j]_q[j+1]_q$. The matrix elements of the generators are precisely those given by Lemma 2.16, showing that the generators given above by Lemma 2.55 are in fact Hermitean with respect to the inner product for which the states $|jm\rangle'$ are orthonormal.

We can write the realization given in Lemma 2.55 entirely in terms of q-boson operators, and can verify directly that the commutation relations of $\mathcal{U}_q(\mathfrak{su}(2))$ are satisfied:

Lemma 2.56 *The following operators, expressed in terms of a single q-boson operator a^q and its conjugate $\overline{a}^q$, satisfy the commutation relations of $\mathcal{U}_q(\mathfrak{su}(2))$:*

$$J_+^q = \overline{a}^q, \quad J_-^q = a^q[2j - N^q]_q, \quad J_z^q = j - N^q.$$

Proof: It suffices to use the properties (2.27,2.33) of q-boson operators, and the identity (2.12) with $a = 2j - N^q, b = N^q, c = -1$. □

Evidently, with the q-boson inner product these generators do not satisfy the Hermiticity properties appropriate to unitary irreps of $\mathcal{U}_q(\mathfrak{su}(2))$ (hence the Hermitean conjugate operators also provide a realization of $\mathcal{U}_q(\mathfrak{su}(2))$). In Chapter 7 we will see that this lack of unitarity may be remedied by a simple redefinition of inner product, however, let us note that we can rewrite this realization of the generators such as to allow us to retain the q-boson inner product of the Fock space $\mathfrak{F}$:

LEMMA **2.57** *A realization of $\mathcal{U}_q(\mathfrak{su}(2))$ which is unitary with respect to the q-boson Fock space is*

$$J_+^q = \sqrt{[2j - N^q]_q}\, \overline{a}^q, \quad J_-^q = a^q \sqrt{[2j - N^q]_q}, \quad J_z^q = j - N^q.$$

This realization is the q-analog of the Holstein-Primakoff realization [96] for $\mathfrak{su}(2)$, and acts in a Fock space for which the number of quanta does not exceed $2j$. The orthonormal basis vectors in the Fock space are $|jm\rangle = ([j-m]_q!)^{-\frac{1}{2}}(a^q)^{j-m}|0\rangle$. This generalization of the Holstein-Primakoff realization has previously been noted by a number of authors, see for example Chaichian et al. [100], also Solomon [2, p. 705], [101].

Let us also record the realization obtained by substituting the expression for a q-boson operator in terms of an ordinary boson operator; we may take advantage of the unspecified parameter α in (2.43) to obtain a realization of $\mathcal{U}_q(\mathfrak{su}(2))$ in terms of a single boson operator a and its conjugate $\overline{a}$, with $N = a\overline{a}$:

$$\begin{aligned} J_+^q &= \sqrt{\frac{[2j - N + \alpha]_q [N - \alpha + 1]_q}{N+1}}\, \overline{a} \\ J_-^q &= a\sqrt{\frac{[2j - N + \alpha]_q [N - \alpha + 1]_q}{N+1}} \\ J_z^q &= j + \alpha - N, \end{aligned}$$

where $\alpha \in \mathbb{R}$.

2.4.4 Mixed Symmetry States and Irreps of $\mathcal{U}_q(\mathfrak{u}(2))$

$\mathcal{U}_q(\mathfrak{su}(2))$ is not typical in one respect: all unitary irreps are constructible from a single (two component) q-boson, since only totally symmetric states are required, that is, the corresponding Weyl patterns consist of only one row. The general $U(2)$ irrep, and hence also $\mathcal{U}_q(\mathfrak{u}(2))$ irreps, are labelled by two integers m_{12}, m_{22} in which mixed symmetries can occur.

Consider first the $q = 1$ case. A general $U(2)$ state in the Gel'fand-Weyl basis (reviewed briefly in §2.6, p. 54) is denoted by $\left|\begin{smallmatrix} m_{12} & & m_{22} \\ & m_{11} & \end{smallmatrix}\right\rangle$ with $m_{12} \geqslant m_{11} \geqslant m_{22}$. The state $|jm\rangle$ corresponds to putting $m_{22} = 0$, $m_{12} = 2j$, $m_{11} = j + m$, as indicated in (2.83). In order to realize such states we must introduce a second, commuting, set of boson operators which we denote here for convenience by $\{b_1, b_2\}$ (instead of $\{a_1^2, a_2^2\}$ as in §2.6). By using the ($q = 1$) boson calculus we find

$$\left|\begin{smallmatrix} m_{12} & & m_{22} \\ & m_{11} & \end{smallmatrix}\right\rangle = \mathcal{N}^{-\frac{1}{2}} (a_{12})^{m_{22}} a_1^{m_{11}-m_{22}} a_2^{m_{12}-m_{11}} |0\rangle \tag{2.58}$$

where $\mathcal{N}$ is the normalization and a_{12} is a determinant:

$$a_{12} \stackrel{\text{def}}{=} a_1 b_2 - a_2 b_1,$$

which corresponds to an antisymmetric tensor in the Young frame, as explained in §2.6.

What are the q-analogs of the states (2.58) and, more particularly, what is the q-analog to the antisymmetric combination a_{12}? We can answer this question explicitly using q-boson operators, and so we now introduce a second commuting set of q-bosons[4] $\{b_1, b_2\}$, and form the generators by co-multiplication:

$$\begin{aligned} J_+ &= q^{\frac{1}{4}(N_1^b - N_2^b)} a_1 \bar{a}_2 + q^{-\frac{1}{4}(N_1^a - N_2^a)} b_1 \bar{b}_2 \\ J_- &= q^{\frac{1}{4}(N_1^b - N_2^b)} a_2 \bar{a}_1 + q^{-\frac{1}{4}(N_1^a - N_2^a)} b_2 \bar{b}_1 \\ J_z &= \tfrac{1}{2}(N_1^a - N_2^a) + \tfrac{1}{2}\left(N_1^b - N_2^b\right), \end{aligned} \tag{2.59}$$

where the superscripts a, b refer to the q-boson sets $\{a_1, a_2\}$, $\{b_1, b_2\}$ respectively, that is, $[N_1^a]_q = a_1 \bar{a}_1$, $[N_1^b]_q = b_1 \bar{b}_1$, etc. These generators act in the Fock space $\mathfrak{F}^4$ of polynomials in a_1, a_2, b_1, b_2 terminated by the vacuum ket, and we seek an irreducible subspace of $\mathfrak{F}^4$ in which the states carry the labels m_{ij}.

One way to proceed, in order to find the q-analog a_{12}^q of a_{12}, is to calculate the state of highest weight in $\mathfrak{F}^4$, which is annihilated by J_+, that is,

$$J_+ \left| \begin{smallmatrix} m_{12} & & m_{22} \\ & m_{11} & \end{smallmatrix} \right\rangle = 0, \tag{2.60}$$

and which is also an eigenfunction of $N_1 = N_1^a + N_1^b$ and $N_2 = N_2^a + N_2^b$. Hence this state must be a linear combination of the polynomials

$$a_1^{m_{11}-s} a_2^{m_{12}-m_{11}+s} b_1^s b_2^{m_{22}-s} |0\rangle,$$

where $0 \leqslant s \leqslant m_{22}$, and the specific linear combination is found by solving (2.60); the general state is then found by applying the lowering operator J_-. By direct calculation we get:

$$\left| \begin{smallmatrix} m_{12} & & m_{22} \\ & m_{11} & \end{smallmatrix} \right\rangle = \mathcal{N}_q^{-\frac{1}{2}} \sum_{s=0}^{m_{22}} \frac{(-)^s q^{m_s}}{[s]_q! [m_{22} - s]_q!} a_1^{m_{11}-s} a_2^{s+m_{12}-m_{11}} b_1^s b_2^{m_{22}-s} |0\rangle, \tag{2.61}$$

where

$$m_s = \tfrac{1}{4}\left((s + m_{12} - m_{11} + 1)m_{22} - s m_{12} - 2s\right),$$

and

$$\mathcal{N}_q = \frac{[m_{12}+1]_q! [m_{12} - m_{11}]_q! [m_{11} - m_{22}]_q! [m_{22}]_q!}{[m_{12} - m_{22} + 1]_q!}.$$

[4]We omit now the affix q for the remainder of this section.

Remarkably, the unwieldy expression (2.61) can be compactly written in the following operator form, an evident q-analog of (2.58):

$$\left|\begin{smallmatrix} m_{12} & & m_{22} \\ & m_{11} & \end{smallmatrix}\right\rangle = \mathcal{N}_q^{-\frac{1}{2}} (a_{12}^q)^{m_{22}} a_1^{m_{11}-m_{22}} a_2^{m_{12}-m_{11}} |0\rangle, \tag{2.62}$$

where a_{12}^q is the operator (first introduced in [102]) defined by

$$a_{12}^q \stackrel{\text{def}}{=} q^{\frac{1}{4}(N_2^a+N_1^b+1)} a_1 b_2 - q^{-\frac{1}{4}(N_1^a+N_2^b+1)} a_2 b_1, \tag{2.63}$$

and is the q-analog to the antisymmetric boson pair operator a_{12}. (The expansion of (2.62) to the expression (2.61) uses the q-binomial theorem and the fact that the summands in a_{12}^q q-commute. See (4.45) below, p. 143, for the explicit expansion). There are numerous significant properties which a_{12}^q satisfies and although these will be developed and proved in §4.4 we observe the following:

REMARK **2.64** 1. The operator a_{12}^q is invariant under each of the interchanges

$$a_1 \longleftrightarrow b_2, \qquad a_2 \longleftrightarrow b_1,$$

(together with a similar interchange of number operators), and changes sign under both of

$$a_1 \longleftrightarrow a_2, \qquad b_1 \longleftrightarrow b_2,$$

provided also $q \leftrightarrow q^{-1}$, and similarly under both of

$$a_1 \longleftrightarrow b_1, \qquad a_2 \longleftrightarrow b_2.$$

2. The operator a_{12}^q does not commute with a_1, a_2, b_1, b_2; unlike the $q = 1$ case, the ordering in (2.62) is essential. This can be understood better once the concept of tensor operators has been introduced in §3.3, where we will see that the q-boson operators $\{a_1, a_2\}$ do not themselves comprise a tensor operator of rank $\frac{1}{2}$, rather, the pair $(a_1 q^{-\frac{N_2^a}{4}}, a_2 q^{\frac{N_1^a}{4}})$ forms a tensor operator, and each component commutes with a_{12}^q (and q-commute with each other). Hence, we can rewrite (2.62) in terms of commuting, or q-commuting, operators.

3. The operator a_{12}^q commutes with the $\mathcal{U}_q(\mathfrak{su}(2))$ generators (2.59). With this knowledge, the basis vectors (2.62) are easily derived, since the action of the generators (2.59) on those q-boson states which are independent of b_1 and b_2 reduces to that of the generators (2.47) obtained from the Jordan map.

4. The operators a_{12}^q and $\overline{a}_{12}^q$ (the Hermitean conjugate of a_{12}^q) generate, together with a diagonal operator, the noncompact quantum group $\mathcal{U}_q(\mathfrak{su}(1,1))$ and these generators commute with $\mathcal{U}_q(\mathfrak{su}(2))$ defined by (2.59). Furthermore, there exists a *third* commuting quantum group! This is generated by another set of bilinear operators $\{K_\pm, K_z\}$ formed from the q-boson operators a_i, b_i $(i = 1, 2)$ and their

conjugates and generates, together with $\{J_\pm, J_z\}$, a direct product quantum group which we denote $\mathcal{U}_q(\mathfrak{su}(2)*\mathfrak{su}(2))$. We return to this quantum factor algebra in §4.4, where we develop the properties in detail. The existence of these commuting algebras generalizes very elegantly the results found by Schwinger [71, p. 241], who used the properties of these commuting algebras to investigate the addition of two angular momenta (see also Biedenharn and Louck [103, p. 124]).

5. The states (2.62) span a subspace of $\mathfrak{F}^4$ which is irreducible under $\{J_\pm, J_z\}$, but in fact are of highest weight with respect to the second $\mathcal{U}_q(\mathfrak{su}(2))$ generated by $\{J'_\pm, J'_z\}$ mentioned in the previous remark. The basis vectors for $\mathcal{U}_q(\mathfrak{su}(2)*\mathfrak{su}(2))$ are calculated in §4.4.1 and are related in form to the representation matrices of the quantum group $SU_q(2)$ (Lemma 4.58, p. 152).

We repeat that these important properties are proved and discussed in detail in Chapter 4.

2.5 Irreducible Unitary Representations of $\mathcal{U}_q(\mathrm{u}(n))$

We have already seen that $\mathcal{U}_q(\mathfrak{su}(2))$ is a deformation of the universal enveloping algebra of $\mathfrak{su}(2)$ and that the irreps of $\mathcal{U}_q(\mathfrak{su}(2))$ are in one-to-one correspondence with those of $\mathfrak{su}(2)$; moreover, the matrix elements of the generators of $\mathcal{U}_q(\mathfrak{su}(2))$ can be obtained from those of $\mathfrak{su}(2)$, as we saw in Lemma 2.16, merely by replacing integer factors with corresponding q-integer factors. As we shall discuss shortly, this remarkable result generalizes to all unitary quantum groups, but first, let us define $\mathcal{U}_q(\mathrm{u}(n))$.

Consider the Lie group $U(n)$ having the Lie algebra $\mathfrak{g}$. The Cartan-Weyl approach divides the generators into three sets, writing the Lie algebra as the union

$$\mathfrak{g} = \mathfrak{n}_+ \cup \mathfrak{h} \cup \mathfrak{n}_-$$

where the elements of $\mathfrak{n}_\pm$, considered as matrix operators, are strictly upper (lower) triangular raising (lowering) operators and $\mathfrak{h}$ comprises the diagonal Cartan subgroup generators. In terms of the Cartan-Weyl generators $\{E_{ij}, i, j = 1, \ldots, n\}$ we have

$$\mathfrak{n}_+ = \{E_{ij} \mid i < j\}, \quad \mathfrak{n}_- = \{E_{ij} \mid i > j\},$$

where these generators satisfy

$$[E_{ij}, E_{kl}] = \delta_{jk} E_{il} - \delta_{il} E_{kj}.$$

The Hermiticity conditions appropriate to unitary irreps of $U(n)$ are $(E_{ij})^\dagger = E_{ji}$.

The Chevalley modification of the Cartan-Weyl procedure uses a subset of the Cartan-Weyl generators as a more concise formulation of the commutation relations, for both the unitary group and its quantum deformation. In this approach, only the $n-1$ raising operators $e_i \stackrel{\text{def}}{=} E_{i,i+1}$ and the $n-1$ lowering operators $f_i \stackrel{\text{def}}{=} E_{i+1,i}$ are used as the basic set of generators together with, for $U(n)$, n diagonal generators which we may take to be $h'_i = E_{ii}$. For $SU(n)$ and hence also for $\mathcal{U}_q(\mathfrak{su}(n))$ we define instead the $n-1$ diagonal generators $h_i \stackrel{\text{def}}{=} E_{ii} - E_{i+1,i+1}$; for $n = 2$ we have $e_1 = J_+ = E_{12}$, $f_1 = J_- = E_{21}$ and $h_1 = 2J_z = E_{11} - E_{22}$. The Hermitean conjugate relations for unitary irreps are $f_i = (e_i)^\dagger, h_i = h_i^\dagger$.

DEFINITION **2.65** *The* $3(n-1)$ *generators of* $\mathcal{U}_q(\mathfrak{su}(n))$ *satisfy the following commutation relations (Jimbo [23, 104], 1985):*

$$\begin{aligned}
[h_i, h_j] &= 0 \\
[h_i, e_j] &= \begin{cases} 2e_j & j = i \\ -e_j & j = i \pm 1 \\ 0 & \textit{otherwise}, \end{cases} \\
[h_i, f_j] &= \begin{cases} -2f_j & j = i \\ f_j & j = i \pm 1 \\ 0 & \textit{otherwise}, \end{cases} \\
[e_i, f_j] &= \delta_{ij}[h_i]_q,
\end{aligned} \tag{2.66}$$

where $i, j = 1, \ldots, n-1$. *In addition we have the* **quadratic Serre relations***:*

$$[e_i, e_j] = [f_i, f_j] = 0, \quad |i - j| \geqslant 2,$$

and the remaining relations for the generators in the Chevalley set are the following **cubic Serre relations***:*

$$\begin{aligned}
e_i^2 e_{i\pm1} - [2]_q e_i e_{i\pm1} e_i + e_{i\pm1} e_i^2 &= 0 \\
f_i^2 f_{i\pm1} - [2]_q f_i f_{i\pm1} f_i + f_{i\pm1} f_i^2 &= 0,
\end{aligned} \tag{2.67}$$

for all possible $i = 1, \ldots, n-1$.

Jimbo [23] in fact stated the defining relations more generally for the q-analog $\mathcal{U}_q(\mathfrak{g})$ of the universal enveloping algebra for any Lie algebra $\mathfrak{g}$, however, we restrict our development here to the unitary quantum groups and refer to Jimbo for the general definition.

The Hopf algebra operations take the form, firstly, for co-multiplication:

$$\begin{aligned}
\Delta(h_i) &= h_i \otimes \mathbb{I} + \mathbb{I} \otimes h_i \\
\Delta(e_i) &= e_i \otimes q^{\frac{h_i}{4}} + q^{-\frac{h_i}{4}} \otimes e_i,
\end{aligned} \tag{2.68}$$

for $i = 1, \ldots, n-1$ and similarly for f_i, by replacing e_i by f_i. The co-unit and antipode are defined by

$$\begin{gathered}\varepsilon(\mathbb{I}) = 1, \quad \varepsilon(e_i) = \varepsilon(f_i) = \varepsilon(h_i) = 0, \\ \gamma(e_i) = -q^{-\frac{1}{2}} e_i, \quad \gamma(f_i) = -q^{\frac{1}{2}} f_i, \quad \gamma(h_i) = -h_i, \end{gathered} \tag{2.69}$$

and with these definitions $\mathcal{U}_q(\mathfrak{su}(n))$ becomes a Hopf algebra. The quantum group $\mathcal{U}_q(\mathfrak{u}(n))$ is defined by the generators above plus the additional generator h_n which commutes with all the other generators. In the Cartan-Weyl notation for the generators E_{ij} of $U(n)$ we have $h_n = E_{11} + \ldots + E_{nn}$. The Hopf algebra operations are the same as those for the other h_i. An alternative definition of the commutation relations of $\mathcal{U}_q(\mathfrak{u}(n))$, in which the diagonal generators are $h'_i \overset{\text{def}}{=} E_{ii}$, is

$$\begin{aligned} \left[h'_i, h'_j\right] &= 0 \\ \left[h'_i, e_j\right] &= (\delta_{ij} - \delta_{i,j+1}) e_j \\ \left[h'_i, f_j\right] &= (\delta_{i,j+1} - \delta_{ij}) f_j \\ \left[e_i, f_j\right] &= \delta_{ij} [h'_i - h'_{i+1}]_q. \end{aligned}$$

These equations replace (2.66), and the Serre relations remain unchanged.

We can specify the precise relation between the Chevalley set of generators and the Cartan-Weyl generators E_{ij}, $(i, j = 1, \ldots, n)$ for $\mathcal{U}_q(\mathfrak{u}(n))$, independent of any representation, with the help of the following definitions:

$$E_{ij} \overset{\text{def}}{=} q^{\frac{1}{2}E_{j-1,j-1}} (E_{i,j-1} E_{j-1,j} - q^{\frac{1}{2}} E_{j-1,j} E_{i,j-1}), \qquad \text{for } j > i+1, \tag{2.70a}$$

$$E_{ji} \overset{\text{def}}{=} q^{-\frac{1}{2}E_{j-1,j-1}} (E_{j,j-1} E_{j-1,i} - q^{-\frac{1}{2}} E_{j-1,i} E_{j,j-1}), \qquad \text{for } j > i+1. \tag{2.70b}$$

Given the raising generators $e_i = E_{i,i+1}$ $(i = 1, \ldots, n-1)$ and the diagonal generators E_{ii} $(i = 1, \ldots, n)$, we define the generators E_{ij} for all $j > i+1$ by putting first $j = i+2$ in (2.70a), which defines $E_{i,i+2}$ in terms of the Chevalley generators $E_{i,i+1}$, $E_{i+1,i+2}$ and the diagonal generator $E_{i+1,i+1}$. Next, we put $j = i+3$ in (2.70a) to define $E_{i,i+3}$ in terms of the known generators $E_{i,i+2}$, $E_{i+2,i+3}$, and by proceeding recursively in this way we define E_{ij} for all $j > i+1$. A similar procedure defines E_{ji} for all $j > i+1$ using (2.70b), and so we obtain the complete set of Cartan-Weyl generators E_{ij}, $(i, j = 1, \ldots, n)$. Note that the lowering generators E_{ji} in (2.70b) are not necessarily the Hermitean conjugates of the raising generators E_{ij} in (2.70a), for $|i - j| > 1$, because we have replaced q by q^{-1} in order to obtain more convenient commutators between the raising and lowering generators. For the case of symmetric representations, however, the Hermiticity conditions $E^{\dagger}_{ij} = E_{ji}$ are satisfied as we shall see when we realize these generators in terms of q-boson operators, which are invariant under $q \leftrightarrow q^{-1}$.

The Serre relations (2.67) can be restated in the simple form

$$[E_{i,i+1}, E_{i,i+2}] = 0, \quad [E_{i,i+1}, E_{i-1,i+1}] = 0,$$
$$[E_{i+1,i}, E_{i+2,i}] = 0, \quad [E_{i+1,i}, E_{i+1,i-1}] = 0. \tag{2.71}$$

Our definition of the Cartan-Weyl generators is guided by a wish to maintain the validity of the Jordan map for all generators other than those of the Cartan subalgebra; we may, however, define Cartan-Weyl generators in ways other than (2.70a,2.70b), and one such way has been used by several authors, including Jimbo [104], and developed by Quesne [105]. We define generators $\mathcal{E}_{ij}$ recursively by q-commutators:

$$\mathcal{E}_{ij} \stackrel{\text{def}}{=} \mathcal{E}_{i,i+1}\mathcal{E}_{i+1,j} - q^{\frac{1}{2}}\mathcal{E}_{i+1,j}\mathcal{E}_{i,i+1} \quad \text{for} \quad i = 1, \ldots, n-2,\ j = i+2, \ldots, n$$
$$\mathcal{E}_{ji} \stackrel{\text{def}}{=} \mathcal{E}_{j,i+1}\mathcal{E}_{i+1,i} - q^{-\frac{1}{2}}\mathcal{E}_{i+1,i}\mathcal{E}_{j,i+1} \quad \text{for} \quad i = 1, \ldots, n-2,\ j = i+2, \ldots, n.$$

In this case the Serre relations are equivalent to

$$\mathcal{E}_{i,i+1}\mathcal{E}_{i,i+2} - q^{-\frac{1}{2}}\mathcal{E}_{i,i+2}\mathcal{E}_{i,i+1} = 0 = \mathcal{E}_{i+1,i+2}\mathcal{E}_{i,i+2} - q^{\frac{1}{2}}\mathcal{E}_{i,i+2}\mathcal{E}_{i+1,i+2},$$
$$\mathcal{E}_{i+2,i}\mathcal{E}_{i+1,i} - q^{\frac{1}{2}}\mathcal{E}_{i+1,i}\mathcal{E}_{i+2,i} = 0 = \mathcal{E}_{i+2,i}\mathcal{E}_{i+1,i} - q^{-\frac{1}{2}}\mathcal{E}_{i+2,i}\mathcal{E}_{i+2,i+1}.$$

Further properties of these generators may be found in [105]. They are related to the definitions (2.70a,2.70b) by

$$\mathcal{E}_{ij} = q^{-\frac{1}{2}(E_{i+1,i+1}+\ldots+E_{j-1,j-1})}E_{ij}, \quad j > i+1$$
$$\mathcal{E}_{ji} = q^{\frac{1}{2}(E_{i+1,i+1}+\ldots+E_{j-1,j-1})}E_{ji}, \quad j > i+1.$$

It was shown by Jimbo [104] that irreps of $\mathcal{U}_q(\mathfrak{gl}(n))$ are deformations of the classical ($q = 1$) irreps, and also [93] that a q-analog of the Gel'fand-Weyl basis exists for $\mathcal{U}_q(\mathfrak{gl}(n))$; explicit matrix elements were given by Ueno et al. [106], which were seen to be deformations of those for $\mathfrak{gl}(n)$. Rosso [70] and (independently) Lusztig [69] showed that these were the only finite-dimensional irreps of $\mathcal{U}_q(\mathfrak{gl}(n))$. Specifically, and more generally, it was shown that all finite-dimensional representations of the deformation $\mathcal{U}_q(\mathfrak{g})$ of the universal enveloping algebra of $\mathfrak{g}$, where $\mathfrak{g}$ is any complex simple Lie algebra, are completely reducible and that the irreps can be classified in terms of highest weights; in particular they are deformations of the irreps of the classical irreps of $\mathcal{U}(\mathfrak{g})$. We do not prove these very general results here, but refer to the original papers [70, 69]. Recall that a representation is **completely reducible** if it is equivalent to the direct sum of irreducible representations.

It is useful to state the explicit matrix elements in any irrep of $\mathcal{U}_q(\mathfrak{u}(n))$. For $U(n)$ these matrix elements were first written down by Gel'fand and Zetlin [107] in 1950, and a description and discussion may be found in the review article by Louck [108] (see also [109, 110]). From these matrix elements we may deduce those of the Chevalley generators of $\mathcal{U}_q(\mathrm{u}(n))$. Let us denote the orthonormal Gel'fand-Weyl basis vectors by $|(m)\rangle$, with irrep labels $[\mathbf{m}] = [m_{1n}, \ldots, m_{nn}]$, where the Gel'fand-Weyl notation is explained in §2.6. The dimension of this irrep is given by the Weyl dimension formula:

$$D([\mathbf{m}]) = \frac{\prod_{i<j}(m_{in} - m_{jn} - i + j)}{1!2!\ldots(n-1)!}. \tag{2.72}$$

The Chevalley generators $e_i = E_{i,i+1}$, corresponding to the simple roots, effect the following transformation of the basis:

$$e_i|(m)\rangle = \sum_{(m')} \langle (m')|\, e_i\, |(m)\rangle\, |(m')\rangle,$$

with the matrix elements being zero, $\langle (m')|\, e_i\, |(m)\rangle = 0$, unless

$$[m'_{1i}, \ldots, m'_{ii}] = [m_{1i} + \delta_{1\tau_i}, \ldots, m_{ii} + \delta_{i\tau_i}] \tag{2.73}$$

where τ_i is any one of $1, \ldots, i$. In other words, the matrix elements vanish unless the final labels (m') agree with the initial labels (m) in all rows except row i, where the allowed final labels are those which can be obtained from the initial labels by adding 1 in all possible positions.

Let us also define the **partial hook length**:

$$p_{ij} \stackrel{\text{def}}{=} m_{ij} + j - i,$$

for each $1 \leqslant i \leqslant j \leqslant n$, and now we may state:

THEOREM **2.74** *In an irreducible representation of $\mathcal{U}_q(\mathrm{u}(n))$ the commuting diagonal generators E_{ii} have eigenvalues given by*

$$E_{ii}|(m)\rangle = \left(\sum_{j=1}^{i} m_{ji} - \sum_{j=1}^{i-1} m_{j,i-1}\right)|(m)\rangle,$$

for $i = 1, \ldots, n$ (with $m_{10} \equiv 0$). The matrix elements of the raising generators e_i, where $1 \leqslant i \leqslant n$, are given by

$$\langle (m')|\, e_i\, |(m)\rangle = \left(\frac{-\prod_{s=1}^{i+1}[p_{s,i+1} - p_{\tau_i i} - 1]_q \prod_{s=1}^{i-1}[p_{s,i-1} - p_{\tau_i i}]_q}{\prod_{\substack{s=1\\ s\neq\tau_i}}^{i}[p_{si} - p_{\tau_i i} - 1]_q[p_{si} - p_{\tau_i i}]_q} \right)^{\frac{1}{2}} \tag{2.75}$$

for some specific assignment of one of the integers $1, \ldots, i$ to τ_i, and where the pattern (m') satisfies the conditions (2.73).

As in the classical case, each factor in these rather complicated matrix elements can be accounted for by general structural properties. The matrix elements of f_i, the lowering generators, are obtained by using $f_i = (e_i)^\dagger$, to give

$$\langle (m)|\, f_i\, |(m')\rangle = \langle (m')|\, e_i\, |(m)\rangle .$$

The matrix elements of the remaining Cartan-Weyl generators can be deduced from their definition, choosing for example (2.70a,2.70b), and can also be determined from the classical matrix elements, given by Gel'fand and Zetlin [107], by replacing integer factors with corresponding q-integers together with explicit q-factors which depend on the precise definition of these generators. Hence, the matrix elements of the Cartan-Weyl generators are not in general invariant under $q \leftrightarrow q^{-1}$, except for the diagonal generators and the Chevalley generators e_i and f_i.

EXAMPLE **2.76** The simplest example is for $\mathcal{U}_q(\mathfrak{u}(2))$ for which the matrix elements of $e_1 = E_{12}, f_1 = E_{21}$ and E_{11}, E_{22} are given by

$$\begin{aligned}
E_{12}\left|\begin{smallmatrix} m_{12} & & m_{22} \\ & m_{11} & \end{smallmatrix}\right\rangle &= \sqrt{[m_{12}-m_{11}]_q[m_{11}+1-m_{22}]_q}\left|\begin{smallmatrix} m_{12} & & m_{22} \\ & m_{11}+1 & \end{smallmatrix}\right\rangle \\
E_{21}\left|\begin{smallmatrix} m_{12} & & m_{22} \\ & m_{11} & \end{smallmatrix}\right\rangle &= \sqrt{[m_{11}-m_{22}]_q[m_{12}+1-m_{11}]_q}\left|\begin{smallmatrix} m_{12} & & m_{22} \\ & m_{11}-1 & \end{smallmatrix}\right\rangle \\
E_{11}\left|\begin{smallmatrix} m_{12} & & m_{22} \\ & m_{11} & \end{smallmatrix}\right\rangle &= m_{11}\left|\begin{smallmatrix} m_{12} & & m_{22} \\ & m_{11} & \end{smallmatrix}\right\rangle \\
E_{22}\left|\begin{smallmatrix} m_{12} & & m_{22} \\ & m_{11} & \end{smallmatrix}\right\rangle &= (m_{12}+m_{22}-m_{11})\left|\begin{smallmatrix} m_{12} & & m_{22} \\ & m_{11} & \end{smallmatrix}\right\rangle .
\end{aligned}$$

The matrix elements of e_2 and f_2 occur in our study of the Borel-Weil construction of irreps for $\mathcal{U}_q(\mathfrak{u}(3))$ in Chapter 7, where we give explicit formulas; see Eqn. (7.63), p. 237, in which the matrix elements of E_{23} are given for $q = 1$, from which the q-analog matrix elements may be deduced by replacing integer factors by corresponding q-integer factors.

The R-matrix in the fundamental representation is given for $\mathcal{U}_q(\mathfrak{u}(n))$ by $R = P\mathcal{R}$ where P is the permutation operator and

$$\mathcal{R} = q^{\frac{1}{2}} \sum_i e_{ii} \otimes e_{ii} + \sum_{i \neq j} e_{ij} \otimes e_{ji} + (q^{\frac{1}{2}} - q^{-\frac{1}{2}}) \sum_{i<j} e_{jj} \otimes e_{ii},$$

where e_{ij} is the $n \times n$ matrix with the only nonzero entry being 1 in the i,j position. The matrix R satisfies $R^2 = (q^{\frac{1}{2}} - q^{-\frac{1}{2}})R + I$, a property satisfied by elements of the Hecke algebra $H_n(q)$ defined in (2.79), indeed, one can determine from R that the centralizer of $\mathcal{U}_q(\mathfrak{su}(n))$ in an n-fold tensor product space obeys the Hecke algebra [30].

2.5.1 The q-Boson Construction for $\mathcal{U}_q(\mathfrak{u}(n))$

We recall now from §2.4 that the Jordan map using boson operators is a completely general construction valid for constructing all unitary irreps of all compact Lie groups. Moreover this construction extends to all unit tensor operators, so that this procedure is a very powerful one indeed. It is natural to pose the question as to whether the Jordan construction can encompass all compact quantum groups as well. The answer is, as yet, not fully known (to our knowledge) but for the quantum unitary group $\mathcal{U}_q(\mathfrak{u}(n))$ the answer is *yes*. The q-boson Jordan map for the fundamental $\mathcal{U}_q(\mathfrak{u}(n))$ irrep succeeds fully.

Let us indicate how this construction works. The Jordan map using the n-component q-boson operators a_i^q (creation) and $\overline{a}_i^q$ (destruction), for $i = 1,\ldots,n$, yields immediately the correct quantum group generators for the raising/lowering generators $\mathfrak{n}_\pm$, that is,

$$E_{ij} = a_i^q \overline{a}_j^q, \quad i,j = 1,\ldots,n,\ i \neq j. \tag{2.77}$$

Just as for $\mathcal{U}_q(\mathfrak{su}(2))$ in §2.4, the Jordan map for the diagonal quantum group generators (comprising the Cartan algebra $\mathfrak{h}$) is different, and uses the number operators N_i^q, $i = 1,\ldots,n$, instead of the q-boson operators $a_i^q \overline{a}_i^q$ which yield now the q-integer valued operators $[N_i^q]_q$. For $\mathcal{U}_q(\mathfrak{su}(n))$, we put

$$h_i = \tfrac{1}{2}(N_i^q - N_{i+1}^q), \quad i = 1,\ldots,n-1$$

and for $\mathcal{U}_q(\mathfrak{u}(n))$ we have $E_{ii} = N_i^q$ for $1 = 1,\ldots,n$.

It is easily verified that the q-boson operator realization satisfies the defining relations (2.66, 2.67). In addition, the relations (2.70a) and (2.70b) defining the Cartan-Weyl generators are also satisfied as are, by inspection, the Serre relations in the form (2.71). As a result, just as in the prototype example $\mathcal{U}_q(\mathfrak{su}(2))$, we can now obtain all totally symmetric unitary irreps (symmetric because they correspond to a Young pattern of one row) as polynomials in the q-boson creation operators acting on the vacuum ket $|0\rangle$. The special properties of symmetric irreps can be ascribed to an extra property satisfied by the realization (2.77), additional to the $\mathcal{U}_q(\mathfrak{u}(n))$ relations, namely $E_{i,i+1}E_{i+1,i} = [N_i^q]_q[N_{i+1}^q]_q = [E_{ii}]_q[E_{i+1,i+1}]_q$.

The orthogonal polynomials in $\mathfrak{F}^n$ are in fact monomials which we can write out explicitly; in the Gel'fand-Weyl notation the symmetric irreps of $\mathcal{U}_q(\mathfrak{u}(n))$ are labelled by one integer m_{1n}, with all other labels zero: $m_{2n} = 0 = \ldots = m_{nn}$. The orthonormal states are labelled by integers $0 \leqslant m_{11} \leqslant m_{12} \leqslant \cdots \leqslant m_{1n}$ and have the explicit form

$$|m_{11}\, m_{12} \ldots m_{1n}\rangle = \prod_{i=1}^{n} \frac{(a_i^q)^{m_{1i}-m_{1,i-1}}}{([m_{1i} - m_{1,i-1}]_q!)^{\frac{1}{2}}} |0\rangle \tag{2.78}$$

where $m_{10} \equiv 0$. These states are eigenvectors of the diagonal generators or, equivalently, of the number operators N_i^q. The irreps are of dimension $\frac{(n+m_{1n}-1)!}{(n-1)!m_{1n}!}$, which is the dimension of the subspace of $\mathfrak{F}^n$ consisting of homogeneous polynomials in $a_1^q, \ldots, a_n^q$ of total degree m_{1n}.

Matrix elements of the generators are readily calculated, and we find

$$\begin{aligned} E_{ii}|m_{11}\, m_{12} \ldots m_{1n}\rangle &= (m_{1i} - m_{1,i-1})|m_{11}\, m_{12} \ldots m_{1n}\rangle \\ E_{i,i+1}|m_{11} \ldots m_{1n}\rangle &= \sqrt{[m_{1,i+1} - m_{1i}]_q [m_{1i} - m_{1,i-1} + 1]_q}\, |m_{11} \ldots, m_{1i}+1, \ldots m_{1n}\rangle \\ E_{i+1,i}|m_{11} \ldots m_{1n}\rangle &= \sqrt{[m_{1,i+1} - m_{1i} + 1]_q [m_{1i} - m_{1,i-1}]_q}\, |m_{11} \ldots, m_{1i}-1, \ldots m_{1n}\rangle, \end{aligned}$$

which are special cases of the matrix elements given in Theorem 2.74. This construction of the symmetric irreps of $\mathcal{U}_q(\mathfrak{u}(n))$ is a straightforward generalization of similar results for $U(n)$, which have been investigated by Louck [109] in connection with harmonic oscillators in n-dimensional space, and also by Moshinsky [111]. The interpretation as q-harmonic oscillator states is considered in §8.1.

2.6 Appendix: Gel'fand-Weyl States and Young Frames

To label the individual vectors in a $U(2)$ irrep, and more generally in a $U(n)$ irrep or a quantum group irrep, requires a systematic notation such as that devised by Gel'fand based on the Weyl branching theorem. This notation greatly simplifies the task of comprehending the relationship of the theory to that of the symmetric group and the generalization to $U(n)$ and, accordingly, also to quantum groups. The role of the symmetric group is replaced for quantum groups, however, by that of the Hecke algebra, as was shown by Jimbo [104]. The Hecke algebra $H_n(q)$ is generated by the $n-1$ elements $h_1, \ldots h_{n-1}$ satisfying

$$\begin{aligned} h_i h_{i+1} h_i = h_{i+1} h_i h_{i+1}, \qquad h_i^2 = (q-1)h_i + q, \\ h_i h_j = h_j h_i, \quad |i-j| > 1. \end{aligned} \tag{2.79}$$

We can compare these relations with those defining the braid group in (1.3). For $q = 1$ these relations are appropriate to the symmetric group with h_i being the transposition $(i, i+1)$.

We review briefly in this section the Gel'fand-Weyl notation, and refer to Biedenharn and Louck [66, §5.9, p. 230] for a more complete description. We begin by defining Young frames, Weyl patterns and then Gel'fand patterns.

A **Young frame** $Y_{[\lambda]}$ of **shape** $[\lambda] = [\lambda_1\, \lambda_2 \cdots \lambda_n]$, where the λ_i are nonnegative integers satisfying $\lambda_1 \geqslant \cdots \geqslant \lambda_n$, is a diagram consisting of λ_1 boxes in row $1, \ldots, \lambda_n$ boxes in row n, arranged as illustrated in Figure 2.80.

A **Weyl pattern** is a Young frame in which the boxes have been "filled in" with integers selected from $1, 2, \ldots, n$. A Weyl pattern is **standard** if the sequence of integers appearing in each row of $Y_{[\lambda]}$ is nondecreasing as read from left to right

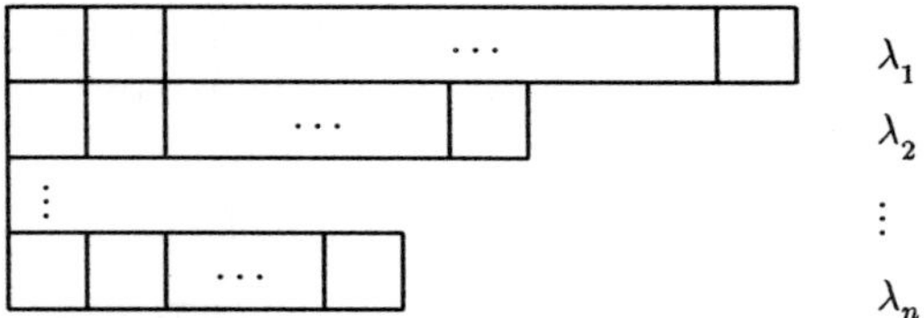

Figure 2.80: A Young frame of shape $[\lambda_1, \lambda_2, \ldots, \lambda_n]$.

and the sequence of integers appearing in each column is strictly increasing as read from top to bottom. The **weight** of a Weyl pattern is defined to be the row vector $(w_1, w_2, \ldots, w_n)$, where w_k equals the number of times the integer k appears in the pattern. If $\lambda_1 + \lambda_2 + \cdots + \lambda_n = N$, then also $w_1 + w_2 + \cdots + w_n = N$. We shall call $[\lambda]$ a **partition** of N into n parts or, more often, a partition when N is unspecified. We generally count the zeros in determining the parts of a partition. For example, the partitions of 4 into 3 parts are $[4\ 0\ 0], [3\,1\,0], [2\ 2\ 0]$, and $[2\ 1\ 1]$.

A **Gel'fand pattern**[5] is a triangular array of n rows of integers, there being one entry in the first row, two entries in the second row, ... and n entries in the nth row. The entries in each row $1, 2, \ldots, n-1$, are arranged so as to fall between the entries in the row above, as illustrated below:

$$(m) = \begin{pmatrix} m_{1n} \quad m_{2n} \quad \cdots \qquad\qquad m_{nn} \\ \ddots \quad \ddots \\ m_{13} \quad m_{23} \quad m_{33} \\ m_{12} \quad m_{22} \\ m_{11} \end{pmatrix}.$$

The entries $m_{ij}, i \leqslant j = 1, 2 \ldots, n$ in this array are integers which satisfy the inequalities

$$m_{ij} \geqslant m_{i,j-1} \geqslant m_{i+1,j}\,, \tag{2.81}$$

which are expressions of the Weyl [112] branching law for the general linear group. An (integral) unitary irrep in $U(n)$ (and therefore also in $\mathcal{U}_q(\mathrm{u}(n))$) is labelled by n integers $m_{1n} \geqslant m_{2n} \geqslant \ldots \geqslant m_{nn}$, not necessarily positive. A vector in an orthonormal basis of the space carrying an irrep of $U(n)$ may be identified uniquely by the triangular Gel'fand pattern and is denoted $|(m)\rangle$, where (m) is a Gel'fand pattern. For $U(n)$ the inequalities (2.81) specify that the vector $|(m)\rangle$ belongs to the irrep $[m_{1j}, \ldots, m_{jj}]$ in $U(j)$ for $j = n, n-1, \ldots, 1$, which uniquely identifies the vector $|(m)\rangle$ using the subgroup chain $U(n) \supset U(n-1) \supset \ldots \supset U(1)$.

[5]These patterns were first introduced by Gel'fand and Zetlin [107] to enumerate basis vectors in the carrier space of an irrep of the general linear group.

For irreps of $SU(n)$ we have $m_{nn} = 0$, however, it will be helpful to introduce a simplifying technical device in this case: we drop the unimodular restriction $m_{nn} = 0$ and instead let m_{nn} be any integer and consider $U(n)$. To return to $SU(n)$ we impose the equivalence relation

$$[m_{in} + k] \simeq [m_{in}],\ i = 1, \ldots n,\ k \in \mathbb{Z},$$

which allows one to formally use $U(n)$ for various manipulations while actually discussing $SU(n)$. We refer to the invariance of matrix elements of $SU(n)$ tensor operators with respect to this equivalence relation as **shift invariance**.

There is a one-to-one correspondence between the set of Gel'fand patterns (m) with an nth row $[m_{1n} m_{2n} \cdots m_{nn}]$ (with $m_{nn} \geqslant 0$) and the set of standard Weyl patterns of this shape, which can be described as follows. The shape of the frame is $[m_{1n} m_{2n} \cdots m_{nn}]$, and the rows of the frame are filled in according to the following rules, read along the diagonals of the Gel'fand pattern:

$$\begin{array}{lll}
\text{Row 1:} & m_{11}\ 1's,\ m_{12} - m_{11}\ 2's,\quad \ldots, & m_{1n} - m_{1n-1}\ n's \\
\text{Row 2:} & m_{22}\ 2's,\ m_{23} - m_{22}\ 3's,\quad \ldots, & m_{2n} - m_{2n-1}\ n's \\
\vdots & & \\
\text{Row } j: & m_{jj}\ j's,\ m_{jj+1} - m_{jj}\ (j+1)'s, \ldots, & m_{jn} - m_{jn-1}\ n's \\
\vdots & & \\
\text{Row } n: & & m_{nn}\ n's.
\end{array}$$

The **weight** of a Gel'fand pattern (m) is the row vector $(w_1, w_2, \ldots, w_n)$, where

$$w_j \stackrel{\text{def}}{=} \sum_{i=1}^{j} m_{ij} - \sum_{i=1}^{j-1} m_{ij-1}.$$

Clearly, this definition of weight coincides with that given earlier for a standard Weyl pattern. The constraint in a standard Weyl pattern that each row (column) should comprise a set of nondecreasing (strictly increasing) nonnegative integers is realized in a Gel'fand pattern by the "geometric" rule that the integers (m_{ij}) satisfy the inequalities (2.81).

Now let us relate this notation to properties of the symmetric group S_n. The important pattern results for S_n are as follows:

1. The set of irreps of S_n is in one-to-one correspondence with the set of partitions $\{[\lambda]\}$ of n into n parts;

2. The set of basis vectors of the carrier space of irrep $[\lambda]$ of S_n is in one-to-one correspondence with the set of standard Young patterns of shape $[\lambda]$ having weight $(1, 1, \ldots, 1)$ (the number of basis vectors or the number of standard patterns is then the dimension of the irrep).

This result may, of course, also be expressed in terms of Gel'fand patterns. For example, the irreps of S_3 are enumerated by the partitions of 3 into 3 parts $[3\,0\,0], [2\,1\,0]$, and $[1\,1\,1]$. The standard Young patterns of weight (1, 1, 1) having these shapes, respectively, are shown in Figure 2.82.

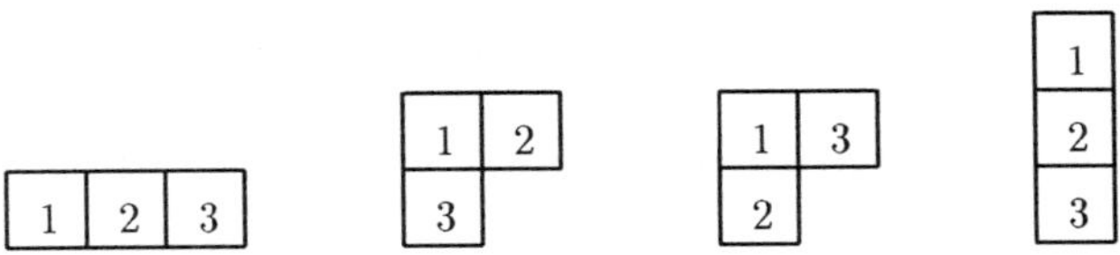

Figure 2.82: Young patterns of weight $(1, 1, 1)$.

Thus, the irreps $[3\,0\,0], [2\,1\,0]$, and $[1\,1\,1]$ are of dimensions 1, 2, and 1, respectively. These same results are enumerated by the Gel'fand patterns:

$$\begin{pmatrix} 3 & & 0 & & 0 \\ & 2 & & 0 & \\ & & 1 & & \end{pmatrix}, \quad \begin{pmatrix} 2 & & 1 & & 0 \\ & 2 & & 0 & \\ & & 1 & & \end{pmatrix}, \quad \begin{pmatrix} 2 & & 1 & & 0 \\ & 1 & & 1 & \\ & & 1 & & \end{pmatrix}, \quad \begin{pmatrix} 1 & & 1 & & 1 \\ & 1 & & 1 & \\ & & 1 & & \end{pmatrix}.$$

This notation applies to irreps of $\mathfrak{su}(2)$ and $\mathcal{U}_q(\mathfrak{su}(2))$ using the fact that:

1. The set of irreps is in one-to-one correspondence with the set of partitions $[2j\ 0]$, $j = 0, \frac{1}{2}, 1 \ldots$;
2. The set of basis vectors of the carrier space of irrep $[2j\ \ 0]$ is in one-to-one correspondence with the set of Gel'fand patterns having the partition $[2j\ 0]$:

$$\begin{pmatrix} 2j & & 0 \\ & j+m & \end{pmatrix}, \quad m = -j, -j+1, \ldots, j. \tag{2.83}$$

Observe that the inequality $2j \geqslant j + m \geqslant 0$ embodies in a natural way the fact that the projection quantum number m runs over the values $m = -j, \ldots, j$. The Weyl pattern corresponding to the Gel'fand pattern (2.83) is the one-row pattern shown in Figure 2.84.

One sees at once that the boson operator form of the basis vector,

$$|jm\rangle = \frac{(a_1)^{j+m}(a_2)^{j-m}}{\sqrt{(j+m)!(j-m)!}}\,|0\rangle,$$

is a direct transcription of this pattern, replacing "1" by a_1, "2" by a_2, and normalizing using the boson inner product. The $U(n)$ generalization of this normalization is discussed in [66], using the concept of "entanglement" of bosons.

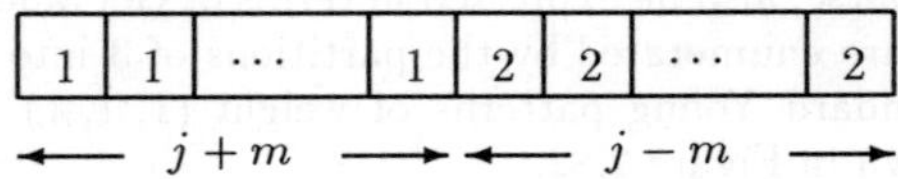

Figure 2.84: The Weyl pattern corresponding to $\mathfrak{su}(2)$ boson states.

Consider next the standard Weyl pattern of two rows, corresponding to the Gel'fand pattern

$$\begin{pmatrix} m_{12} & & m_{22} \\ & m_{11} & \end{pmatrix}, \tag{2.85}$$

where $m_{12} \geqslant m_{11} \geqslant m_{22}$, shown in Figure 2.86.

$\longleftarrow m_{22} \longrightarrow \longleftarrow m_{11}-m_{22} \longrightarrow \longleftarrow m_{12}-m_{11} \longrightarrow$

1	1	⋯	1	1	1	⋯	1	2	2	⋯	2
2	2	⋯	2								

Figure 2.86: The Weyl pattern corresponding to $\mathfrak{u}(2)$ boson states.

The mapping from Weyl patterns to Gel'fand patterns to boson operators is given by:

$$\boxed{1} \rightarrow \begin{pmatrix} 1 & & 0 \\ & 1 & \end{pmatrix} \rightarrow a_1^1, \qquad \boxed{2} \rightarrow \begin{pmatrix} 1 & & 0 \\ & 0 & \end{pmatrix} \rightarrow a_2^1,$$

$$\begin{array}{|c|}\hline 1 \\ \hline 2 \\ \hline \end{array} \rightarrow \begin{pmatrix} 1 & & 1 \\ & 1 & \end{pmatrix} \rightarrow \det \begin{pmatrix} a_1^1 & a_1^2 \\ a_2^1 & a_2^2 \end{pmatrix} \stackrel{\text{def}}{=} a_{12}. \tag{2.87}$$

The last Weyl pattern corresponds to antisymmetrized bosons made up of two independent bosons a_i^1 and a_i^2 $(i=1,2)$.

By using the correspondence (2.87), we obtain the following boson state vector, corresponding to the Gel'fand pattern $\left| \begin{smallmatrix} m_{12} & & m_{22} \\ & m_{11} & \end{smallmatrix} \right\rangle$ and the Weyl pattern in Figure 2.86:

$$\left| \begin{matrix} m_{12} & & m_{22} \\ & m_{11} & \end{matrix} \right\rangle = \mathcal{N}^{-\frac{1}{2}} (a_{12})^{m_{22}} (a_1^1)^{m_{11}-m_{22}} (a_2^1)^{m_{12}-m_{11}} |0\rangle, \tag{2.88}$$

where the normalization factor is given by

$$\mathcal{N} = \frac{(m_{12}+1)!(m_{11}-m_{22})!(m_{12}-m_{11})!(m_{22})!}{(m_{12}-m_{22}+1)!}$$

The angular momentum labels for the states (2.88) are

$$j = \tfrac{1}{2}(m_{12} - m_{22}), \quad m = m_{11} - \tfrac{1}{2}(m_{12} + m_{22}). \tag{2.89}$$

The $2m_{22}$ antisymmetric paired bosons commute with the angular momentum generators, that is, a_{12} is invariant under unitary unimodular transformations.

It must be emphasized that the orthonormal Gel'fand-Weyl states in $U(n)$ do not correspond for general n to the mapping shown in (2.87); indeed, such a mapping from Weyl patterns leads always to monomials in the boson operators, whereas basis vectors in $U(n)$ are generally special polynomial functions of the boson operators. For example the basis vectors for $U(3)$ can be expressed as terminating hypergeometric functions [110].

Consider now the generalization to quantum groups of the results reviewed in this section. The Gel'fand-Weyl notation for basis vectors applies without change, because the irreps of $\mathcal{U}_q(\mathfrak{u}(n))$ are smooth deformations of those of $U(n)$, as explained in §2.5. Moreover, we see that the monomial basis states for $\mathcal{U}_q(\mathfrak{su}(2))$, given in (2.48), can be obtained directly from the corresponding Weyl patterns, just as for $\mathfrak{su}(2)$, by mapping "i" in the Weyl pattern to a_i^q in the q-boson polynomial for $i = 1, 2$. Indeed, this mapping is valid for general n for the "symmetric" irreps of $\mathcal{U}_q(\mathfrak{u}(n))$, as our discussion in §2.5.1 shows. In particular, the monomial basis states (2.78) can be obtained immediately from the corresponding Weyl patterns. This mapping does not, however, extend to irreps for which the Weyl pattern has two or more rows. We consider the $\mathcal{U}_q(\mathfrak{u}(2))$ case in detail in §2.4.4, where we determine the q-analog of the determinant a_{12} and find an expression in terms of q-boson operators with explicit q-dependent factors. We will see that the appropriate operators which correspond to the boxes of the Weyl patterns are elementary q-tensor operators which do not commute, and a_{12} generalizes to a quantum determinant (see Theorem 4.55, p. 150). The symmetric group is replaced by the Hecke algebra, which acts on non-commuting operators, and generalizes the role of the symmetric group; the carrier space is reduced using the q-analog of the Young symmetrizers of the symmetric group (Jimbo [104]). In terms of q-boson operators this requires the construction of the q-analog of antisymmetric tensors such as $a_{12\ldots n}$ with which we project into the irreducible subspaces, and uses properties of elementary tensor operators, as discussed in Chapters 3 and 4.

2.7 Appendix: Properties of q-Numbers

We have introduced q-integers in §2.1, Eqn. (2.3) and have observed that they play a role which extends beyond mere notational convenience; indeed, the properties of q-integers, which appear for example in the matrix elements shown in Lemma 2.16, are fundamental in the derivation of the representation matrices. It is remarkable that many, although not all, matrix elements of operators (such as generators) and coupling coefficients (such as Racah coefficients) for quantum groups can be obtained by the simple recipe of replacing integer factors (n) by the corresponding q-integers $[n]_q$. We investigate now in more detail therefore the properties of q-integers, particularly those which can be considered fundamental in determining quantum group properties.

Firstly we point out that q-integers, as defined by (2.3), form an additive group:

LEMMA **2.90** *Under the addition law*

$$[m]_q \oplus [n]_q \stackrel{\text{def}}{=} q^{-\frac{n}{2}}[m]_q + q^{\frac{m}{2}}[n]_q = [m+n]_q, \quad m,n \in \mathbb{Z}$$

we have the isomorphism $\{\mathbb{Z}, +\} \cong \{[n]_q, \oplus \mid n \in \mathbb{Z}\}$.

The set of q-integers does not form a ring, although we can perform a multiplication in the following sense:

$$[n]_q \lim_{q \to q^n} [n']_q = [nn']_q.$$

Nevertheless, we do have several useful properties, some already mentioned such as $[-n]_q = -[n]_q$ from which follows $[0]_q = 0$. Conversely, $[n]_q = 0$ implies $n = 0$, provided $q \in \mathbb{R}^+$. The set of q-integers has no zero divisors for $q \in \mathbb{R}^+$, since $[n]_q[m]_q = 0$ implies $n = 0$ or $m = 0$.

There are numerous identities satisfied by q-integers which, however, can be stated more generally. In order to do this we extend the concept of a q-integer to a q-number $[x]_q$, where $x \in \mathbb{R}$, by the formula

$$[x]_q \stackrel{\text{def}}{=} \frac{q^{\frac{x}{2}} - q^{-\frac{x}{2}}}{q^{\frac{1}{2}} - q^{-\frac{1}{2}}}. \tag{2.91}$$

At $q = 1$ we have $[x]_q = x$. We will find that this extension to real arguments is necessary in Chapter 6 when considering complex values of q such that q is a root of unity, but let us consider now $q \in \mathbb{R}^+$. By means of the identity

$$[x+y]_q = q^{\frac{x}{2}}[y]_q + [x]_q q^{-\frac{y}{2}} = q^{-\frac{x}{2}}[y]_q + [x]_q q^{\frac{y}{2}}, \tag{2.92}$$

we deduce

LEMMA **2.93** *Under the addition law* $[x]_q \oplus [y]_q \stackrel{\text{def}}{=} [x+y]_q$ *the set of q-numbers* $\{[x]_q \mid x \in \mathbb{R}\}$ *forms an additive group which is isomorphic to* $\mathbb{R}$.

It is useful to write $\theta = \frac{1}{2}\log q$, for then we can express $[x]_q$ in terms of a hyperbolic sin function:

$$[x]_q = \frac{\sinh x\theta}{\sinh \theta},$$

from which several properties follow immediately. Firstly, we see that $[x]_q$ is unbounded above and below as a function of x. Secondly, since sinh is a strictly increasing function on $\mathbb{R}$, we find that the mapping $[\ ]_q : \mathbb{R} \to \mathbb{R}$ is surjective and, furthermore,

$$x < y \iff [x]_q < [y]_q,$$

showing that q-numbers (and hence also q-integers) preserve the order of the real numbers, that is, q-numbers form an ordered group. Thirdly, from the inequality $|\sinh x/x| \geqslant 1$ we determine that

$$|[x]_q| = \left|\frac{\sinh x\theta}{\sinh\theta}\right| \geqslant c|x|$$

for all $x \in \mathbb{R}$, where $c = \theta/\sinh\theta$ is a constant which depends on q. For q-integers, however, we may determine a more precise bound:

LEMMA **2.94** *$[n]_q \geqslant n$ for all $n \in \mathbb{N}$, with equality for all $n \in \mathbb{N}$ iff $q = 1$.*

PROOF: We differentiate $[n]_q = \sum_{r=1}^{n} q^{\frac{1}{2}(n-2r+1)}$ with respect to q to find

$$\frac{d[n]_q}{dq} = \tfrac{1}{2}\sum_{r=1}^{n}(n-2r+1)q^{\frac{1}{2}(n-2r-1)} = \tfrac{1}{2}(q-1)\sum_{r=1}^{n-1} r(n-r)q^{\frac{1}{2}(n-2r-3)},$$

and, since each term under the last sum is positive (for $q > 0$), we find that $[n]_q$ has a minimum as a function of q at $q = 1$.□

This result can be used to demonstrate the convergence of q-analog functions formed by replacing n by $[n]_q$ in the series representation of the function; an example is the q-exponential function.

2.7.1 Symmetries and Identities of q-Numbers

We will frequently use the many identities and symmetries of q-numbers, including identities such as the q-binomial theorem which are best expressed in terms of q-commuting quantities; we shall introduce elements x, y of a noncommutative algebra satisfying $q\,xy = yx$. Such noncommuting "coordinates" play a central role in the formulation of quantum groups, and will appear naturally in our development in Chapter 4 as components of tensor operators. In the following we consider general q-numbers $[a]_q$ with $a \in \mathbb{R}$, $q \in \mathbb{R}^+$. For completeness, let us point out that in the literature many of these results are expressed in terms of related q-numbers, or q-integers, defined as follows:

$$[n;q] \stackrel{\text{def}}{=} \frac{1-q^n}{1-q} = q^{\frac{1}{2}(n-1)}[n]_q$$

$$(q^n;q)_1 \stackrel{\text{def}}{=} (1-q)q^{\frac{1}{2}(n-1)}[n]_q.$$

The first notation $[n;q]$ is used by Exton [113], and is often shortened to $[n]$, (Jackson, Cigler [114]) or q_n (Feinsilver [115]). The second notation is a special case of the notation introduced in (2.120) and (2.121) below, and is used in connection with basic hypergeometric functions.

The symmetry $[a]_q = [a]_{q^{-1}}$ implies various identities, for example (2.92) has two forms which reflect this invariance. Rational functions $f(q)$ of q which are invariant under $q \leftrightarrow q^{-1}$, such as functions of q-integers without explicit q-factors, satisfy $f'(1) = 0$, which provides a means, in some cases, of relating various $3n-j$ symbols. We will also encounter functions which are either even or odd as functions of $q^{\frac{1}{2}}$, and invariance under $q^{\frac{1}{4}} \leftrightarrow iq^{-\frac{1}{4}}$ has been considered in the development of the q-symplecton [116].

Now let us mention several useful identities, the first of which is the following, in which products of q-numbers are added:

$$[a]_q[b-c]_q + [b]_q[c-a]_q + [c]_q[a-b]_q = 0. \tag{2.95}$$

In the case when $c = a + b$, this cyclic relation becomes

$$[a]_q^2 - [b]_q^2 = [a+b]_q[a-b]_q.$$

We may express a product of q-numbers as a sum of q-numbers:

$$[2a+1]_q[2b+1]_q = [2|a-b|+1]_q + [2(|a-b|+1)+1]_q + \cdots + [2(a+b)+1]_q.$$

Some other useful relations are:

$$[a]_q[b]_q[b-a]_q + [b]_q[c]_q[c-b]_q + [c]_q[a]_q[a-c]_q + [b-a]_q[c-b]_q[a-c]_q = 0,$$
$$[a+b+c]_q - [a+b-c]_q - [b+c-a]_q - [c+a-b]_q = [a]_q[b]_q[c]_q\left(q^{\frac{1}{2}} - q^{-\frac{1}{2}}\right)^2.$$

Next, recall from (2.3) that q-integers are rational functions in $q^{\frac{1}{2}}$. We can be more precise:

LEMMA **2.96** *Any q-integer $[n]_q$ is expressible as a polynomial in $[2]_q$ of degree $n-1$ with integer coefficients.*

For example,

$$[3]_q = [2]_q^2 - 1, \quad [4]_q = [2]_q^3 - 2[2]_q, \quad [5]_q = [2]_q^4 - 3[2]_q^2 + 1.$$

For a proof of the lemma we use the property that every q-integer may be generated from the recursion relation

$$[n]_q = [2]_q[n-1]_q - [n-2]_q, \quad [0]_q = 0, \quad [1]_q = 1, \tag{2.97}$$

which follows from (2.95) with $a = n, b = 2, c = 1$. Hence, every q-integer is a polynomial in $q^{\frac{1}{2}} + q^{-\frac{1}{2}}$ with integer coefficients.

Bacry [117] has discussed the polynomial f_n defined by $f_{n+1}(t) = t\, f_n(t) - f_{n-1}(t)$ with $f_0(t) = 0$ and $f_1(t) = 1$. If $t = [2]_q$, this recurrence relation coincides with (2.97), implying $f_n([2]_q) = [n]_q$. We also have

$$\sum_{n=0}^{\infty} [n]_q z^n = \frac{z}{1 - [2]_q z + z^2}, \tag{2.98}$$

which gives a generating function for q-integers. This formula can be proved directly by summing the geometric series which results from substituting the form (2.91) for a q-integer into the left hand side of (2.98), or by applying the finite difference operator D^q defined by (2.113) to the geometric series for z. The series (2.98) converges for $|z| < \min(q^{\frac{1}{2}}, q^{-\frac{1}{2}})$.

Using $[n]_q^2 + [2n+1]_q = [n+1]_q^2$, which is obtained from (2.95) with $a = n$, $b = n+1$ and $c = 2n+1$, we derive (as in [116]) $\sum_{r=0}^{n-1} [2r+1]_q = [n]_q^2$. Similarly,

$$\sum_{r=1}^{n} [2r]_q = [n]_q [n+1]_q. \tag{2.99}$$

In the q-symplecton algebra [116] a special role is played by the series

$$F(n) \stackrel{\text{def}}{=} \sum_{r=1}^{n} [r]_q,$$

which may be summed to give

$$F(n) = \frac{[n]_q [n+1]_q}{1 - [n]_q + [n+1]_q} = \frac{\left[\frac{n}{2}\right]_q \left[\frac{n+1}{2}\right]_q}{\left[\frac{1}{2}\right]_q}. \tag{2.100}$$

The left expression for $F(n)$ can be proved directly by induction on n using (2.95) with $a = -n, b = -n-1, c = 1$. The equality of the two expressions for $F(n)$ follows with the help of the identity

$$[a]_q - [b]_q = \frac{[a+b]_q [\frac{1}{2}(a-b)]_q}{[\frac{1}{2}(a+b)]_q},$$

where $a, b \in \mathbb{R}$. Each of (2.99) and (2.100) is a q-analog of the well-known sum $\sum_{r=0}^{n} r = \frac{1}{2} n(n+1)$.

A generalization of (2.99) is

$$\sum_{r=0}^{n} [a+2r]_q = [a+n]_q [n+1]_q$$

for any $a \in \mathbb{R}$ and also, for (2.100),

$$\sum_{r=0}^{n} [a+r]_q = \frac{\left[a + \frac{n}{2}\right]_q \left[\frac{n+1}{2}\right]_q}{\left[\frac{1}{2}\right]_q} \tag{2.101}$$

again for any $a \in \mathbb{R}$. These two identities may be proved by induction on n using the now familiar identity (2.95) with $c = 1$ and $c = \frac{1}{2}$ respectively, and are related by the identity $[2n]_q = [2]_q [n]_{q^2}$.

2.7.2 The q-Binomial Theorem

We will find the q-analog of the binomial theorem to be particularly useful. This theorem was used by Cauchy and Heine[6] and is well-known as the expansion of certain finite products in terms of Gaussian polynomials (see Andrews [118] and Gasper and Rahman [64, §1.3]). The most elegant formulation is in terms of elements of the quantum plane, that is, we introduce "coordinates" x, y satisfying[7]

$$xy = q^{-1}yx,$$

and we say that the elements x, y q-commute. More precisely, as discussed in §4.2.2, we consider the associative algebra generated by x, y factored by the relation $xy - q^{-1}yx = 0$. Now we may state the **q-binomial theorem**:

THEOREM **2.102** *If $xy = q^{-1}yx$ then*

$$(x+y)^n = \sum_{r=0}^{n} q^{\frac{1}{2}r(n-r)} \frac{[n]_q!}{[r]_q![n-r]_q!} x^r y^{n-r} \tag{2.103}$$

for all $n \in \mathbb{N}$.

The proof is by induction on n. A discussion of this theorem and applications using q-calculus have been given by Feinsilver [115], see also Exton [113, §1.3.1] and [119, p. 793]). We will extend the theorem to all $n \in \mathbb{Z}$ below, when the sum becomes infinite, but we observe that for $n \in \mathbb{N}$ the sum terminates as shown. (As commonly occurs in such sums, the range of summation is determined by the denominator factors.)

It is convenient to define the **q-binomial coefficients** for $m \in \mathbb{N}$ and for all $a \in \mathbb{R}$ by

$$\begin{bmatrix} a \\ m \end{bmatrix} \stackrel{\text{def}}{=} \frac{[a]_q[a-1]_q \cdots [a-m+1]_q}{[m]_q!}, \qquad \begin{bmatrix} a \\ 0 \end{bmatrix} \stackrel{\text{def}}{=} 1, \tag{2.104}$$

and rewrite the q-binomial theorem as

$$(x+y)^n = \sum_r q^{\frac{1}{2}r(n-r)} \begin{bmatrix} n \\ r \end{bmatrix} x^r y^{n-r}. \tag{2.105}$$

The combination $q^{\frac{1}{2}m(n-m)} \begin{bmatrix} n \\ m \end{bmatrix}$ defines a **Gaussian polynomial**, which for $n, m \in \mathbb{N}$ is a polynomial in q of degree $m(n-m)$ and is denoted here by

$$\begin{aligned} \begin{bmatrix} n \\ m \end{bmatrix}_{\text{Gauss}} &\stackrel{\text{def}}{=} q^{\frac{1}{2}m(n-m)} \begin{bmatrix} n \\ m \end{bmatrix} \\ &= \frac{(1-q^n)(1-q^{n-1})\cdots(1-q^{n-m+1})}{(1-q^m)(q-q^{m-1})\cdots(1-q)} = \frac{(q;q)_n}{(q;q)_m(q;q)_{n-m}}, \end{aligned}$$

[6] Dr. R. Askey has kindly informed us that the q-binomial theorem was first stated by Rothe in 1812, but that special cases were known to Euler.

[7] In §4.2.2 (p. 121) elements of the quantum plane are defined to satisfy $xy = q^{-\frac{1}{2}}yx$, but in applications of the q-binomial theorem the q-commutator involves q, not $q^{\frac{1}{2}}$.

where $0 \leqslant m \leqslant n$, and where in the last equation we have used the notation of (2.120) and (2.121) below. These polynomials have been used extensively in the theory of partitions, see Andrews [118, Chapter 3], and by Gasper and Rahman [64, p. 20], where they are defined to be the q-binomial coefficients themselves.

We can use the q-binomial theorem to derive properties of the q-binomial coefficients. Firstly, we observe that

$$\begin{bmatrix} -a \\ m \end{bmatrix} = (-1)^m \begin{bmatrix} a+m-1 \\ m \end{bmatrix}.$$

Now we apply the q-binomial theorem to the identity $(x+y)^{n+r} = (x+y)^n(x+y)^r$ to obtain

$$\begin{bmatrix} n+r \\ m \end{bmatrix} = \sum_k q^{\frac{1}{2}(nk+rk-nm)} \begin{bmatrix} n \\ k \end{bmatrix} \begin{bmatrix} r \\ m-k \end{bmatrix}. \tag{2.106}$$

This relation, with $r=1$ and n replaced by $n-1$, gives

$$\begin{bmatrix} n \\ m \end{bmatrix} = q^{\frac{m}{2}} \begin{bmatrix} n-1 \\ m \end{bmatrix} + q^{\frac{1}{2}(m-n)} \begin{bmatrix} n-1 \\ m-1 \end{bmatrix}, \tag{2.107}$$

which is a recursive formula for these coefficients (note that by replacing q by q^{-1}, which leaves the coefficient invariant, we obtain a similar formula). Another way to derive (2.106) is via the following equation, which may be regarded as determining a generating function for the q-binomial coefficient, and demonstrates the role of these coefficients in expanding finite products:

$$\sum_{i=0}^{n} q^{\frac{1}{2}i(n-1)} \begin{bmatrix} n \\ i \end{bmatrix} z^i = \prod_{i=0}^{n-1} (1+q^i z).$$

By applying this equation to the identity

$$\prod_{i=0}^{n+r-1} (1+q^i z) = \prod_{i=0}^{r-1} (1+q^i z) \prod_{j=0}^{n-1} (1+q^{j+r} z)$$

we again obtain (2.106). Yet another way to derive (2.106) is to use the following identity (following Nomura [120]):

$$\sum_{r=0}^{n} \begin{bmatrix} n \\ r \end{bmatrix} q^{\frac{1}{4}r^2} a^{n-r} b^r = \prod_{r=0}^{n-1} \left(a + q^{-\frac{1}{4}(2r+1)} b\right),$$

where a, b are q-commuting coordinates satisfying $ab = q^{-\frac{1}{2}} ba$. Again, we see that noncommuting coordinates appear naturally in connection with q-number identities.

The q-binomial theorem can be extended to negative values of n, provided that we assume the existence of the inverse elements x^{-1}, y^{-1} of x, y respectively. Indeed, the definition of the q-binomial coefficient (2.104) is valid for all $a \in \mathbb{R}$, in particular for negative integers $a = -n+1$. Hence, the statement (2.105) of the q-binomial theorem can be extended to all integers n, with the finite sum now replaced by an infinite sum over all nonnegative integers r (here we ignore questions of convergence in the algebra of the elements x, y). We may prove this generalized q-binomial theorem by induction on $m = 1 - n \in \mathbb{N}$ in the following way. Firstly, we rewrite the right hand side of (2.105) as

$$(x+y)\sum_r q^{-\frac{1}{2}r(m+r)} \begin{bmatrix} -m \\ r \end{bmatrix} x^r y^{-m-r}, \tag{2.108}$$

which follows from a modified version of (2.107):

$$\begin{bmatrix} -m+1 \\ r \end{bmatrix} = q^{-\frac{r}{2}} \begin{bmatrix} -m \\ r \end{bmatrix} + q^{\frac{1}{2}(r+m-1)} \begin{bmatrix} -m \\ r-1 \end{bmatrix}.$$

The expression (2.108) is equal to $(x+y)$ times the right hand side of (2.105) with n being replaced by $-m$, which is sufficient to complete the induction argument.

Now let us extend the identities for the q-binomial coefficients to all integers, obtaining in the process some orthogonality relations. Firstly, by applying the generalized q-binomial theorem to the identity

$$(x+y)^{n-r-1} = (x+y)^n (x+y)^{-r-1} \tag{2.109}$$

($r \in \mathbb{N}$, $n \in \mathbb{Z}$) we derive the relation

$$\begin{bmatrix} n-r-1 \\ m \end{bmatrix} = \sum_k (-1)^k q^{\frac{1}{2}k(r-n+1)-\frac{1}{2}m(r+1)} \begin{bmatrix} n \\ m-k \end{bmatrix} \begin{bmatrix} r+k \\ r \end{bmatrix}, \tag{2.110}$$

where n is any integer and $m, r \in \mathbb{N}$. From this follows the orthogonality relation:

$$\sum_k (-1)^k q^{\frac{1}{2}k(n-m+1)} \begin{bmatrix} m \\ k \end{bmatrix} \begin{bmatrix} k \\ n \end{bmatrix} = \delta_{m,n}(-1)^m q^{\frac{m}{2}},$$

where both $n, m \in \mathbb{N}$. By substituting $m \to m - r$ and summing over $\ell = k + r$ in (2.110) we get

$$\begin{bmatrix} n-r-1 \\ m-r \end{bmatrix} = \sum_\ell (-1)^{\ell-r} q^{\frac{1}{2}(\ell-m)(r+1)+\frac{1}{2}(r-\ell)n} \begin{bmatrix} n \\ m-\ell \end{bmatrix} \begin{bmatrix} \ell \\ r \end{bmatrix}.$$

The case $n = r + 1$ reduces to another orthogonality relation:

$$\sum_\ell (-1)^{\ell-m} \begin{bmatrix} r+1 \\ m-\ell \end{bmatrix} \begin{bmatrix} \ell \\ r \end{bmatrix} = \delta_{rm},$$

which in this case is invariant under $q \leftrightarrow q^{-1}$. By replacing $n \to -(n+1)$ in (2.109) we obtain yet another relation,

$$\begin{bmatrix} m+r+1 \\ m-n \end{bmatrix} = \sum_{k=0}^{m-n} q^{\frac{1}{2}k(n+r+2)+\frac{1}{2}(n-m)(r+1)} \begin{bmatrix} m-k \\ n \end{bmatrix} \begin{bmatrix} r+k \\ r \end{bmatrix}.$$

In the case of $n = r = 0$, this identity reduces to the expression (2.3) (with $n = m+1$) for the q-integer $[n]_q$.

We may summarize the main properties of the q-binomial coefficients as follows:

LEMMA 2.111 *The q-binomial coefficients* $\begin{bmatrix} n \\ m \end{bmatrix}$, $0 \leqslant m \leqslant n$, *are invariant under* $q \leftrightarrow q^{-1}$ *and satisfy the following relations:*

$$\begin{aligned} \begin{bmatrix} n \\ 0 \end{bmatrix} &= \begin{bmatrix} n \\ n \end{bmatrix} = 1, \\ \begin{bmatrix} n \\ m \end{bmatrix} &= \begin{bmatrix} n \\ n-m \end{bmatrix}, \\ \begin{bmatrix} n \\ m \end{bmatrix} &= q^{\frac{m}{2}} \begin{bmatrix} n-1 \\ m \end{bmatrix} + q^{\frac{1}{2}(m-n)} \begin{bmatrix} n-1 \\ m-1 \end{bmatrix}, \end{aligned} \tag{2.112}$$

and $q^{\frac{1}{2}m(n-m)} \begin{bmatrix} n \\ m \end{bmatrix}$ *is a polynomial in q of degree $m(n-m)$.*

Further properties of these coefficients may be found in [118] (in the form of Gaussian polynomials), including summation formulas which are related to the q-analog of Saalschütz's formula, considered below in §2.8.3.

2.8 Appendix: q-Calculus and q-Functions

We have already seen in §2.4.1 that q-boson operators, which are fundamental entities in our constructive approach to quantum groups, can be represented as finite difference operators. Let us consider therefore in more detail properties of these operators and the corresponding q-calculus of finite sum and difference operators.

These q-analog concepts each play a role in our development of quantum groups, although most of them were understood well before the advent of quantum groups (see for example Feinsilver [115], where an operator calculus based on commutation rules equivalent to (2.31) is developed; also Cigler [114], where the same commutation rules are postulated and q-identities derived). In our development we point out particularly how noncommuting coordinates, which are fundamental to the theory of q-tensor operators for quantum groups, either help to simplify expressions involving q-analog functions, as in the q-binomial theorem, or enter in an apparently essential way, as with the addition theorem of q-exponential functions (Lemma 2.116 and Eqn. (2.118) below).

2.8.1 q-Derivation and Integration

Many of the basic properties of q-derivatives were stated by Jackson (1908) [121], using the definition

$$\nabla_q f(x) \stackrel{\text{def}}{=} \frac{f(qx)-f(x)}{x(q-1)}$$

of q-derivative, which reduces to the usual operation of derivation when $q \to 1$. If now we let X denote the operator of multiplication, that is, $Xf(x) = xf(x)$, then the commutation relations

$$\nabla_q X - qX\nabla_q = I$$

are satisfied, which we recognize as the q-boson relations shown in the form (2.31). We find a more convenient definition of the finite difference operator to be the following:

$$D^q f(x) \stackrel{\text{def}}{=} \frac{f(xq^{\frac{1}{2}}) - f(xq^{-\frac{1}{2}})}{x(q^{\frac{1}{2}} - q^{-\frac{1}{2}})}, \tag{2.113}$$

which is invariant under $q \leftrightarrow q^{-1}$. As stated in Lemma 2.40 this operator, together with the operator X of multiplication and N, generates the q-boson algebra. The two definitions of q-derivative are related by the operator equation $\nabla_q = q^{\frac{N}{2}} D^q$. A useful property is

$$D^q x^n = [n]_q x^{n-1}$$

which is equivalent to the operator equation (2.34). We refer to Feinsilver [115] and also Bracken et al. [88] for the further development of q-differentiation, including properties such as the q-Leibniz rule.

We have already defined q-integration in §2.4.1, again in a way that is symmetric under $q \leftrightarrow q^{-1}$:

$$(I^q f)(x) = \int_0^x f(t)\, d_q t \stackrel{\text{def}}{=} \sum_{r=0}^{\infty} f\left(xq^{r+\frac{1}{2}}\right)\left(xq^r - xq^{r+1}\right).$$

This definition is closely related to that introduced by Thomae [92] and Jackson [87], and discussed by Andrews [57], and limits to the Riemann integral $\int_0^x f(x)dx$ as $q \to 1$. In the operator form which we used in §2.4.1, namely,

$$D^q = x^{-1}[N]_q, \quad I^q = ([N]_q)^{-1} x$$

we see easily that $I^q D^q f(x) = D^q I^q f(x) = f(x)$, that is, these operations are inverse to each other.

For further discussion of q-integration we refer to those articles already cited, and in particular to [64, §1.11].

2.8.2 The q-Exponential Function

We may introduce a q-analog of the exponential function as an eigenfunction of D^q and with a termwise application of D^q we are lead to the following series representation:

DEFINITION **2.114** *The q-exponential function is defined and denoted by*

$$\exp_q(z) \stackrel{\text{def}}{=} \sum_{n=0}^{\infty} \frac{z^n}{[n]_q!}.$$

This reduces to the usual exponential function as $q \to 1$, and is invariant under $q \leftrightarrow q^{-1}$. Some of the fundamental properties of the q-exponential may be summarized as follows:

Lemma 2.115 *(i) The q-exponential series converges absolutely for all z; (ii) the q-exponential function is holomorphic on the complex plane; (iii) the q-exponential function satisfies*

$$D^q \exp_q(\alpha z) = \alpha \exp_q(\alpha z)$$

where $\alpha \in \mathbb{C}$.

Proof: Part (i) follows by comparison with e^z using the inequality $[n]_q \geqslant n$ of Lemma 2.94, and it follows that $\exp_q$ is a holomorphic function of z. Part (iii) follows by term-by-term q-differentiation of the series definition □.

Conversely to this lemma, $f(z) = \exp_q(\alpha z)$ is the unique solution to the difference equation $D^q f(z) = \alpha f(z)$ with $f(0) = 1$.

Another important property of the exponential function which we seek to generalize is the addition formula $e^{x+y} = e^x e^y$. The generalization is best expressed in operator form by introducing an operator representation of noncommuting coordinates. Let

$$x = q^{\frac{N_2}{2}} z_1, \quad y = q^{-\frac{N_1}{2}} z_2,$$

where $z_1, z_2 \in \mathbb{C}$ and, as usual, $N_i = z_i \partial_i$. Then we have $yx = q\,xy$. Hence, we may apply the q-binomial theorem to the series expansion of $\exp_q(x+y)\bullet$, where we use the symbol "$\bullet$" to indicate that the operators x, y act on all factors to their right, up to "$\bullet$" which stands for the constant 1 (this notation also serves to emphasize that the identity to be obtained is real-valued, not operator-valued). Equivalently, we could rewrite these expressions in terms of boson or q-boson operators, with associated number operators N_1, N_2, acting on the vacuum, using the realization described in §2.4.1. Next, we use $x^r y^{n-r} \bullet = q^{\frac{1}{2}r(n-r)} z_1^r z_2^{n-r}$ and, after a change of summation order, we obtain:

Lemma 2.116 *The q-exponential function satisfies the addition formula*

$$\exp_q(z_1)\exp_q(z_2) = \exp_q(q^{-\frac{N_2}{2}} z_1 + q^{\frac{N_1}{2}} z_2)\bullet$$

where the operators $N_i = z_i \partial_i$ $(i = 1, 2)$ act to the right on the constant 1, indicated by "$\bullet$".

Because of the operator form of this addition theorem, we cannot substitute specific values for z_1, z_2 until we have evaluated the action of the right hand side on the terminating constant. Hence, $\exp_q(z)\exp_q(-z) \neq 1$, in contrast to the $q = 1$ case. The generalization of this lemma to n non-commuting variables is stated in Lemma 7.39 (p. 229), where it is used in the Borel-Weil construction of irreps of $\mathcal{U}_q(\mathfrak{u}(n))$.

The combination of operators which appears in the exponent in Lemma 2.116, namely $q^{-\frac{N_2^q}{2}} a_1^q + q^{\frac{N_1^q}{2}} a_2^q$ in q-boson notation, is similar to that obtained by the co-multiplication introduced in (2.5) for the generators $J_\pm$ of $\mathcal{U}_q(\mathfrak{su}(2))$, with J_z

represented by essentially the number operator. This combination is in fact co-multiplication for the Hermitean conjugate of the realization of $\mathcal{U}_q(\mathfrak{su}(2))$ described in Lemma 2.56 (p. 39), for which $J_- = a^q$, $J_z = j - N^q$ and

$$\Delta(J_-) = q^{-\frac{N_2^q}{2}} q^{\frac{j}{2}} a_1^q + q^{\frac{N_1^q}{2}} q^{-\frac{j}{2}} a_2^q$$

(the numerical q-factors may be absorbed into the q-boson operators).

Although Definition 2.114 is the one which is most useful for our purposes because of its invariance under $q \leftrightarrow q^{-1}$, there is no unique way of generalizing the ordinary exponential, for we could include arbitrary q-factors under the sum in the series definition. We refer to Gasper and Rahman [64, §1.3], Exton [113, §4.5] and Feinsilver [115] for further discussion, however let us mention briefly some properties of the function $E_q(z)$ defined by

$$E_q(z) \stackrel{\text{def}}{=} \sum_{n=0}^{\infty} \frac{q^{-\frac{1}{4}n(n-1)} z^n}{[n]_q!} \tag{2.117}$$

which is precisely the q-analog exponential defined by Exton and Feinsilver, where it is denoted $E(z)$. By contrast, the function denoted $E_q(z)$ by Gasper and Rahman is equal to $E_{q^{-1}}(z(1-q)^{-1})$ in our notation. The definition (2.117) is related to $\exp_q$ by the formula

$$E_q(z) = \exp_q(zq^{-\frac{N}{2}})\bullet$$

where $N = z\frac{\partial}{\partial z}$, as follows from the operator identity $\left(zq^{-\frac{N}{2}}\right)^n = z^n q^{-\frac{n}{4}(2N+n-1)}$. This series representation of $E_q(z)$ has a finite radius of convergence provided that $0 < q < 1$. E_q satisfies an addition formula which may be elegantly expressed in terms of noncommuting coordinates:

$$E_q(x+y) = E_q(x)E_q(y) = E_{q^{-1}}(y)E_{q^{-1}}(x), \quad yx = q\,xy \tag{2.118}$$

which can be generalized to

$$E_q(x_1 + x_2 + \cdots + x_n) = E_q(x_1)E_q(x_2)\cdots E_q(x_n), \quad x_i x_j = q x_j x_i \quad \text{for} \quad i > j.$$

Further properties of E_q are discussed in [113] and [114], see also [88] for further discussion of $\exp_q$.

We have defined

$$[n]_q! \stackrel{\text{def}}{=} [n]_q[n-1]_q \ldots [1]_q \tag{2.119}$$

and it is natural to investigate generalizations to the q-gamma function, which we may denote Γ_q. One definition is

$$\Gamma_q(x) \stackrel{\text{def}}{=} \frac{(q;q)_\infty}{(q^x;q)_\infty}(1-q)^{1-x}, \quad 0 < q < 1,$$

(using the notation of (2.120) below) which reduces for $x = n+1$ to $\Gamma_q(n+1) = q^{\frac{1}{4}n(n-1)}[n]_q!$. Again, further details may be found in [64, §1.10].

2.8.3 Basic Hypergeometric Functions

The q-analogs of hypergeometric functions were introduced by Heine in 1846, and developed systematically by him and others over many years; see particularly the books by Bailey [58] and Slater [59], and more recently the detailed account, which summarizes modern developments, by Gasper and Rahman [64]. We will find that basic hypergeometric functions occur naturally in quantum groups in a variety of contexts, including that of basis functions (described in §4.4.1), and as q-analogs of Wigner-Clebsch-Gordan and Racah coefficients which may be related to ${}_3\phi_2$ and ${}_4\phi_3$ functions (see §3.5.2,3.6). In §7.7.4 we show how ${}_8\phi_7$ functions appear in the method of algebraic induction applied to $\mathcal{U}_q(\mathfrak{u}(3))$, and how classical identities such as Watson's formula are equivalent to quantum group properties. Using the q-symplecton construction we show in §8.3.2 that ${}_2\phi_1$ functions also appear as components of tensor operators in $\mathcal{U}_q(\mathfrak{su}(2))$, leading to summation formulas that can be related directly to quantum group properties. Similar connections have been observed by several authors, including Kirillov and Reshitikhin [65], Floreanini and Vinet [2, p. 235] and [4, p. 264], and Groza et al. [60], indeed there has been extensive discussion in the literature on this topic.

Generally, q-analogs of classical functions are obtained by replacing parameters $a, b \ldots$ by $[a]_q, [b]_q \ldots$, although such a generalization is not unique, for we could include arbitrary q-factors in such a definition. A standard definition is the symbol $(a;q)_n$ for $n \in \mathbb{N}$:

$$(a;q)_n \stackrel{\text{def}}{=} \begin{cases} 1 & n=0, \\ (1-a)(1-aq)\ \ldots(1-aq^{n-1}) & n>0. \end{cases} \tag{2.120}$$

A useful identity for the limiting case $n \to \infty$ (valid for $|q|<1$) is

$$(q^\alpha;q)_n = \frac{(q^\alpha;q)_\infty}{(q^{n+\alpha};q)_\infty}.$$

In terms of the q-numbers defined by (2.91) we have

$$(a;q)_n = (q^\alpha;q)_n = (1-q)^n\, q^{\frac{n}{4}(n+2\alpha-3)}([\alpha]_q)_n \tag{2.121}$$

where we have put $a = q^\alpha$, and where the **rising product** (or **shifted factorial** [64]) $([\alpha]_q)_n$ is defined by

$$([\alpha]_q)_n \stackrel{\text{def}}{=} \begin{cases} [\alpha]_q[\alpha+1]_q \ldots [\alpha+n-1]_q & \text{for } n>0 \\ 1 & \text{for } n=0 \\ \dfrac{1}{[\alpha-1]_q[\alpha-2]_q\ldots[\alpha+n]_q} & \text{for } n<0. \end{cases} \tag{2.122}$$

Although we will generally use this definition for positive n only, it is helpful to keep in mind the relation

$$([\alpha]_q)_{-n}([\alpha-n]_q)_n = 1,$$

for any integers n. For $\alpha = 1$ we have $([1]_q)_n = [n]_q!$, where the q-factorial $[n]_q!$ is defined in (2.119). It will be useful to record the identities

$$[m-n]_q! = \frac{(-1)^n[m]_q!}{([-m]_q)_n}, \quad [m+n]_q! = [m]_q!([m+1]_q)_n$$

where $n, m \in \mathbb{N}$.

The basic hypergeometric function ${}_{p+1}\phi_p$ is defined by

$${}_{p+1}\phi_p\begin{pmatrix} a_1 a_2 \dots a_{p+1} \\ b_1 b_2 \dots b_p \end{pmatrix}; q, z \Bigg) \stackrel{\text{def}}{=} \sum_{n=0}^{\infty} \frac{(a_1;q)_n(a_2;q)_n \dots (a_{p+1};q)_n}{(q;q)_n(b_1;q)_n \dots (b_p;q)_n} z^n.$$

In the notation of q-numbers we can write

$$\begin{aligned} {}_{p+1}\phi_p\left(\begin{matrix} q^{\alpha_1}q^{\alpha_2}\dots q^{\alpha_{p+1}} \\ q^{\beta_1}q^{\beta_2}\dots q^{\beta_p}\end{matrix} ;q,z\right) &= \sum_{n=0}^{\infty} \frac{([\alpha_1]_q)_n([\alpha_2]_q)_n \dots ([\alpha_{p+1}]_q)_n q^{\frac{n\sigma}{2}}}{[n]_q!([\beta_1]_q)_n([\beta_2]_q)_n \dots ([\beta_p]_q)_n} z^n \\ &= {}_{p+1}\phi_p\left(\begin{matrix} q^{-\alpha_1}q^{-\alpha_2}\dots q^{-\alpha_{p+1}} \\ q^{-\beta_1}q^{-\beta_2}\dots q^{-\beta_p}\end{matrix} ;q^{-1},zq^{\sigma}\right) \end{aligned} \tag{2.123}$$

where

$$\sigma = \sum_{i=1}^{p+1}\alpha_i - \sum_{i=1}^{p}\beta_i - 1.$$

For $q=1$ the basic hypergeometric function reduces to

$${}_{p+1}F_p\left(\begin{matrix}\alpha_1 \dots \alpha_{p+1} \\ \beta_1 \dots \beta_p\end{matrix} ;z\right),$$

the generalized Gauss hypergeometric function. Of course, we could more generally consider functions ${}_{p'}\phi_p$ for any p, p', but we find that usually only the case $p' = p+1$ occurs in the study of quantum groups. If one of the numerator parameters $\alpha_1, \dots \alpha_{p+1}$ is a negative integer $-m$, the series terminates and in this case any one of the denominator parameters $\beta_1 \dots \beta_p$ can also be negative provided it is less than $-m$.

When $\sigma = -2$ and $z = q$ the function ${}_{p+1}\phi_p$ is called **balanced** or **Saalschützian**. If

$$1+\alpha_1 = \alpha_2+\beta_1 = \alpha_3+\beta_2 = \dots = \alpha_{p+1}+\beta_p,$$

the basic hypergeometric function is said to be **well poised**. In terms of the parameters a_i, b_i, well poised means

$$qa_1 = a_2b_1 = a_3b_2 = \dots = a_{p+1}b_p.$$

If in addition $a_2 = qa_1^{\frac{1}{2}}$ and $a_3 = -qa_1^{\frac{1}{2}}$ the function is said to be **very well poised**.

We will find that the many identities, mostly well-known, satisfied by basic hypergeometric functions play a significant role in the study of quantum groups. An example of such an identity is the q-analog of Saalschütz's formula (see [64, §1.7]), which reads

$$ {}_3\phi_2\left(\begin{matrix} q^a, & q^b, & q^{-n} \\ q^c, & q^{a+b-c+1-n} \end{matrix}; q, q\right) = \frac{(q^{c-a};q)_n(q^{c-b};q)_n}{(q^c;q)_n(q^{c-a-b};q)_n} \tag{2.124} $$

for $n = 0, 1, 2, \ldots$, and which sums a terminating Saalschützian basic hypergeometric function. (Observe that since $\sigma = -2$ and $z = q$ this function is indeed Saalschützian). This formula, first derived by Jackson [122] in 1910, reduces to Saalschütz's formula on letting $q \to 1$. In the notation of q-numbers it reads

$$ \sum_{m=0}^{n} \frac{([a]_q)_m([b]_q)_m([-n]_q)_m}{[m]_q!([c]_q)_m([a+b-c+1-n]_q)_m} = \frac{([c-a]_q)_n([c-b]_q)_n}{([c]_q)_n([c-a-b]_q)_n} \tag{2.125} $$

which we see is an identity in which q enters only through q-numbers, and so is symmetric under $q \leftrightarrow q^{-1}$. The special case of $n = 1$ reduces to

$$ [a]_q[b]_q + [c]_q[c-a-b]_q - [c-a]_q[c-b]_q = 0, $$

which is equivalent to the frequently used identity (2.95).

Two further special cases of (2.124) are the limits $a \to \pm\infty$. To investigate these we use

$$ ([x+a]_q)_n \to \frac{q^{\frac{n}{4}(2x+2a+n-1)}}{(q^{\frac{1}{2}} - q^{-\frac{1}{2}})^n}, \quad ([x-a]_q)_n \to (-1)^n \frac{q^{-\frac{n}{4}(2x-2a+n-1)}}{(q^{\frac{1}{2}} - q^{-\frac{1}{2}})^n}, $$

for $a \to \infty$ and assume $q > 1$. Hence from (2.125) we obtain, by letting $a \to \infty$ and for $q > 1$,

$$ \sum_{m=0}^{n} \frac{([b]_q)_m([-n]_q)_m}{[m]_q!([c]_q)_m} q^{\frac{m}{2}(-b+c-1+n)} = q^{-\frac{nb}{2}} \frac{([c-b]_q)_n}{([c]_q)_n} \tag{2.126} $$

which can also be written

$$ {}_2\phi_1\left(\begin{matrix} q^b, q^{-n} \\ q^c \end{matrix}; q, q^{c-b+n}\right) = \frac{(q^{c-b};q)_n}{(q^c;q)_n}. $$

This formula, which is valid for all $q \in \mathbb{R}^+$, is the q-analog of Gauss' summation formula in the terminating case. By reversing the order of summation, we obtain the identity

$$ {}_2\phi_1\left(\begin{matrix} q^b, q^{-n} \\ q^c \end{matrix}; q, q\right) = q^{nb} \frac{(q^{c-b};q)_n}{(q^c;q)_n}, \tag{2.127} $$

which can also be derived by taking the limit $a \to -\infty$ of (2.125).

The special case of (2.126) for $n = 1$ is merely an expression of the identity (2.92). In general, by this method of taking limits we obtain identities involving explicit q-factors which are not invariant under $q \leftrightarrow q^{-1}$; this situation will occur when we investigate the limit of the Racah coefficients, expressible in terms of a ${}_4\phi_3$ function, which become q-WCG coefficients which have an explicit dependence on q (see Lemma 3.74, p. 106).

We have in fact already encountered the form (2.127) of the q-analog of Gauss' summation formula in Eqn. (2.106) involving q-binomial coefficients since, by putting $b = -m, c = r - m + 1$, $m \in \mathbb{N}$ in (2.127), we obtain the relation (2.106). The q-binomial theorem itself may be stated as a basic hypergeometric function identity; it reads [64, §1.3]

$$ {}_1\phi_0(a; -; q, z) = \frac{(az; q)_\infty}{(z; q)_\infty}, \quad |z| < 1, \ |q| < 1. $$

Finally, we mention that the q-exponential function defined by (2.117) can be expressed in terms of basic hypergeometric functions, specifically,

$$ {}_1\phi_0(0; -; q, z) = \frac{1}{(z; q)_\infty} = E_q\left(\frac{z}{1-q}\right) $$

and

$$ {}_0\phi_0(-; -; q, z) = (-z; q)_\infty = E_{q^{-1}}\left(\frac{z}{1-q}\right). $$

The first of these functions is denoted $e_q(z)$ and the second $E_q(z)$ in [64].

Chapter 3

Tensor Operators in Quantum Groups

In this chapter we introduce the tensor operators associated with a quantum group, considering mainly the unitary quantum groups. The algebra of tensor operators, whether for the classical or the quantum groups, provides a far-reaching generalization of the concept of the enveloping algebra of a Lie group, capable of unifying and extending the representation theory of Lie groups and quantum groups, and especially the physical applications of the theory. A typical application is the classification as tensor operators, with respect to a symmetry group, of explicit symmetry breaking terms in quantum Hamiltonians. The algebra of tensor operators, and its q-tensor operator extension, is well developed for $SU(2)$ and $\mathcal{U}_q(\mathfrak{su}(2))$ and is almost as fully developed for $SU(3)$ but less so for $\mathcal{U}_q(\mathfrak{u}(3))$. These operator algebras are characterized by a well-defined and fully explicit operator product expansion which generalizes the universal enveloping algebra.

3.1 Introduction

The concept of tensor operators stems from quantum physics. The development and applications of quantum physics since its inception have been strongly influenced by symmetry considerations and, directly or indirectly, by group theory. The paradigm for applying symmetry techniques in quantum physics has been the exploitation of rotational symmetry in the form of the quantum theory of angular momentum involving the Lie group $SU(2)$. A key concept in such applications has been that of tensor operators, a concept developed specifically for the quantum theory of angular momentum. A basic problem has been to extend, as far as possible, to an arbitrary Lie group the tensor operator concept. The developments in physics stimulated by this problem — originating in the work of Weyl, Heisenberg, Wigner and Racah — have led to results (in the physics literature) anticipating and/or independently obtaining similar results by mathematicians in Lie group theory (such as

Harish-Chandra, Kostant, Weil, Borel).

The applications of symmetry techniques in physics are widespread and varied. There is no universal way to apply symmetries, and the many applications depend strongly on the specific way in which the symmetry is realized. In the Wigner approach it is realized linearly and the operator algebras are *equivariant* (defined in §3.2). In this approach, a given symmetry group has the implication for physics not only of classifying possible state vectors and associated quantum numbers but also of classifying and partially determining physical transitions between these symmetry-classified allowed states. This latter structure is determined by the algebra of tensor operators.

Let us briefly mention two well-known examples of the physical application of tensor operators, and refer to [66, Chapter 7] for an extensive discussion. Elementary examples of tensor operators are the momentum and position operators $\boldsymbol{P}, \boldsymbol{Q}$ satisfying the canonical commutation relations $[Q_i, P_j] = i\delta_{ij}$ and which are spin 1 tensor operators (vectors) under the orbital angular momentum generators $\boldsymbol{L} = \boldsymbol{Q} \times \boldsymbol{P}$. This property is useful in solving the nonrelativistic hydrogen atom for which the Hamiltonian is

$$H = |\boldsymbol{P}|^2 - \frac{2}{|\boldsymbol{Q}|}$$

(in dimensionless form), and is an invariant under $\boldsymbol{L}$. The Runge-Lenz operator

$$\boldsymbol{A} = \frac{\boldsymbol{Q}}{|\boldsymbol{Q}|} + \tfrac{1}{2}(\boldsymbol{L} \times \boldsymbol{P} - \boldsymbol{P} \times \boldsymbol{L})$$

is also a vector operator under $\boldsymbol{L}$, and also commutes with H. The two vector operators $\boldsymbol{L}, \boldsymbol{A}$ together generate the group $SU(2) \times SU(2)$ and this fact, together with their tensor operator properties, leads to an algebraic solution of the hydrogen atom. A second, spectacular, example is that of $SU(3)$ symmetry and the classification of hadrons (strongly interacting elementary particles). Multiplets of hadrons, such as the octet, can be classified according to irreps of $SU(3)$ and would have equal masses within a multiplet if the symmetry were exact. Since this is not the case, the Lagrangian which models the hadrons is not an $SU(3)$ invariant but must include a symmetry breaking term, which is assumed to be a tensor operator of a certain type (transforming like the eighth component of the octet). This term is responsible for the mass splitting and leads to a mass formula which is well satisfied. For details we refer to the collection of articles [123] edited by Gell-Mann and Ne'eman.

One can fully expect the importance of tensor operators for classical unitary groups to extend to quantum groups in at least some physical applications, and so we are lead naturally to investigate tensor operators in quantum groups. We begin this chapter by reviewing the theory of tensor operators in the classical groups, emphasizing the concepts which we seek to generalize to quantum groups, in particular equivariance and the derivative property. Then we extend these concepts to quantum groups, discussing initially general quantum groups but then specializing to the unitary quantum groups and finally to $\mathcal{U}_q(\mathfrak{su}(2))$. We give specific examples of tensor operators in $\mathcal{U}_q(\mathfrak{su}(2))$ and use them to derive the fundamental q-Wigner-Clebsch-Gordan (q-WCG) coefficients. Then we develop general $\mathcal{U}_q(\mathfrak{su}(2))$ q-WCG coefficients in detail followed by a discussion of the q-$6j$ symbols, and finally we describe the pattern calculus which enables us to write down matrix elements of all elementary unit tensor operators in $\mathcal{U}_q(\mathfrak{u}(n))$.

3.2 Classical Theory of Tensor Operators

Consider now a compact Lie group G which is a physical symmetry, that is, the action by G leaves some Hamiltonian or Lagrangian invariant (as well as the vacuum). For definiteness, let us take G to be $U(n)$ or $SU(n)$. This symmetry allows one to classify a Hilbert space basis for physical states using the irreps of G. To discuss the problem of interactions, which may break the symmetry and induce transitions within and between irreps, it is useful to focus on the group theoretically defined properties of the transition operators. The most expeditious way to accomplish this separation is to construct a **model space** following Gel'fand [124, 125]:

DEFINITION **3.1** *A model space $\mathfrak{M}$ is defined to be the direct sum of vector spaces carrying unitary irreps of the group G, each equivalence class of irreps occurring once and only once, that is,*

$$\mathfrak{M} = \sum_{\lambda} \oplus \mathfrak{V}_{\lambda}, \tag{3.2}$$

where λ denotes the irrep labels of G, and $\mathfrak{V}_{\lambda}$ denotes the vector space carrying the irrep λ.

This definition generalizes without change to quantum groups, because the irreps of quantum groups, for real positive q, are in one-to-one correspondence with those of the classical groups. For unitary groups the irrep labels λ are given by $[\mathbf{m}] \stackrel{\text{def}}{=} [m_{1n} \cdots m_{nn}]$, as explained in §2.6. As an example of a model space we have, for both $\mathfrak{su}(2)$ and $\mathcal{U}_q(\mathfrak{su}(2))$, $\mathfrak{M} = \sum_j \oplus \mathfrak{V}_j$, where $\mathfrak{V}_j$ is the $2j+1$ dimensional space carrying the irrep labelled by j, spanned by the vectors $|jm\rangle$; that is, $\mathfrak{M}$ is the infinite-dimensional space consisting of all vectors $|jm\rangle$ with all possible values of m and j. In the realization using boson or q-boson operators, discussed in §2.4.2, $\mathfrak{M}$ is simply the Fock space $\mathfrak{F}^2$.

Let us denote by $\mathfrak{T}$ the linear space comprising all linear operators on $\mathfrak{M}$, that is $\mathbf{T} : \mathfrak{M} \to \mathfrak{M}$ for all elements $\mathbf{T} \in \mathfrak{T}$. The symmetry G can be exploited to give a partial, group theoretic, classification of the operators belonging to $\mathfrak{T}$, provided these operators are **equivariant**, a concept which we now define. Under the symmetry $g \in G$, elements of the model space transform according to $\mathfrak{M} \to \mathbf{U}(g)\,\mathfrak{M}$, where $\mathbf{U}(g)$ is an operator whose action on the vectors in $\mathfrak{M}$ corresponds to matrix multiplication by irrep matrices.

DEFINITION **3.3** *Equivariance is the property that the action by $g \in G$ on $\mathfrak{T}$ obeys*

$$\mathbf{U}(g)\,\mathbf{T}_i\,\mathbf{U}^{-1}(g) \;=\; \sum_j \mathbf{T}_j\;\mathfrak{D}_{ji}(g),$$

where $\mathfrak{D}(g)$ is a representation matrix of G and $\mathbf{T}_i$ denotes a component of $\mathbf{T} \in \mathfrak{T}$.

We now define a **tensor operator T** to be the set of operators $\{\mathbf{T}_i\}$. An **irreducible tensor operator** is a tensor operator whose components $\{\mathbf{T}_i\}$ satisfy equivariance with respect to an an irrep $\lambda = [\mathbf{m}]$ of $G = U(n)$. The individual components $\mathbf{T}_i$ of the irreducible tensor operator $\mathbf{T}$ are labelled by the Gel'fand-Weyl patterns (m) belonging to the irrep $[\mathbf{m}]$.

For $SU(2)$, irreducible tensor operators $\mathbf{T}_j$ are labelled by the irrep label j, with $2j+1$ components $\mathbf{T}_{jm}$, where $-j \leqslant m \leqslant j$, and the equivariance condition states

$$\mathbf{U}(g)\,\mathbf{T}_{jm}\;\mathbf{U}^{-1}(g) = \sum_{m'=-j}^{j} \mathbf{T}_{jm'}\;\mathfrak{D}^j_{m'm}(g), \quad g \in SU(2),$$

where $\mathfrak{D}^j$ are irrep matrices of $SU(2)$.

At the level of the Lie algebra $\mathfrak{g}$ the action of the generator $E_\alpha \in \mathfrak{g}$ on the components $\mathbf{T}_{(m)}$ of the irreducible tensor operator $\mathbf{T}$, is the **adjoint action**, given by

$$[E_\alpha, \mathbf{T}_{(m)}] \;=\; \sum_{(m')} \langle (m')|E_\alpha|(m)\rangle \mathbf{T}_{(m')},$$

where $\langle (m')|E_\alpha|(m)\rangle$ is the matrix of the generator E_α for the irrep $[\mathbf{m}]$. For $\mathfrak{su}(2)$ this adjoint action is given by the equations

$$\begin{aligned} \left[J_\pm, \mathbf{T}_{jm}\right] &= \sqrt{(j \pm m+1)(j \mp m)}\;\mathbf{T}_{j,m\pm 1} \\ \left[J_z, \mathbf{T}_{jm}\right] &= m\mathbf{T}_{jm}, \end{aligned} \tag{3.4}$$

where $J_\pm, J_z$ are the usual $\mathfrak{su}(2)$ generators. We refer to this property of $\mathfrak{su}(2)$ tensor operators as the **derivative** property.

The tensor operator classification is not categoric, as is clear from noting that multiplying a given irreducible tensor operator by an operator invariant under G leaves the classification unchanged. To eliminate this freedom one can define unit tensor operators whose norm (defined below in Corollary 3.24) is unity. For $G = U(n)$ with $n \geqslant 3$, distinct unit tensor operators with the same irrep labels $[m_{1n} \cdots m_{nn}]$ exist, but it has been shown, as stated below in Theorem 3.10, that a canonical, or natural, labelling splitting this multiplicity exists for $n = 3$, and very likely for all $n > 3$.

Applications of symmetries in quantum physics are heavily dependent on the fundamental theorem for tensor operators, namely the generalized Wigner-Eckart theorem, and the corresponding generalized WCG coefficients. Such a theorem need not exist for a given symmetry, but depends rather upon the specific way in which the symmetry is realized [66]. The two required properties of the realization, as mentioned already, are equivariance and an adjoint action on the tensor operators (or more generally the derivation property). It is a consequence of these two properties that the generalized WCG coefficients for a given symmetry group G occur in two logically distinct ways:

1. as *coupling coefficients* for the Kronecker product of irreps carried by kinematically independent constituent systems (this is the Clebsch-Gordan problem) and,

2. as *matrix elements*, up to a rotationally invariant scale factor, of physical transition operators (this is the Wigner-Eckart problem).

If equivariance and derivation are not valid for a given realization, then property (2) does not hold.

Let us now summarize the main properties of tensor operators and then discuss these properties in greater detail for quantum groups:

1. Tensor operators $\mathbf{T}$ form an algebra with a composition, denoted Ⓦ since the composition is constructed with WCG coefficients, that carries two tensor operators into a sum of tensor operators. This algebra is associative and unital, with an involution; the conjugate tensor operator is denoted $\mathbf{T}^\dagger$ or $\overline{\mathbf{T}}$.

2. A positive semi-definite norm can be defined which maps each irreducible tensor operator $\mathbf{T}_\lambda$ into $\mathbf{T}_\lambda^\dagger \boldsymbol{\cdot} \mathbf{T}_\lambda$ which is an invariant positive operator. The zeros of the semi-definite norm determine the null space of the tensor operator.

3. Unit tensor operators $\widehat{\mathbf{T}}_\lambda$ can now be defined as normalized tensor operators such that $\widehat{\mathbf{T}}_\lambda^\dagger \boldsymbol{\cdot} \widehat{\mathbf{T}}_\lambda$ is a projection operator with eigenvalues 1 or 0 on irrep subspaces of $\mathfrak{M}$. Since unit tensor operators are bounded, it follows from the Gel'fand-Naimark theorem that the components of unit tensor operators belong to a C*-algebra.

4. The unit tensor operators form a linear basis for all tensor operators with invariant operators as scalar multipliers.

5. **Elementary unit tensor operators** are those with an irrep having a Young frame of the form: $[\underbrace{1\cdots1}_{k}\underbrace{0\cdots0}_{n-k}]$, $(k = 1, 2, \cdots n-1.)$ All elementary unit tensor operators for $SU(n)$, $n \geqslant 2$, are explicitly constructible by an algorithm known as the pattern calculus (due to Biedenharn and Louck [126]), and described for $\mathcal{U}_q(\mathrm{u}(n))$ in §3.7 (p. 107). The elementary unit tensor operators constitute an algebraic generating set for all tensor operators.

6. The universal enveloping algebra is a subalgebra of the algebra of tensor operators characterized by, firstly, an action on $\mathfrak{M}$ that is diagonal on the irrep labels in subspaces of $\mathfrak{M}$ and, secondly, by a semi-definite norm that is polynomial in the generalized Casimir invariants.

3.2.1 The Classification Problem for Tensor Operators

The problem of classifying all tensor operators is both fundamental and difficult. The equivariance condition (Definition 3.3) greatly helps in reducing the problem to more manageable proportions. For irreducible tensor operators we have already seen that equivariance allows a partial classification, the assignment of the irrep label λ.

In addition, components of irreducible tensor operators in $U(n)$ are labelled by a Gel'fand-Weyl pattern which provides a canonical labelling for vectors in the carrier space of the irrep $\lambda = [\mathbf{m}]$ and hence a labelling for an individual operator component in the set $\mathbf{T}_\lambda$. The Gel'fand-Weyl pattern (m) encodes the information that the component $\mathbf{T}_{\lambda,(m)}$ of the tensor operator $\mathbf{T}_\lambda$ is, firstly, a component of a tensor operator belonging to the $U(n)$ irrep $[m_{1n} \cdots m_{nn}]$; secondly, a component of a tensor operator belonging to the $U(n-1)$ irrep $[m_{1,n-1} \cdots m_{n-1,n-1}]$, acting in a subspace of $\mathfrak{M}$; and so on, and finally, a component of a tensor operator belonging to the $U(1)$ irrep m_{11} acting in a subspace of $\mathfrak{M}$. That is, the components of the tensor operator $\mathbf{T}_\lambda$ are classified by irrep labels in the canonical subgroup chain $U(n) \supset U(n-1) \supset \cdots \supset U(1)$ using the Weyl branching theorem.

As noted in §2.6, the Gel'fand-Weyl pattern determines the weights (w_i) assigned to vectors carrying the irrep $[m]$ by the relation: $w_i = \sum_j m_{j,i} - \sum_j m_{j,i-1}$, where $m_{ij} \equiv 0$ if $i > j$. The labelling by a Gel'fand-Weyl pattern uniquely and canonically splits all multiplicity in the weight space labelling.

Having classified irreducible tensor operators by their irreps, and their components by Gel'fand-Weyl patterns, we must also normalize to obtain unit tensor operators in order to eliminate arbitrary multiplicative invariant factors. Still, the classification is not complete. To refine the classification further one introduces a labelling based on the change in the irrep labels of subspaces of $\mathfrak{M}$ induced by the action of the irreducible unit tensor operator $\hat{\mathbf{T}}_\lambda$. These changes are given by the **shifts**, defined by

$$\Delta_i \stackrel{\text{def}}{=} m_{in}^{\text{final}} - m_{in}^{\text{initial}},$$

where $[m^{\text{final}}]$ and $[m^{\text{initial}}]$ are $U(n)$ irrep labels determined by the action of $\hat{\mathbf{T}}_\lambda$ on irrep subspaces $\mathfrak{V}_{([m^{\text{initial}}])}$ and $\mathfrak{V}_{([m^{\text{final}}])}$ of $\mathfrak{M}$, that is, $\hat{\mathbf{T}}_{\lambda,(m)} : \mathfrak{V}_{([m^{\text{initial}}])} \to \mathfrak{V}_{([m^{\text{final}}])}$.

For a tensor operator belonging to the irrep $\lambda = [m_{1n} \cdots m_{nn}]$, the shift values are constrained by group theory, that is, by the decomposition of the Kronecker product into irreducible constituents. It is well-known that the *possible* (as opposed to actual) shift values for a tensor operator $\mathbf{T}_\lambda$ acting in $\mathfrak{M}$ are precisely the weights of the irrep λ, including multiplicity. The multiplicities are the Kostka numbers [127]. The *actual* shift values are determined by the triple of irreps, $[m^{\text{initial}}]$, λ and $[m^{\text{final}}]$, with multiplicities given by the Littlewood-Richardson numbers [127].

The shifts $\Delta = \{\Delta_i\}$ are the operator space analog, for an irreducible tensor operator $\mathbf{T}_\lambda$, of the weights of a representation in group theory. Just as the weights of a vector in an irrep space do not always characterize a vector uniquely, so too the shifts of the unit tensor operator component $\widehat{\mathbf{T}}_{\lambda,(m)}$ do not in general categorize an operator uniquely. Two concepts which are of assistance in order to further refine the classification are those of operator patterns and the characteristic nullspace, considered in the following section. However, in special cases, in particular for $SU(2)$ and $\mathcal{U}_q(\mathfrak{su}(2))$, the shifts are sufficient in order to complete the classification of all unit tensor operators.

A canonical classification of *all* irreducible unit tensor operator components has, so far, been proven to exist only for $U(2)$ and $U(3)$ but the result is conjectured to be true for all $U(n)$ and has been verified in a large number of individual cases. If this conjecture is valid, the canonical construction of $3j, 6j, \cdots, 3nj$ tensor operators [103, 66] can be carried out, developing extended tensor operator algebras [103], and for quantum groups as well. This would have many significant consequences, for example it would imply the existence of a very large number of generalized hypergeometric and basic hypergeometric identities.

3.2.2 Operator Patterns and the Characteristic Null Space

Significant progress toward fully resolving the classification problem of the previous section can be made by exploiting further the analogy that exists between the set of weights $\{w_i\}$ and the shifts $\{\Delta_i\}$ of operator space, and by taking advantage of an important result, due to Baird and Biedenharn [128]:

LEMMA **3.5** *The number of linearly independent irreducible tensor operators with irrep labels* $[\mathbf{m}]$ *equals the dimension* $D([\mathbf{m}])$ *of the irrep of* $U(n)$ *with labels* $[\mathbf{m}]$.

Hence, we can enumerate irreducible tensor operators using patterns similar to the Gel'fand-Weyl patterns. Let us therefore introduce the notion of an **operator pattern**:

DEFINITION **3.6** *An operator pattern* (Γ) *is a triangular array of integers* Γ_{ij}*:*

$$(\Gamma) \stackrel{\text{def}}{=} \begin{pmatrix} \Gamma_{1n} & & \Gamma_{2n} & \cdots & \Gamma_{n-1,n} & & \Gamma_{nn} \\ & \Gamma_{1,n-1} & & \cdots & & \Gamma_{n-1,n-1} & \\ & & & \vdots & & & \\ & & & \Gamma_{11} & & & \end{pmatrix}$$

where (a) the integers Γ_{ij} *obey the inequalities (betweenness relations)* $\Gamma_{i,j} \geqslant \Gamma_{i,j-1} \geqslant \Gamma_{i+1,j}$ *and (b) for the unit tensor operator* $\hat{\mathbf{T}}_\lambda$ *with* $\lambda = [m_{1n}, \cdots, m_{nn}]$ *we have* $\Gamma_{i,n} = m_{i,n}$.

We use these patterns (Γ) to enumerate irreducible tensor operators, now denoted $\hat{\mathbf{T}}_{(\Gamma)}$, noting the following:

Remark 3.7 1. The shifts Δ_i associated with the unit tensor operator $\hat{\mathbf{T}}_{(\Gamma)}$ are given by:

$$\Delta_i = \sum_j \Gamma_{j,i} - \sum_j \Gamma_{j,i-1} \qquad (\text{with } \Gamma_{ij} \equiv 0 \text{ if } i > j).$$

This shift rule for operators patterns is in precise analogy to the weight rule for Gel'fand-Weyl patterns.

2. An operator pattern is structurally a Gel'fand-Weyl pattern of integers since both obey similar inequalities and determine 'weights' but, unlike a Gel'fand-Weyl pattern, for an operator pattern there is no group-subgroup implication. Put differently, there is no analog for an operator pattern of the Weyl branching theorem encoded in the structure of a Gel'fand-Weyl pattern.

3. For a unit tensor operator component of $\hat{\mathbf{T}}_\lambda$, classified both by a Gel'fand-Weyl pattern (m) and an operator pattern (Γ), we have $\lambda = [m_{1n}, m_{2n}, \cdots, m_{nn}] = [\Gamma_{1n}, \Gamma_{2n} \cdots, \Gamma_{nn}]$ with $\Gamma_{in} = m_{in}$. That is, (Γ) and (m) share a common top row, the Young frame λ. Since inverting a triangular pattern of integers leaves the betweenness relations unchanged, it is economical notationally to join the patterns (m) and (Γ) inverted as a single diamond-shaped pattern sharing a common largest row, and denote the tensor operator by

$$\left\langle \begin{matrix} (\Gamma) \\ [M] \\ (m) \end{matrix} \right\rangle,$$

where $[M]$ denotes the $U(n)$ irrep.

4. A tensor operator in $U(n)$ is also a tensor operator in $U(n-1)$ and since unit tensor operators in $U(n-1)$ form a basis for all $U(n-1)$ tensor operators, we may expand the $U(n)$ tensor operator as a linear combination:

$$\left\langle \begin{matrix} (\Gamma)_{n-1} \\ [M]_n \\ (m)_{n-1} \end{matrix} \right\rangle = \sum_{(\gamma)_{n-2}} \left[\begin{matrix} (\Gamma)_{n-1} \\ [M]_n \\ (\gamma)_{n-1} \end{matrix} \right] \left\langle \begin{matrix} (\gamma)_{n-2} \\ [M]_{n-1} \\ (m)_{n-2} \end{matrix} \right\rangle, \tag{3.8}$$

where the pattern subscripts indicate to which group the pattern refers, and where the coefficients, denoted $[\vdots]$, are $U(n-1)$ invariant operators which form a **unit projective operator** in $U(n)/U(n-1)$.

As an example of operator patterns consider $U(2)$, for which a component of an irreducible tensor operator is denoted

$$\left\langle \begin{matrix} & \Gamma_{11} & \\ m_{12} & & m_{22} \\ & m_{11} & \end{matrix} \right\rangle$$

with $m_{12} \geqslant \Gamma_{11} \geqslant m_{22}$, $m_{12} \geqslant m_{11} \geqslant m_{22}$, and where the operator pattern entry Γ_{11} coincides with the weight Δ_1. For $SU(2)$, and hence for $\mathcal{U}_q(\mathfrak{su}(2))$, the components are labelled

$$\left\langle \begin{matrix} & J+\Delta & \\ 2J & & 0 \\ & J+M & \end{matrix} \right\rangle \tag{3.9}$$

where $-J \leqslant \Delta \leqslant J$, $-J \leqslant M \leqslant J$, and J is the irrep label.

The space of operator patterns is strikingly similar to the space of Gel'fand-Weyl patterns but, unlike Gel'fand-Weyl patterns, one can show that there is no underlying group structure. Something more than a group seems to be involved in which some group-like properties are restored in the limit of large m_{nn}.

Next, let us consider the null space properties of tensor operators which enable us to further classify these operators, in particular, to distinguish tensor operators with identical weights Δ_i. The invariant semi-definite norm associated with a unit tensor operator $\hat{\mathbf{T}}_\lambda$ is a projection operator on $\mathfrak{M}$. Let us define the **characteristic null space** $\mathfrak{N}$ of the unit tensor operator $\hat{\mathbf{T}}_\lambda$ to be the direct sum of all irrep carrier spaces $\mathfrak{V}_{([\mu])}$, where μ are $U(n)$ irreps, such that all vectors in $\mathfrak{V}_{([\mu])}$ are annihilated by all components of $\hat{\mathbf{T}}_\lambda$. It follows by construction that the characteristic null space of the unit tensor operator $\hat{\mathbf{T}}_\lambda$ coincides with the null space of the invariant norm.

Consider now the set of distinct irreducible unit tensor operators $\hat{\mathbf{T}}_\lambda$ having the same shift labels Δ. It can be shown [127] that the characteristic null spaces $\{\mathfrak{N}_i\}$ associated with this set are simply ordered by inclusion: $\mathfrak{N}_1 \supset \mathfrak{N}_2 \supset \cdots$. If the number of distinct $\mathfrak{N}_i$ equals the Kostka number for the multiplicity of the set $\{\hat{\mathbf{T}}_\lambda(\Delta)\}$ then it follows that a labelling by characteristic null spaces canonically splits the multiplicity for the set $\{\hat{\mathbf{T}}_\lambda(\Delta)\}$.

For $SU(2)$, the characteristic null space of the unit tensor operator

$$\left\langle \begin{matrix} & J+\Delta & \\ 2J & & 0 \\ & \bullet & \end{matrix} \right\rangle,$$

which has components as shown in (3.9), is the subspace $\mathfrak{N}_{\frac{1}{2}(J-\Delta-1)}$ of $\mathfrak{M}$ consisting of the direct sum

$$\mathfrak{N}_{\frac{1}{2}(J-\Delta-1)} \stackrel{\text{def}}{=} \mathfrak{V}_0 \oplus \mathfrak{V}_{\frac{1}{2}} \oplus \mathfrak{V}_1 \oplus \ldots \oplus \mathfrak{V}_{\frac{1}{2}(J-\Delta-1)},$$

where $\mathfrak{V}_j$ is the vector space spanned by $|jm\rangle$, $-j \leqslant m \leqslant j$. Clearly. these null spaces are simply ordered (for a detailed discussion see [103, Chapter 2, p. 21]).

For $SU(3)$ we may state [129]:

THEOREM **3.10** *For irreducible unit tensor operators in $SU(3)$ the characteristic null spaces are all distinct and ordered by inclusion. There exists a canonical splitting of all multiplicity in $SU(3)$ classifying uniquely all components of all irreducible unit tensor operators.*

This is a very satisfactory result for $SU(3)$, and *a posteriori* for $SU(2)$. Unfortunately, a general classification by characteristic null space alone is not possible, as a theorem due to Baclawski [130] shows.

Further information on the meaning of the operator patterns, and their numerical values, which classify irreducible tensor operators can be obtained by consideration of limit relations, to which we now turn. Progress in the analysis of tensor operator algebras has been achieved [131, 132] by exploiting a variant of the Borel-Weil construction of irreps, which we describe in Chapter 7 together with its quantum group extension, and which can be used to determine recursive properties such as limit relations.

The Borel-Weil construction induces an irreducible representation from a one-dimensional representation of the Cartan maximal toroid. The variant which we shall exploit, following Hecht [133], uses the Borel-Weil procedure recursively, to generate a $U(n)$ irreducible representation from one of the subgroup $U(n-1) \times U(1)$. The irreps generated by this procedure are, to within an easily determined normalization, the tensor product of a symmetric $U(n-1)$ irrep and a generic $U(n-1) \times U(1)$ irrep, so that by assuming properties of $U(n-1)$ (which is the recursion hypothesis) we may determine $U(n)$ properties in terms of those of $U(n-1)$. The tensor product is effected by $U(n-1)$ $3j$ coefficients, which are matrix elements of $U(n-1)$ irreducible unit tensor operators. The significance of this construction can be seen by remarking that, firstly, these $U(n)$ irreps are constructed algebraically entirely using properties of $U(n-1)$ tensor operators and, secondly, the method is well-adapted to investigating, order-by-order, the limit in which $m_{nn} \to -\infty$. We return to this inductive approach again in Chapter 7 where we consider algebraic induction in detail and generalize the methods to quantum groups. However, let us state a theorem which may be used to assign numerical Γ patterns in $U(n)$:

THEOREM **3.11** *Matrix elements of the $U(n)$ irreducible unit tensor operator $\hat{\mathbf{T}}_{(m),(\Gamma)}$ labelled by the Gel'fand pattern (m) and the operator pattern (Γ) obey the limit relation:*

$$\lim_{m_{nn} \to -\infty} \hat{\mathbf{T}}_{(m),(\Gamma)} = \prod_{i=1}^{n-1} \delta_{m_{i,n-1},\Gamma_{i,n-1}} \mathcal{I}\, \hat{\mathbf{T}}_{(m'),(\Gamma')}$$

where it is understood that (a) matrix elements are to be taken in $\mathfrak{M}$ (so that m_{nn} is well defined) (b) the operator $\mathcal{I}$ is an invariant of $U(n)$ and is uniquely defined, and (c) the (m') and (Γ') patterns have row n (the top row) deleted.

For a more complete statement we refer to [131, 132]. This result is essential to the classification problem since it establishes recursively the value of all operator pattern integers Γ_{ij}. (Recall that by contrast the Gel'fand-Weyl pattern integers m_{ij} are eigenvalues of well-defined operators [134]).

3.3 Tensor Operators in Quantum Groups

Now let us consider the generalization of the results, reviewed and discussed in the previous sections, to quantum groups. The concept of a model space extends without modification. However, it is not obvious that one can extend the two fundamental properties of equivariance and derivation to quantum groups, particularly when one realizes that the derivative property corresponds to a co-product which is commutative, and therefore not relevant to a quantum group. Moreover, the equivariance condition is problematic as well. For a Lie algebra, equivariance is effected by the adjoint action. This cannot work in any simple way for a quantum group since the quantum group irrep corresponding to the adjoint representation is finite dimensional in contrast to the infinite number of linearly independent algebra elements that may be obtained under commutation. The resolution of the Clebsch-Gordan problem for $\mathcal{U}_q(\mathfrak{su}(2))$ has been published by many authors, including Kirillov and Reshetikhin [65], Nomura [76], Vaksman [136], Ruegg [137], Groza et al. [60], and for the Wigner-Eckart problem, the resolution is given by Biedenharn and Tarlini [138], see also [65]. We now follow [138].

Let the Hilbert space on which the operators act be the model space $\mathfrak{M}$ for the quantum group and let us now introduce the key concept of an **induced action**, which is a mapping

$$\mathfrak{T} \otimes \mathfrak{T} \longrightarrow \mathfrak{T}$$

which we will denote by $A(B) = C$, that is, A acts on B to give C, where $A, B, C \in \mathfrak{T}$. One must distinguish this action from the natural product action of operators on a Hilbert space, which we denote as usual by juxtaposition: AB. Using these concepts, we can now recognize that the standard setting for the generalized Wigner-Eckart theorem for classical Lie algebras consists first of the induced co-product Δ:

$$\Delta(E_\alpha)(A \otimes |m\rangle) = E_\alpha(A) \otimes \mathbb{I}\,|m\rangle + \mathbb{I}(A) \otimes E_\alpha\,|m\rangle, \tag{3.12}$$

(where E_α is a generator of $\mathfrak{g}$, $A \in \mathfrak{T}$ and $|m\rangle \in \mathfrak{M}$), followed by a mapping $c : \mathfrak{T} \otimes \mathfrak{M} \to \mathfrak{M}$. Applying this operation in the standard case ($q = 1$) yields

$$E_\alpha(A|m\rangle) \;=\; E_\alpha(A)|m\rangle + AE_\alpha|m\rangle,$$

for which one uses the induced action $E_\alpha(A) \overset{\text{def}}{=} [E_\alpha, A]$, that is, the adjoint action.

The distinctive features of this construction are that the co-product Δ acts, via (3.12), on the tensor product $\mathfrak{T} \otimes \mathfrak{M}$ of different vector spaces and in addition there is a further operation, the mapping c, which requires that the induced co-product and the induced action be compatible. The compatibility of these two structures is guaranteed for the standard Lie group case since the matrix action of operators in the Hilbert space induces, by commutation, a commutative co-product (which is a derivation) and the induced commutator action of the generators realizes, by equivariance, an irrep carried by the tensor operators. This compatibility is the content of the tensor operator theorem, or generalized Wigner-Eckart theorem.

The compatibility requirement can be expressed most succinctly by using the language of diagrams. Consider the following diagram (where $E \in \mathfrak{g}$):

$$\begin{array}{ccc} \mathfrak{T} \otimes \mathfrak{M} & \xrightarrow{\Delta(E)} & \mathfrak{T} \otimes \mathfrak{M} \\ c \big\downarrow & & \big\downarrow c \\ \mathfrak{M} & \xrightarrow{E} & \mathfrak{M} \end{array} \tag{3.13}$$

The requirement of compatibility is that the diagram above be commutative. Assuming a given co-product determines a compatible action (or vice-versa). Thus, for example, for a Lie group one uses the diagonal, commutative, co-product and determines the commutator to be the compatible induced action.

We can now extend this structure to tensor operators for quantum groups (which we generally refer to as q-tensor operators). Let us formalize these considerations, following [138]:

Definition **3.14** *Let $\mathfrak{T}$ denote the vector space of operators $\mathbf{T}$ mapping the model space $\mathfrak{M}$ of the compact quantum group G_q into itself, that is, $\mathbf{T} : \mathfrak{M} \longrightarrow \mathfrak{M}$. An irreducible q-tensor operator is a set of operators $\{\mathbf{T}_{[m],(m)}\} \in \mathfrak{T}$ which carries a finite-dimensional irrep $[m]$, with vectors labelled (m), of the quantum group G_q. That is:*

$$E_\alpha\left(\mathbf{T}_{[m],(m)}\right) = \sum_{(m')} \langle [m],(m')|\, E_\alpha\, |[m],(m)\rangle\, \mathbf{T}_{[m],(m')},$$

where E_α is a generator of G_q, $E_\alpha\left(\mathbf{T}_{[m],(m)}\right)$ denotes the action of E_α on $\mathbf{T}_{[m],(m)} \in \mathfrak{T}$, and $\left\langle \cdots \right\rangle$ denotes the matrix elements of the generators for the irrep $[m]$. A q-tensor operator is accordingly a linear combination of irreducible q-tensor operators, with coefficients invariant under the action of the quantum group.

Next, we note that the direct product of irreps of a quantum group is completely reducible (Jimbo [23]), that is, the direct product may be decomposed into a sum of irreducible constituents. Hence, we may define the q-analog of the Wigner-Clebsch-Gordan coefficients, q-WCG coefficients for short, in the usual way:

Definition **3.15** *The q-WCG coefficients of the quantum group G_q are matrix elements of the unitary transformation which diagonalizes the direct product of two unitary irreps of G_q.*

As an example, consider $\mathcal{U}_q(\mathfrak{su}(2))$ and denote by $\mathfrak{V}_j$ the space carrying an irreducible representation parametrized by the half-integer j. We may decompose the direct product of the irrep spaces $\mathfrak{V}_{j_1}$ and $\mathfrak{V}_{j_2}$, which is of dimension $(2j_1+1)(2j_2+1)$, as follows:

$$\mathfrak{V}_{j_1} \otimes \mathfrak{V}_{j_2} = \sum_j \oplus \mathfrak{V}_j,$$

where the sum is over all values of j such that $|j_1 - j_2| \leqslant j \leqslant j_1 + j_2$ (this follows in the same way as for $SU(2)$ [66, §3.11]). The requirements that

$$j \in \{j_1 + j_2, j_1 + j_2 - 1, \ldots, |j_1 - j_2|\} \tag{3.16}$$

are known as the **triangle conditions**. Let $|jm\rangle$ be a weight basis in the irreducible component $\mathfrak{V}_j$ and denote by $|(j_1 j_2) jm\rangle$ a basis in $\mathfrak{V}_{j_1} \otimes \mathfrak{V}_{j_2}$. The q-WCG coefficients ${}_qC^{j_1\ j_2\ j}_{m_1\ m_2\ m}$, which comprise the unitary transformation which diagonalizes the reducible direct product (co-multiplied) representation, are given by

$$|(j_1 j_2) jm\rangle = \sum_{m_1, m_2} {}_qC^{j_1\ j_2\ j}_{m_1\ m_2\ m}\, |j_1 m_1\rangle \otimes |j_2 m_2\rangle. \tag{3.17}$$

We say that these q-WCG coefficients couple the irreps $(j_1) \otimes (j_2) \to (j)$. Since the diagonalizing transformation is unitary, and since we may choose the q-WCG coefficients to be real, these coefficients satisfy orthogonality conditions corresponding to orthogonality of the rows, and the columns, of the diagonalizing matrix. Hence, we may rewrite (3.17) as

$$|j_1 m_1\rangle \otimes |j_2 m_2\rangle = \sum_{j,m} \varepsilon_{j_1 j_2 j}\, {}_qC^{j_1\ j_2\ j}_{m_1\ m_2\ m}\, |(j_1 j_2) jm\rangle,$$

where $\varepsilon_{j_1 j_2 j} = 1$ if the triangle conditions are satisfied, and zero otherwise. We consider detailed properties of the q-WCG coefficients for $\mathcal{U}_q(\mathfrak{su}(2))$, including orthogonality, recurrence relations, special cases, and symmetries in §3.5,3.5.1,3.5.2.

We now state and prove a theorem which shows, as in the classical case, that the q-WCG coefficients are also matrix elements of irreducible tensor operators:

THEOREM **3.18** *If $\{\mathbf{T}_{[m],(m)}\}$ is a q-tensor operator of the compact quantum group G_q such that the co-product of G_q is compatible with the action $E_\alpha\left(\mathbf{T}_{[m],(m)}\right)$, that is, diagram (3.13) is commutative, then the matrix elements of $\{\mathbf{T}_{[m],(m)}\}$ with respect to a basis in $\mathfrak{M}$ are proportional to the q-WCG coefficients of G_q, with the constant of proportionality an invariant.*

Conversely, if $\{\mathbf{T}_{[m],(m)}\}$ is a q-tensor operator and if the matrix elements of $\{\mathbf{T}_{[m],(m)}\}$ are proportional to the q-WCG coefficients, then diagram (3.13) is commutative.

PROOF: Since the q-WCG coefficients of G_q are matrix elements of the unitary transformation which diagonalizes the direct product of two unitary irreps of G_q, these matrix elements accordingly obey the defining relation:

$$\begin{aligned}
&\Delta(E_{\pm i})\left(\left|\begin{matrix}a\\ \alpha'\end{matrix}\right\rangle \otimes \left|\begin{matrix}b\\ \beta'\end{matrix}\right\rangle\right)\\
&= \left(q^{-\frac{h_i}{4}}\left|\begin{matrix}a\\ \alpha'\end{matrix}\right\rangle\right) \otimes \left(E_{\pm i}\left|\begin{matrix}b\\ \beta'\end{matrix}\right\rangle\right) + \left(E_{\pm i}\left|\begin{matrix}a\\ \alpha'\end{matrix}\right\rangle\right) \otimes \left(q^{\frac{h_i}{4}}\left|\begin{matrix}b\\ \beta'\end{matrix}\right\rangle\right)\\
&= \sum_{c,\gamma,\gamma',\Gamma} {}_{q^{-1}}C^{b\,\binom{\Gamma}{a}\,c}_{\beta'\,\alpha'\,\gamma'} \left\langle\begin{matrix}c\\ \gamma\end{matrix}\right| E_{\pm i} \left|\begin{matrix}c\\ \gamma'\end{matrix}\right\rangle \sum_{\alpha,\beta} {}_{q^{-1}}C^{b\,\binom{\Gamma}{a}\,c}_{\beta\,\alpha\,\gamma} \left|\begin{matrix}a\\ \alpha\end{matrix}\right\rangle \otimes \left|\begin{matrix}b\\ \beta\end{matrix}\right\rangle,
\end{aligned} \tag{3.19}$$

where $E_{\pm i}$ denotes a raising/lowering generator of G_q and h_i is the associated Cartan basis element, $|{}^{a}_{\alpha}\rangle$ is the vector labelled α in the irrep a, and ${}_{q^{-1}}C^{b\,\binom{\Gamma}{a}\,c}_{\beta\,\alpha\,\gamma}$ denotes the q-WCG coefficients of G_q coupling the irreps $a \otimes b \to c$ (in this order, see the following Remark 3.23). The operator pattern given by the label Γ is determined by a unique orthogonal splitting of any multiplicity. As explained in §3.2.1 and §3.2.2, this labelling is known to be free of arbitrariness, to within equivalence, for all simply reducible groups (which includes $\mathcal{U}_q(\mathfrak{su}(2))$), and also for $\mathcal{U}_q(\mathfrak{u}(3))$ as shown in [129]. It is very likely to be canonical for all $\mathcal{U}_q(\mathfrak{su}(n))$.

Now, because of the way the action has been defined for the q-tensor operator $\{\mathbf{T}_{[m],(m)}\}$, Eqn. (3.19) is valid for $\mathbf{T}_{a,\alpha}$ acting on $|{}^{b}_{\beta}\rangle$. That is,

$$\begin{aligned}
&\Delta(E_{\pm i})\left(\mathbf{T}_{a,\alpha'} \otimes \left|\begin{matrix}b\\ \beta'\end{matrix}\right\rangle\right)\\
&= \sum_{c,\gamma,\gamma',\Gamma} {}_{q^{-1}}C^{b\,\binom{\Gamma}{a}\,c}_{\beta'\,\alpha'\,\gamma'} \left\langle\begin{matrix}c\\ \gamma\end{matrix}\right| E_{\pm i} \left|\begin{matrix}c\\ \gamma'\end{matrix}\right\rangle \sum_{\alpha,\beta} {}_{q^{-1}}C^{b\,\binom{\Gamma}{a}\,c}_{\beta\,\alpha\,\gamma} \mathbf{T}_{a,\alpha} \otimes \left|\begin{matrix}b\\ \beta\end{matrix}\right\rangle.
\end{aligned} \tag{3.20}$$

If we assume that diagram (3.13) is commutative, then we can contract and take matrix elements with the bra-vector $\langle{}^{c''}_{\gamma''}|$ belonging to the dual space of $\mathfrak{M}$; then we have for the left hand side of (3.20):

$$\sum_{\gamma'''} \left\langle\begin{matrix}c''\\ \gamma''\end{matrix}\right| E_{\pm i} \left|\begin{matrix}c''\\ \gamma'''\end{matrix}\right\rangle \left\langle\begin{matrix}c''\\ \gamma'''\end{matrix}\right| \mathbf{T}_{a,\alpha'} \left|\begin{matrix}b\\ \beta'\end{matrix}\right\rangle,$$

where we used the completeness relation to write the sum as shown. The right hand side of (3.20) becomes

$$\sum_{c,\gamma,\gamma',\Gamma} \left\langle\begin{matrix}c\\ \gamma\end{matrix}\right| E_{\pm i} \left|\begin{matrix}c\\ \gamma'\end{matrix}\right\rangle {}_{q^{-1}}C^{b\,\binom{\Gamma}{a}\,c}_{\beta'\,\alpha'\,\gamma'} \left(\sum_{\alpha,\beta} {}_{q^{-1}}C^{b\,\binom{\Gamma}{a}\,c}_{\beta\,\alpha\,\gamma} \left\langle\begin{matrix}c''\\ \gamma''\end{matrix}\right| \mathbf{T}_{a,\alpha} \left|\begin{matrix}b\\ \beta\end{matrix}\right\rangle\right).$$

Consider the term $(\cdots)$ appearing in this equation. The q-WCG coefficients, by definition, couple to a unique vector in an irrep of G_q, and hence it follows that the term $(\cdots)$ is equal to

$$\sum_{\alpha,\beta} {}_{q^{-1}}C^{b\ \binom{\Gamma}{a}\ c}_{\beta\ \alpha\ \gamma} \left\langle \begin{matrix} c'' \\ \gamma'' \end{matrix} \right| \mathbf{T}_{a,\alpha} \left| \begin{matrix} b \\ \beta \end{matrix} \right\rangle = \delta_c^{c''} \delta_\gamma^{\gamma''}\, \mathcal{I}(c, \mathbf{T}_a, b; \Gamma),$$

where $\mathcal{I}(c, \mathbf{T}_a, b; \Gamma)$ denotes an invariant proportionality constant.

By combining these results, and re-labelling dummy indices, we find

$$\sum_{\gamma'} \left\langle \begin{matrix} c \\ \gamma \end{matrix} \right| E_{\pm i} \left| \begin{matrix} c \\ \gamma' \end{matrix} \right\rangle \left(\left\langle \begin{matrix} c \\ \gamma' \end{matrix} \right| \mathbf{T}_{a,\alpha'} \left| \begin{matrix} b \\ \beta' \end{matrix} \right\rangle - \sum_{\Gamma} {}_{q^{-1}}C^{b\ \binom{\Gamma}{a}\ c}_{\beta'\ \alpha'\ \gamma'}\, \mathcal{I}(c, \mathbf{T}_a, b; \Gamma) \right) = 0. \tag{3.21}$$

By multiplying with the ket-vector $\left|\begin{smallmatrix} c \\ \gamma \end{smallmatrix}\right\rangle$, and summing over all γ, the completeness relation implies that (3.21) is valid for all generators of G_q and all vectors in the irrep $c \in \mathfrak{M}$. This implies (dropping the primes) that

$$\left\langle \begin{matrix} c \\ \gamma \end{matrix} \right| \mathbf{T}_{a,\alpha} \left| \begin{matrix} b \\ \beta \end{matrix} \right\rangle = \sum_{\Gamma} {}_{q^{-1}}C^{b\ \binom{\Gamma}{a}\ c}_{\beta\ \alpha\ \gamma}\, \mathcal{I}(c, \mathbf{T}_a, b; \Gamma), \tag{3.22}$$

which is the assertion of the theorem. (The occurrence of the operator pattern (Γ) shows that q-tensor operators can be classified further: in addition to the irrep label a and the state vector pattern α there exists an operator pattern label Γ.)

The converse, in which we assume (3.22) to be valid, is easily seen to imply compatibility. □

REMARK **3.23** We have defined the q-WCG coefficient ${}_{q^{-1}}C^{\cdots}_{\cdots}$ in this theorem with q^{-1} instead of q in order to conform to the standard definition of the q-WCG coefficients in $\mathcal{U}_q(\mathfrak{su}(2))$. For the coupling of irreps in the order $(j_1) \otimes (j_2) \to (j)$ the q-WCG coefficient is given by ${}_qC^{j_1\ j_2\ j}_{m_1\ m_2\ m}$ as shown in (3.17), and if we reverse the order of the coupling then q is replaced by q^{-1}, as the symmetry (3.63b) derived below in Theorem 3.62 shows.

3.4 The Algebra of q-Tensor Operators

We have seen in Theorem 3.18 that the q-WCG coefficients are the matrix elements of a map W from $\mathfrak{M} \otimes \mathfrak{M}$ into $\mathfrak{M}$. Expressing this diagrammatically we have the following commutative diagram:

$$\begin{array}{ccc} \mathfrak{M} \otimes \mathfrak{M} & \xrightarrow{\Delta(E)} & \mathfrak{M} \otimes \mathfrak{M} \\ W \big\downarrow & & \big\downarrow W \\ \mathfrak{M} & \xrightarrow{E} & \mathfrak{M}. \end{array}$$

Now, because of the way the action has been defined on $\{\mathbf{T}_{[m],(m)}\}$, when $\mathbf{T}_{[m],(m)}$ acts on an element $\left|\begin{smallmatrix}[m']\\(m')\end{smallmatrix}\right\rangle$ of $\mathfrak{T}\otimes\mathfrak{M}$, we have a well-defined compatible co-product action $\Delta(E)$, with $\mathfrak{T}\otimes\mathfrak{M}$ replacing $\mathfrak{M}\otimes\mathfrak{M}$ above. It follows from the diagrams that the mapping c can be identified with the mapping W (to within an invariant factor), that is, with the q-WCG coefficients. Moreover, the co-product Δ has a well-defined compatible action that extends to $\mathfrak{T}\otimes\mathfrak{T}$ thus yielding a further commuting diagram. Hence we deduce:

COROLLARY **3.24** *1. There exists an algebra of q-tensor operators,* $\mathfrak{T}\otimes\mathfrak{T}\xrightarrow{W}\mathfrak{T}$*, carrying products of irreducible q-tensor operators into irreducible q-tensor operators.*

2. *There exists a product, having the properties of an inner product, carrying an irreducible q-tensor operator* $\mathbf{T}$ *and its Hermitian conjugate* $\mathbf{T}^\dagger$ *into an invariant. Thus, a norm exists and irreducible unit q-tensor operators, denoted* $\widehat{\mathbf{T}}$*, are well-defined tensor operators whose matrix elements, by Theorem 3.18, are q-WCG coefficients.*

3. *If we denote the product of q-tensor operators by* $\mathbf{T}\circledw\mathbf{T}\in\mathfrak{T}$*, and denote the invariant product by* $\mathbf{T}^\dagger\bullet\mathbf{T}$*, then the invariant 6j operators are defined by* $\widehat{\mathbf{T}}^\dagger\bullet(\widehat{\mathbf{T}}\circledw\widehat{\mathbf{T}})$.

EXAMPLE **3.25** Consider again $\mathcal{U}_q(\mathfrak{su}(2))$. Theorem 3.18 and Corollary 3.24 shows that the q-WCG coefficients are matrix elements of unit tensor operators. Specifically, we have

$$\langle j+\Delta, m+M|\left\langle {}_{2J}\begin{smallmatrix}J+\Delta\\ \\J+M\end{smallmatrix}{}_0\right\rangle|jm\rangle = {}_{q^{-1}}C^{j\ \ J\ \ j+\Delta}_{m\ M\ m+M}.$$

Conjugate tensor operators and invariant $6j$ operators for $\mathcal{U}_q(\mathfrak{su}(2))$ are considered and their properties developed in following sections.

3.4.1 $\mathcal{U}_q(\mathfrak{su}(2))$ q-Tensor Operators and Coupling Coefficients

In order to clarify the meaning of the results in §3.3 and §3.4 let us give completely explicit formulas for $\mathcal{U}_q(\mathfrak{su}(2))$. Tensor operators $\mathbf{T}$ in $\mathcal{U}_q(\mathfrak{su}(2))$ have components $\mathbf{T}_{jm}$ where j is the invariant label and $-j\leqslant m\leqslant j$. Given any two such tensor operators $\mathbf{T}_{j_1m_1}$ and $\mathbf{T}'_{j_2m_2}$ we can form a third tensor operator $\mathbf{T}''_{jm}$, denoted by $\mathbf{T}\circledw\mathbf{T}'$ above, according to the q-WCG coupling

$$\mathbf{T}''_{jm} = \sum_{m_1,m_2} {}_qC^{j_1\ j_2\ j}_{m_1\ m_2\ m}\mathbf{T}_{j_1,m_1}\mathbf{T}'_{j_2,m_2}, \tag{3.26}$$

where the q-WCG coefficients ${}_qC^{j_1\ j_2\ j}_{m_1\ m_2\ m}$ are given explicitly in §3.5, with detailed properties including symmetries discussed in §3.5.2. In particular, we can form an invariant $\mathbf{T}\bullet\mathbf{T}'$ according to

$$\mathbf{T}\bullet\mathbf{T}' = \sum_{m_1+m_2=0} {}_qC^{j_1\ j_2\ 0}_{m_1\ m_2\ 0}\, \mathbf{T}_{j_1,m_1} \mathbf{T}'_{j_2,m_2},$$

and by using the formula[1] (which can be derived directly from the explicit expression for the q-WCG coefficients in the Racah form, see (3.60) or (5.27) below):

$${}_qC^{j_1\ j_2\ 0}_{m_1\ m_2\ 0} = \delta_{j_1,j_2}\delta_{m_1,-m_2} \frac{(-1)^{j_1-m_1} q^{\frac{m_1}{2}}}{\sqrt{[2j_1+1]}} \tag{3.27}$$

we can write the invariant, upon discarding a normalization, as

$$\mathbf{T}\bullet\mathbf{T}' = \sum_m (-1)^{j-m} q^{\frac{m}{2}} \mathbf{T}_{jm} \mathbf{T}'_{j,-m}. \tag{3.28}$$

We may regard the coefficients ${}_qC^{j\ \ j\ \ 0}_{m_1\ m_2\ 0}$ as defining a **metric**; as such they may also be interpreted as 1-j symbols.

The explicit induced action of the generators $\mathbf{J}$ on a tensor operator $\mathbf{T}$ can be determined from the diagram (3.13), which implies that

$$J_\pm(\mathbf{T}\,|jm\rangle) = q^{-\frac{J_z}{2}}(\mathbf{T}) J_\pm\,|jm\rangle + J_\pm(\mathbf{T}) q^{\frac{J_z}{2}}\,|jm\rangle.$$

We may express this as an operator relation:

$$J_\pm(\mathbf{T}) = J_\pm \mathbf{T} q^{-\frac{J_z}{2}} - q^{-\frac{J_z}{2}}(\mathbf{T}) J_\pm q^{-\frac{J_z}{2}}.$$

The action of $q^{-\frac{J_z}{2}}$ on $\mathbf{T}$ is the usual action because $\Delta(J_z)$ is commutative, hence we have

$$q^{-\frac{J_z}{2}}(\mathbf{T}) = q^{-\frac{J_z}{2}} \mathbf{T} q^{\frac{J_z}{2}}.$$

Accordingly the induced action becomes

$$J_\pm(\mathbf{T}) = J_\pm \mathbf{T} q^{-\frac{J_z}{2}} - q^{-\frac{J_z}{2}\pm\frac{1}{2}} \mathbf{T} J_\pm. \tag{3.29}$$

For $q = 1$ we regain the commutator action.

The result of the induced action is the set of matrix elements shown in Definition 3.14, which for $\mathcal{U}_q(\mathfrak{su}(2))$ leads to the explicit results:

$$\begin{aligned} q^{-\frac{J_z}{2}}(\mathbf{T}_{jm}) &= q^{-\frac{m}{2}} \mathbf{T}_{jm}, \\ J_\pm(\mathbf{T}_{jm}) &= \sqrt{[j\pm m+1][j\mp m]}\, \mathbf{T}_{j,m\pm 1}. \end{aligned}$$

Hence, we deduce the q-analog of the derivative property (3.4):

[1] **We denote q-integers by $[n]$.**

THEOREM **3.30** *An irreducible tensor operator* $\mathbf{T}_{jm}$ *in* $\mathcal{U}_q(\mathfrak{su}(2))$ *satisfies the relations:*

$$\begin{aligned}\left(J_\pm \mathbf{T}_{jm} - q^{-\frac{m}{2}} \mathbf{T}_{jm} J_\pm\right) q^{-\frac{J_z}{2}} &= \sqrt{[j \pm m + 1][j \mp m]}\, \mathbf{T}_{j,m\pm 1} \\ \left[J_z, \mathbf{T}_{jm}\right] &= m \mathbf{T}_{jm}.\end{aligned}$$

We now verify directly from this defining property that the construction of tensor operators given by (3.26) is valid:

LEMMA **3.31** *Given two tensor operators* $\mathbf{T}, \mathbf{T}'$ *we can form a third tensor operator* $\mathbf{T}''$ *by the product law (3.26).*

PROOF: We must show that if $\mathbf{T}$ and $\mathbf{T}'$ each satisfy the relations of Theorem 3.30 then $\mathbf{T}''$ also satisfies these same relations. It is sufficient to verify the validity of this coupling for the spin $\frac{1}{2}$ case only, since given any spin j tensor we may construct a spin $j + \frac{1}{2}$ tensor by coupling with an arbitrary spin $\frac{1}{2}$ tensor. Then by coupling recursively we may construct a tensor of arbitrary spin. Therefore we need to verify, given the spin j tensor operator $\mathbf{T}_{jm}$ and the spin $\frac{1}{2}$ tensor operator $\left(\mathbf{T}'_{\frac{1}{2},\frac{1}{2}}, \mathbf{T}'_{\frac{1}{2},-\frac{1}{2}}\right)$, that $\mathbf{T}''$ is also a tensor operator, where

$$\mathbf{T}''_{j+\frac{1}{2},m+\frac{1}{2}} = q^{\frac{1}{4}(j-m)} \sqrt{\frac{[j+m+1]}{[2j+1]}}\, \mathbf{T}_{jm} \mathbf{T}'_{\frac{1}{2},\frac{1}{2}} + q^{-\frac{1}{4}(j+m+1)} \sqrt{\frac{[j-m]}{[2j+1]}}\, \mathbf{T}_{j,m+1} \mathbf{T}'_{\frac{1}{2},-\frac{1}{2}}.$$

Here we have substituted for the spin $\frac{1}{2}$ q-WCG coefficients from (3.43) below. The verification that the relations of Theorem 3.30 are satisfied by $\mathbf{T}''_{j+\frac{1}{2},m+\frac{1}{2}}$ is lengthy, but straightforward. It is sufficient to check the relation for J_+, say, using the identity $[n+1] = q^{\frac{1}{2}}[n] + q^{-\frac{n}{2}}$ with $n = j + m + 1$, and the relation for J_- follows similarly. The commutator involving J_z is seen by inspection to be valid. □

From Theorem 3.30 we determine that the **conjugate irreducible q-tensor operator**, denoted $\overline{\mathbf{T}}_{jm}$, and defined by:

$$\overline{\mathbf{T}}_{jm} \stackrel{\text{def}}{=} (-)^{j-m} q^{\frac{m}{2}} \mathbf{T}_{j,-m} \tag{3.32}$$

satisfies the following relations:

$$\begin{aligned}q^{-\frac{J_z}{2}} \left(q^{-\frac{m}{2}} J_\pm \overline{\mathbf{T}}_{jm} - \overline{\mathbf{T}}_{jm} J_\pm\right) &= -\sqrt{[j \pm m][j \mp m + 1]}\, \overline{\mathbf{T}}_{j,m\mp 1} \\ \left[J_z, \overline{\mathbf{T}}_{jm}\right] &= -m \overline{\mathbf{T}}_{jm}.\end{aligned} \tag{3.33}$$

An example of a conjugate tensor operator is the Hermitean conjugate of $\mathbf{T}$ since, as may be verified directly from Theorem 3.30, $\mathbf{T}^\dagger$ satisfies (3.33).

We can now write the invariant product of two q-tensor operators (3.28) in the form

$$\mathbf{T}' \bullet \mathbf{T} = \sum_m \mathbf{T}'_{jm} \overline{\mathbf{T}}_{jm}. \tag{3.34}$$

Whereas the matrix elements of unit tensor operators are the q-WCG coefficients, as shown in Example 3.25, conversely we may define a unit tensor operator and its conjugate from these coefficients:

DEFINITION **3.35** *The* (J, M, Δ) **Wigner operator** *is the operator in* $\mathfrak{M}$ *defined and denoted by*

$$\left\langle {}_{2J} \begin{smallmatrix} J+\Delta \\ \\ J+M \end{smallmatrix} {}_{0} \right\rangle \stackrel{\text{def}}{=} \sum_{j,m} |j+\Delta, m+M\rangle \; {}_{q^{-1}}C^{j\;\;J\;\;j+\Delta}_{m\;M\;m+M} \langle jm| \,.$$

The **conjugate Wigner operator** *is defined and denoted by*

$$\left\langle {}_{2J} \begin{smallmatrix} J+\Delta \\ \\ J+M \end{smallmatrix} {}_{0} \right\rangle^{\dagger} \stackrel{\text{def}}{=} \sum_{j,m} |j-\Delta, m-M\rangle \; {}_{q^{-1}}C^{j-\Delta\;\;J\;\;j}_{m-M\;M\;m} \langle jm| \,.$$

By means of these definitions properties of q-WCG coefficients, such as orthogonality, may be expressed as operator relations.

3.4.2 Examples of q-Tensor Operators in $\mathcal{U}_q(\mathfrak{su}(2))$

It is useful to consider a variety of examples of tensor operators in $\mathcal{U}_q(\mathfrak{su}(2))$, and demonstrate the construction of invariants and unit tensor operators. The spin 1 example is of particular interest because it shows that the generators themselves do not form components of a spin 1 tensor operator, in contrast to the $q = 1$ case, but that one can form a genuine tensor operator using a certain combination of the generators. We consider several realizations of $\mathcal{U}_q(\mathfrak{su}(2))$ as examples and note that q-boson operators must also be modified in order to form components of a spin $\frac{1}{2}$ tensor operator.

EXAMPLE **3.36 (spin $\frac{1}{2}$)** The induced action (3.29) of the generators $\mathbf{J}$ on a spin $\frac{1}{2}$ q-tensor operator (or q-spinor), denoted now by $\mathbf{t}$, obeys the equivariance condition and the result of this action is the following:

$$J_\pm\left(\mathbf{t}_{\frac{1}{2},\mp\frac{1}{2}}\right) = \mathbf{t}_{\frac{1}{2},\pm\frac{1}{2}}, \quad J_\pm\left(\mathbf{t}_{\frac{1}{2},\pm\frac{1}{2}}\right) = 0, \quad J_z\left(\mathbf{t}_{\frac{1}{2},\pm\frac{1}{2}}\right) = \pm\tfrac{1}{2}\mathbf{t}_{\frac{1}{2},\pm\frac{1}{2}}.$$

We may write these relations in the following commutator form:

$$\begin{aligned} \left[q^{\frac{J_z}{2}} J_\pm \,,\, \mathbf{t}_{\frac{1}{2},\mp\frac{1}{2}}\right] &= q^{J_z \mp \frac{1}{4}} \mathbf{t}_{\frac{1}{2},\pm\frac{1}{2}}, \\ \left[q^{\frac{J_z}{2}} J_\pm \,,\, \mathbf{t}_{\frac{1}{2},\pm\frac{1}{2}}\right] &= 0, \\ \left[J_z \,,\, \mathbf{t}_{\frac{1}{2},\pm\frac{1}{2}}\right] &= \pm\tfrac{1}{2}\mathbf{t}_{\frac{1}{2},\pm\frac{1}{2}} \,. \end{aligned} \tag{3.37}$$

We can form an invariant $\mathbf{t}\,\bullet\mathbf{t}'$ from the components of $\mathbf{t}$ and $\mathbf{t}'$ by using (3.28) to find that

$$\mathbf{t}\bullet\mathbf{t}' = q^{\frac{1}{4}}\mathbf{t}_{\frac{1}{2},\frac{1}{2}}\mathbf{t}'_{\frac{1}{2},-\frac{1}{2}} - q^{-\frac{1}{4}}\mathbf{t}_{\frac{1}{2},-\frac{1}{2}}\mathbf{t}'_{\frac{1}{2},\frac{1}{2}}, \tag{3.38}$$

and we may check explicitly, using (3.37), that indeed this expression commutes with the generators $\mathbf{J}$.

By using the q-WCG operator coupling (3.26), the induced action for the spin $\frac{1}{2}$ q-tensor operators extends to all q-tensor operators in $\mathcal{U}_q(\mathfrak{su}(2))$, which then gives the relations in Theorem 3.30.

EXAMPLE **3.39 (q-boson realization for spin $\frac{1}{2}$)** Define the operator pair by

$$\mathbf{t}_{\frac{1}{2},\frac{1}{2}} \overset{\text{def}}{=} a_1\, q^{-\frac{N_2}{4}}, \quad \mathbf{t}_{\frac{1}{2},-\frac{1}{2}} \overset{\text{def}}{=} a_2\, q^{\frac{N_1}{4}}, \tag{3.40}$$

where a_i denotes a q-boson operator[2]. By direct calculation we verify that this pair is a q-tensor operator of $\mathcal{U}_q(\mathfrak{su}(2))$ with the generators realized as in (2.47), with the explicit induced action as given in (3.37). Note that only for $q = 1$ does the q-boson pair a_1, a_2 itself constitute a spin $\frac{1}{2}$ tensor operator. The invariant $\mathbf{t}\bullet\mathbf{t}$ given explicitly in (3.38) is identically zero. This corresponds to the fact that, for $q = 1$, the inner product for $SU(2)$ is symplectic. (The inner product in Corollary 3.24 uses the conjugate $\mathbf{t}^\dagger$ which accords with the q-analog of symplectic behaviour).

The fact that $\mathbf{t}\bullet\mathbf{t} = 0$ means that the components of $\mathbf{t}$ q-commute, that is,

$$ab = q^{-\frac{1}{2}}ba, \quad \text{where} \quad a = \mathbf{t}_{\frac{1}{2},\frac{1}{2}},\ b = \mathbf{t}_{\frac{1}{2},-\frac{1}{2}}.$$

We will encounter this property frequently in Chapter 4, where non-commuting coordinates arise as q-spinor components for which the invariant $\mathbf{t}\bullet\mathbf{t}$ is zero.

Given a q-spinor $\mathbf{t}$ we may form another q-spinor $\mathbf{t}^\dagger$ by Hermitian conjugation, and we find from (3.37) that the components are given by

$$\mathbf{t}^\dagger = \left(-q^{\frac{1}{4}}\overline{\mathbf{t}}_{\frac{1}{2},-\frac{1}{2}},\, q^{-\frac{1}{4}}\overline{\mathbf{t}}_{\frac{1}{2},\frac{1}{2}}\right), \tag{3.41}$$

which for the realization (3.40) becomes

$$\mathbf{t}^\dagger = (-\overline{a}_2 q^{\frac{1}{4}(N_1+1)},\, \overline{a}_1 q^{-\frac{1}{4}(N_2+1)}),$$

and is also a q-spinor with respect to $\mathcal{U}_q(\mathfrak{su}(2))$. If we form the invariant $\mathbf{t}\bullet\mathbf{t}^\dagger$ according to (3.38), we obtain

$$\mathbf{t}\bullet\mathbf{t}^\dagger = [N_1 + N_2],$$

which is indeed an invariant operator that commutes with the generators $\mathbf{J}$ given in (2.47) by the Jordan map. This invariant is in fact essentially the norm (squared)

[2] In this chapter we omit the affix q for q-boson operators.

of the q-spinor $\mathbf{t}$ and so, by normalizing, we obtain a unit tensor operator which we may identify with the fundamental unit Wigner operator as follows:

$$\begin{aligned}
a_1 q^{-\frac{N_2}{4}}[N_1+N_2+1]^{-\frac{1}{2}} &\longleftrightarrow \left\langle 1 \begin{smallmatrix} 1 \\ \\ 1 \end{smallmatrix} 0 \right\rangle \\
a_2 q^{\frac{N_1}{4}}[N_1+N_2+1]^{-\frac{1}{2}} &\longleftrightarrow \left\langle 1 \begin{smallmatrix} 1 \\ \\ 0 \end{smallmatrix} 0 \right\rangle \\
-\overline{a}_2 q^{\frac{1}{4}(N_1+1)}[N_1+N_2+1]^{-\frac{1}{2}} &\longleftrightarrow \left\langle 1 \begin{smallmatrix} 0 \\ \\ 1 \end{smallmatrix} 0 \right\rangle \\
\overline{a}_1 q^{-\frac{1}{4}(N_2+1)}[N_1+N_2+1]^{-\frac{1}{2}} &\longleftrightarrow \left\langle 1 \begin{smallmatrix} 0 \\ \\ 0 \end{smallmatrix} 0 \right\rangle .
\end{aligned} \tag{3.42}$$

This is the $\mathcal{U}_q(\mathfrak{su}(2))$ generalization of similar relations for ordinary boson operators (see [103, p. 16]).

We can use this realization of the fundamental Wigner operator to calculate a restricted class of q-WCG coefficients, using the formula of Example 3.25 (p. 86). Special cases of this formula are:

$$\begin{aligned}
{}_{q^{-1}}C^{j\ \frac{1}{2}\ j+\frac{1}{2}}_{m\ \frac{1}{2}\ m+\frac{1}{2}} &= \langle j+\tfrac{1}{2}, m+\tfrac{1}{2}| \left\langle 1 \begin{smallmatrix} 1 \\ \\ 1 \end{smallmatrix} 0 \right\rangle |j,m\rangle \\
{}_{q^{-1}}C^{j\ \frac{1}{2}\ j+\frac{1}{2}}_{m\ -\frac{1}{2}\ m-\frac{1}{2}} &= \langle j+\tfrac{1}{2}, m-\tfrac{1}{2}| \left\langle 1 \begin{smallmatrix} 1 \\ \\ 0 \end{smallmatrix} 0 \right\rangle |j,m\rangle \\
{}_{q^{-1}}C^{j\ \frac{1}{2}\ j-\frac{1}{2}}_{m\ \frac{1}{2}\ m+\frac{1}{2}} &= \langle j-\tfrac{1}{2}, m+\tfrac{1}{2}| \left\langle 1 \begin{smallmatrix} 0 \\ \\ 1 \end{smallmatrix} 0 \right\rangle |j,m\rangle \\
{}_{q^{-1}}C^{j\ \frac{1}{2}\ j-\frac{1}{2}}_{m\ -\frac{1}{2}\ m-\frac{1}{2}} &= \langle j-\tfrac{1}{2}, m-\tfrac{1}{2}| \left\langle 1 \begin{smallmatrix} 1 \\ \\ 0 \end{smallmatrix} 0 \right\rangle |j,m\rangle ,
\end{aligned}$$

which, by substituting for the q-boson basis states given in Eqn. (2.48), p. 34, and the Wigner operators in (3.42), leads to:

$$\begin{aligned}
{}_qC^{j\ \frac{1}{2}\ j+\frac{1}{2}}_{m\ \frac{1}{2}\ m+\frac{1}{2}} &= q^{\frac{1}{4}(j-m)}\sqrt{\frac{[j+m+1]}{[2j+1]}} \\
{}_qC^{j\ \frac{1}{2}\ j+\frac{1}{2}}_{m\ -\frac{1}{2}\ m-\frac{1}{2}} &= q^{-\frac{1}{4}(j+m)}\sqrt{\frac{[j-m+1]}{[2j+1]}} \\
{}_qC^{j\ \frac{1}{2}\ j-\frac{1}{2}}_{m\ \frac{1}{2}\ m+\frac{1}{2}} &= -q^{-\frac{1}{4}(j+m+1)}\sqrt{\frac{[j-m]}{[2j+1]}} \\
{}_qC^{j\ \frac{1}{2}\ j-\frac{1}{2}}_{m\ -\frac{1}{2}\ m-\frac{1}{2}} &= q^{\frac{1}{4}(j-m+1)}\sqrt{\frac{[j+m]}{[2j+1]}} .
\end{aligned} \tag{3.43}$$

(We have reversed the sign of the q exponent in the matrix element of the tensor operator to obtain the coefficient ${}_qC^{\cdot\ \cdot\ \cdot}_{\cdot\ \cdot\ \cdot}$, as explained in Remark 3.23). These expressions for the matrix elements may be written down directly using the pattern calculus rules (see §3.7.1, p. 109).

EXAMPLE **3.44 (spin 1)** Consider now the q-tensor operator carrying the three-dimensional (adjoint) irrep, denoted $\mathbf{T}_{1,m}$ with $m = \pm 1, 0$. These components satisfy equations similar to those for the q-spinor in (3.37):

$$\begin{aligned} \left[q^{\frac{J_z}{2}} J_{\pm}\,,\, \mathbf{T}_{1,\pm 1}\right] &= 0 \\ \left[q^{\frac{J_z}{2}} J_{\pm}\,,\, \mathbf{T}_{1,0}\right] &= \sqrt{[2]}\, q^{J_z \mp \frac{1}{2}} \mathbf{T}_{1,\pm 1} \\ \left[q^{\frac{J_z}{2}} J_{\pm}\,,\, \mathbf{T}_{1,\mp 1}\right] &= \sqrt{[2]}\, q^{J_z} \mathbf{T}_{1,0} \\ \left[J_z\,,\, \mathbf{T}_{1,m}\right] &= m \mathbf{T}_{1,m}, \quad (m = 0, \pm 1). \end{aligned} \tag{3.45}$$

Again, we may form an invariant $\mathbf{T} \bullet \mathbf{T}$ of $\mathcal{U}_q(\mathfrak{su}(2))$ from the components of $\mathbf{T}$, and from (3.28) we find that

$$\mathbf{T} \bullet \mathbf{T} = (\mathbf{T}_{1,0})^2 - q^{\frac{1}{2}} \mathbf{T}_{1,1} \mathbf{T}_{1,-1} - q^{-\frac{1}{2}} \mathbf{T}_{1,-1} \mathbf{T}_{1,1}\,, \tag{3.46}$$

which, as we may verify directly, commutes with $\mathbf{J}$.

The generators of $\mathcal{U}_q(\mathfrak{su}(2))$ are *not* irreducible q-tensor operators, contrary to the standard case $q = 1$, as can be seen from the fact that they do not satisfy (3.45). Instead, we find that the following combination can be identified with $j = 1$ tensor operator components:

$$\begin{aligned} \mathbf{T}_{1,\pm 1} &= \mp \frac{1}{\sqrt{[2]}}\, q^{\frac{J_z}{2}} J_{\pm}, \\ \mathbf{T}_{1,0} &= \frac{1}{[2]} \left(q^{-\frac{1}{2}} J_+ J_- - q^{\frac{1}{2}} J_- J_+\right) \end{aligned} \tag{3.47}$$

as one can verify by direct calculation. For $q = 1$ these components reduce to the standard spin 1 tensor operator[3] $(-J_+/\sqrt{2},\, J_z,\, J_-/\sqrt{2})$. The form for $\mathbf{T}_{1,0}$ is interesting, and we note that it is not expressible in terms of J_z alone. One can verify that under the induced action $\mathbf{T}_{1,m}$ transforms correctly as the adjoint q-irrep. The invariant $\mathbf{T} \bullet \mathbf{T}$ formed according to (3.28) and (3.46) is, in fact, fourth order in the generators and may be simplified to:

$$[2]^2 \mathbf{T} \bullet \mathbf{T} = \left(q^{\frac{1}{2}} - q^{-\frac{1}{2}}\right)^2 J_-^2 J_+^2 + \left([2J_z + 4] - [2J_z]\right) J_- J_+ + [2J_z][2J_z + 2].$$

Hence, the eigenvalues are given by

$$[2]^2 \mathbf{T} \bullet \mathbf{T}\, |jm\rangle = [2j][2j+2]\, |jm\rangle \tag{3.48}$$

in contrast to the eigenvalues $[j][j+1]$ given by (2.11). Less systematic approaches to constructing tensor operator invariants yield forms for the Casimir invariant different from that in (3.46), and distinct from each other, as we have seen in §2.2 (see (2.11)

---

[3] **The signs accord with the Wigner-Condon-Shortley phase conventions for irrep vectors labelled by (j, m).**

and (2.14), p. 92). These forms introduce fractional q-numbers in the eigenvalues, in contra-distinction to (3.48), which involves only q-integers. Although q-integers do not form a field, or even a ring, algebraic operations in $\mathcal{U}_q(\mathfrak{su}(2))$ preserve the q-integer form, and fractional q-numbers are never necessary.

EXAMPLE **3.49** We can form a spin 1 tensor operator $\mathbf{T}$ from the spin $\frac{1}{2}$ tensor operators $\mathbf{t} = (a_1 q^{-\frac{N_2}{4}}, a_2 q^{\frac{N_1}{4}})$ and $\mathbf{t}' = (-\bar{a}_2 q^{\frac{1}{4}(N_1+1)}, \bar{a}_1 q^{-\frac{1}{4}(N_2+1)})$ considered in Example 3.39 by using the coupling indicated in (3.26). Explicitly, given any two q-spinors $\mathbf{t}$ and $\mathbf{t}'$, we can form $\mathbf{T}$ according to

$$\begin{aligned} \mathbf{T}_{1,\pm1} &= \mathbf{t}_{\frac{1}{2},\pm\frac{1}{2}} \mathbf{t}'_{\frac{1}{2},\pm\frac{1}{2}}, \\ \mathbf{T}_{1,0} &= \frac{1}{\sqrt{[2]}} \left(q^{-\frac{1}{4}} \mathbf{t}_{\frac{1}{2},\frac{1}{2}} \mathbf{t}'_{\frac{1}{2},-\frac{1}{2}} + q^{\frac{1}{4}} \mathbf{t}_{\frac{1}{2},-\frac{1}{2}} \mathbf{t}'_{\frac{1}{2},\frac{1}{2}} \right) \end{aligned}$$

where we have used the following values of the q-WCG coefficients:

$${}_qC^{\frac{1}{2}\,\frac{1}{2}\,1}_{\frac{1}{2}\,\frac{1}{2}\,1} = 1 = {}_qC^{\frac{1}{2}\,\frac{1}{2}\,1}_{-\frac{1}{2}\,-\frac{1}{2}\,-1}, \qquad {}_qC^{\frac{1}{2}\,\frac{1}{2}\,1}_{\frac{1}{2}\,-\frac{1}{2}\,0} = \frac{q^{-\frac{1}{4}}}{\sqrt{[2]}}, \qquad {}_qC^{\frac{1}{2}\,\frac{1}{2}\,1}_{-\frac{1}{2}\,\frac{1}{2}\,0} = \frac{q^{\frac{1}{4}}}{\sqrt{[2]}}.$$

These values are special cases of those shown in (3.43). Hence we obtain (after normalizing suitably):

$$\begin{aligned} \sqrt{[2]}\,\mathbf{T}_{1,1} &= q^{\frac{1}{4}(N_1-N_2)} a_1 \bar{a}_2 \\ [2]\,\mathbf{T}_{1,0} &= q^{-\frac{1}{2}(N_2+1)}[N_1] - q^{\frac{1}{2}(N_1+1)}[N_2] \\ \sqrt{[2]}\,\mathbf{T}_{1,-1} &= -q^{\frac{1}{4}(N_1-N_2)} a_2 \bar{a}_1, \end{aligned}$$

which accords with (3.47), upon identifying the generators $\mathbf{J}$ as in (2.47).

Hence, we may construct the generators directly from q-spinor components with formulas such as

$$q^{\frac{1}{2}J_z} J_+ = -\mathbf{t}_{\frac{1}{2},\frac{1}{2}} \mathbf{t}'_{\frac{1}{2},\frac{1}{2}}. \tag{3.50}$$

3.4.3 $\mathcal{U}_q(\mathfrak{u}(n))$ q-Tensor Operators

Having discussed in detail the induced action for tensor operators in $\mathcal{U}_q(\mathfrak{su}(2))$, let us now turn briefly to the general case $\mathcal{U}_q(\mathfrak{u}(n))$ and consider in particular the induced action for the elementary unit tensor operators, and generalize Theorem 3.30. We may denote these operator components by $\mathbf{T}_{[1,\dot{0}],i}$ where $[1,\dot{0}]$ is the Gel'fand-Weyl irrep label (for which the irrep is of dimension n), and $i = 1, 2, \ldots, n$ labels the n components, with $i = 1$ being the component of highest weight. The generators of

$\mathcal{U}_q(\mathfrak{u}(n))$ are denoted $E_{i,i+1}$ and $E_{i+1,i}$ for the raising and lowering Chevalley generators, and E_{ii} for the diagonal generators, where $i = 1, \ldots, n$, as discussed in §2.5. We also define $H_i \stackrel{\text{def}}{=} E_{ii} - E_{i+1,i+1}$.

The induced action of the generators on $\mathbf{T}_{[1,\dot{0}],i}$ can be determined as for the $n = 2$ case, and the result of this action is as shown in Definition 3.14 (as also discussed by Quesne [2, p. 601] and Rittenberg and Scheunert [135]). The formulas are as follows:

THEOREM **3.51** *An elementary tensor operator $\mathbf{T}_{[1,\dot{0}],i}$ in $\mathcal{U}_q(\mathfrak{u}(n))$ satisfies the relations:*

$$\begin{aligned}
E_{jj}\left(\mathbf{T}_{[1,\dot{0}],i}\right) &= [E_{jj}, \mathbf{T}_{[1,\dot{0}],i}] = \delta_{ij}\mathbf{T}_{[1,\dot{0}],i} \\
E_{j,j+1}\left(\mathbf{T}_{[1,\dot{0}],i}\right) &= E_{j,j+1}\mathbf{T}_{[1,\dot{0}],i}q^{-\frac{1}{4}H_j} - q^{-\frac{1}{4}H_j+\frac{1}{2}}\mathbf{T}_{[1,\dot{0}],i}E_{j,j+1} = \delta_{i,j+1}\mathbf{T}_{[1,\dot{0}],i-1} \\
E_{j+1,j}\left(\mathbf{T}_{[1,\dot{0}],i}\right) &= E_{j+1,j}\mathbf{T}_{[1,\dot{0}],i}q^{-\frac{1}{4}H_j} - q^{-\frac{1}{4}H_j+\frac{1}{2}}\mathbf{T}_{[1,\dot{0}],i}E_{j+1,j} = \delta_{ij}\mathbf{T}_{[1,\dot{0}],i+1}\,.
\end{aligned}$$

The previous case $n = 2$ shown in Theorem 3.30 is obtained by putting $j = 1$, $H_1 = 2J_z$, $E_{12} = J_+$, $E_{21} = J_-$ and identifying $\mathbf{T}_{[1,\dot{0}],1}$ and $\mathbf{T}_{[1,\dot{0}],2}$ with $\mathbf{T}_{\frac{1}{2},\frac{1}{2}}$ and $\mathbf{T}_{\frac{1}{2},-\frac{1}{2}}$ respectively. As in §3.4.1, we can determine similar relations satisfied by the conjugate tensor operator, which is related to the Hermitean conjugate tensor operator according to (see [2, p. 605]):

$$\mathbf{T}_{[\dot{0},\,-1],-i} = (-)^{-\frac{1}{4}(n-2i+1)}\left(\mathbf{T}_{[1,\dot{0}],i}\right)^{\dagger}.$$

EXAMPLE **3.52** If we realize the symmetric irreps of $\mathcal{U}_q(\mathfrak{u}(n))$ with q-boson operators as described in §2.5.1 (p. 49) then the operators

$$\mathbf{T}_{[1,\dot{0}],i} = a_i q^{\frac{1}{4}(2N_1+\ldots+2N_{i-1}+N_i)}$$

form an elementary unit tensor operator. The case $n = 2$ reduces to (3.40) upon multiplying through by the invariant $q^{-\frac{1}{4}(N_1+N_2)}$.

We return to elementary unit tensor operators in §3.7.1 where we describe the pattern calculus, which enables us to write down matrix elements of all tensor operators in $\mathcal{U}_q(\mathfrak{u}(n))$ with irrep labels of the form $[1\ldots1\,0\ldots0]$.

3.5 q-Wigner-Clebsch-Gordan Coefficients

The explicit decomposition of the tensor product of two finite-dimensional unitary irreps of $\mathcal{U}_q(\mathfrak{su}(2))$ has been carried out by Kirillov and Reshetikhin [65] and also Vaksman [136] beginning with (3.17), and the coupling coefficients ${}_qC^{j_1\ j_2\ j}_{m_1\ m_2\ m}$ have been calculated in several forms, including q-analogs of the Racah and van der Waerden forms. We present an independent derivation of these coefficients in §4.4.2 (p. 141) which leads to the following explicit expression in the van der Waerden form:

$$
\begin{aligned}
{}_qC^{j_1\ j_2\ j}_{m_1\ m_2\ m} &= \delta_{m,m_1+m_2} q^{\frac{1}{4}(j_1+j_2-j)(j_1+j_2+j+1)+\frac{1}{2}(j_1m_2-j_2m_1)}\Delta(j_1j_2j) \\
&\times\Big([j_1+m_1]![j_1-m_1]![j_2+m_2]![j_2-m_2]![j+m]![j-m]![2j+1]\Big)^{\frac{1}{2}} \\
&\times\sum_n \frac{(-)^n q^{-\frac{n}{2}(j_1+j_2+j+1)}}{[n]![j_1+j_2-j-n]![j_1-m_1-n]![j_2+m_2-n]![j-j_2+m_1+n]![j-j_1-m_2+n]!}
\end{aligned}
\tag{3.53}
$$

where the triangle function Δ is given by

$$
\Delta(abc) \stackrel{\text{def}}{=} \left(\frac{[a+b-c]![a-b+c]![-a+b+c]!}{[a+b+c+1]!}\right)^{\frac{1}{2}}.
$$

For $q \to 1$ the q-WCG coefficients reduce to the usual WCG coefficients $C^{j_1\ j_2\ j}_{m_1\ m_2\ m}$ in the van der Waerden form. Kirillov and Reshetikhin have also given q-analog expressions of the Racah and Majumdar formulas, and a different derivation of the van der Waerden form has been given by Ruegg [137] and Nomura [119].

Since the q-WCG coefficients are elements of a unitary matrix and may be chosen to be real, they satisfy orthogonality properties, the first comprising the orthogonality of columns:

$$
\sum_{j,m} {}_qC^{j_1\ j_2\ j}_{m_1\ m_2\ m}\, {}_qC^{j_1\ j_2\ j}_{m_1'\ m_2'\ m} = \delta_{m_1,m_1'}\delta_{m_2,m_2'}, \tag{3.54}
$$

and the second the orthogonality of rows:

$$
\sum_{m_1,m_2} {}_qC^{j_1\ j_2\ j}_{m_1\ m_2\ m}\, {}_qC^{j_1\ j_2\ j'}_{m_1\ m_2\ m'} = \delta_{jj'}\delta_{mm'}\varepsilon_{j_1j_2j}, \tag{3.55}
$$

where we have included the factor $\varepsilon_{j_1j_2j}$ in order to enforce the triangle conditions. An example of these properties is given by the orthogonal matrix P in Example 2.17.

We may express these orthogonality properties as operator relations for the unit Wigner operators defined in Definition 3.35:

$$
\begin{aligned}
\sum_{\Delta} \left\langle 2J \begin{matrix} J+\Delta \\ J+M' \end{matrix}\ 0 \right\rangle^{\dagger} \left\langle 2J \begin{matrix} J+\Delta \\ J+M \end{matrix}\ 0 \right\rangle &= \delta_{M'M} \\
\sum_{M} \left\langle 2J \begin{matrix} J+\Delta' \\ J+M \end{matrix}\ 0 \right\rangle \left\langle 2J \begin{matrix} J+\Delta \\ J+M \end{matrix}\ 0 \right\rangle^{\dagger} &= \delta_{\Delta'\Delta}\varepsilon_{j-\Delta,J,j} \\
\sum_m q^m \langle jm| \left\langle 2J' \begin{matrix} J'+\Delta' \\ J'+M' \end{matrix}\ 0 \right\rangle \left\langle 2J \begin{matrix} J+\Delta \\ J+M \end{matrix}\ 0 \right\rangle^{\dagger} |jm\rangle &= q^M \frac{[2j+1]}{[2J+1]}\delta_{J'J}\delta_{M'M}\delta_{\Delta'\Delta}.
\end{aligned}
$$

The last equation is verified by use of the symmetry (3.63e) below, which also explains the appearance of the explicit q-factors and the ratio of q-integers.

By applying the operator $\Delta(J_\pm)$ to both sides of (3.17) we obtain the following recurrence relations:

$$\begin{aligned}\left([j\mp m][j\pm m+1]\right)^{\frac{1}{2}}\,{}_qC^{j_1\ j_2\ j}_{m_1\ m_2\ m\pm1} &= q^{-\frac{m_1}{2}}\left([j_2\pm m_2][j_2\mp m_2+1]\right)^{\frac{1}{2}}\,{}_qC^{j_1\ \ j_2\ \ j}_{m_1\ m_2\mp1\ m}\\ &+ q^{\frac{m_2}{2}}\left([j_1\pm m_1][j_1\mp m_1+1]\right)^{\frac{1}{2}}\,{}_qC^{\ j_1\ \ j_2\ j}_{m_1\mp1\ m_2\ m},\end{aligned}$$

(obtained by relabelling the summation indices suitably in the sum (3.17)). These relations determine all q-WCG coefficients for fixed j_1, j_2, j from their values when m_1, m_2, m take boundary values. We may explicitly verify a special case of these relations using the spin-$\frac{1}{2}$ coefficients given in (3.43). Further recurrence relations have been discussed by Nomura [139] and Groza et al. [60].

In addition, the q-WCG coefficients satisfy relations, known as the hexagon identities (since they result from the compatibility of commutativity and associativity of a product functor, see Majid [14]), which are obtained by taking matrix elements of the two equations in (1.9), namely, $(\Delta\otimes\mathrm{id})\mathcal{R}=\mathcal{R}_{13}\mathcal{R}_{23}$ and $(\mathrm{id}\otimes\Delta)\mathcal{R}=\mathcal{R}_{13}\mathcal{R}_{12}$. The respective identities read [65]:

$$\begin{aligned}\sum_m\left(\mathcal{R}^{jj_3}\right)^{m',m_3'}_{m,m_3}\,{}_qC^{j_1\ j_2\ j}_{m_1\ m_2\ m} &= \sum_{m_1',m_2',m_3''}{}_qC^{j_1\ j_2\ j}_{m_1'\ m_2'\ m'}\left(\mathcal{R}^{j_1j_3}\right)^{m_1',m_3''}_{m_1,m_3}\left(\mathcal{R}^{j_2j_3}\right)^{m_2',m_3'}_{m_2,m_3''}\\ \sum_m\left(\mathcal{R}^{j_1j}\right)^{m_1',m'}_{m_1,m}\,{}_qC^{j_2\ j_3\ j}_{m_2\ m_3\ m} &= \sum_{m_2',m_3',m_3''}{}_qC^{j_2\ j_3\ j}_{m_2'\ m_3'\ m'}\left(\mathcal{R}^{j_1j_3}\right)^{m_1',m_3''}_{m_1,m_3}\left(\mathcal{R}^{j_1j_2}\right)^{m_2',m_3'}_{m_2,m_3''}.\end{aligned}$$

A derivation is given in Majid [14, p. 77]. For $q=1$ the R-matrix is the unit matrix: $(\mathcal{R}^{j_1j_2})^{m_1,m_2}_{m_1',m_2'}=\delta^{m_1}_{m_1'}\delta^{m_2}_{m_2'}$, in which case the hexagon identities become trivial.

3.5.1 Special Cases of q-Wigner-Clebsch-Gordan Coefficients

The q-WCG coefficients are, from Theorem 3.18, the matrix elements of irreducible unit tensor operators. Thus, as we have done in §3.4.2, we can determine the explicit spin-$\frac{1}{2}$ $\mathcal{U}_q(\mathfrak{su}(2))$ q-WCG coefficients as matrix elements of the fundamental unit Wigner operators between generic boson operator eigenstates $|jm\rangle$. It is useful to express these coefficients, listed in (3.43), in terms of the Gel'fand-Weyl labels m_{ij} described in §2.6 (see (2.85), p. 54), and determined from j, m by $m_{12}=2j$, $m_{11}=j+m$ with $m_{22}=0$. We may shift these labels according to $m_{11}\to m_{11}-m_{22}$, $m_{12}\to m_{12}-m_{22}$ in order to obtain the $\mathcal{U}_q(\mathfrak{u}(2))$ coefficients, for this is a special case of the shift-invariance of the $\mathcal{U}_q(\mathfrak{u}(n))$ q-WCG coefficients under the transformation $m_{ij}\to m_{ij}-k$, for any $k\in\mathbb{Z}$. Conversely, we may obtain the $\mathcal{U}_q(\mathfrak{su}(2))$ coefficients by identifying labels according to

$$j=\tfrac{1}{2}(m_{12}-m_{22}),\quad m=m_{11}-\tfrac{1}{2}(m_{12}+m_{22}).$$

With this procedure we may construct Table 3.56, in which we display the coefficients ${}_qC^{j\ \frac{1}{2}\ j+\Delta j}_{m\ \Delta m\ m+\Delta m}$ in terms of the Gel'fand-Weyl labels.

*	$\Delta m = \frac{1}{2}$	$\Delta m = -\frac{1}{2}$
$\Delta j = \frac{1}{2}$	$q^{\frac{1}{4}(m_{12}-m_{11})}\sqrt{\frac{[m_{11}+1-m_{22}]}{[m_{12}+1-m_{22}]}}$	$q^{-\frac{1}{4}(m_{11}-m_{22})}\sqrt{\frac{[m_{12}+1-m_{11}]}{[m_{12}+1-m_{22}]}}$
$\Delta j = -\frac{1}{2}$	$-q^{-\frac{1}{4}(m_{11}+1-m_{22})}\sqrt{\frac{[m_{12}-m_{11}]}{[m_{12}+1-m_{22}]}}$	$q^{\frac{1}{4}(m_{12}+1-m_{11})}\sqrt{\frac{[m_{11}-m_{22}]}{[m_{12}+1-m_{22}]}}$

Table 3.56: Spin-$\frac{1}{2}$ q-WCG coefficients ${}_qC^{j\ \frac{1}{2}\ j+\Delta j}_{m\ \Delta m\ m+\Delta m}$ for $\mathcal{U}_q(\mathfrak{u}(2))$

More generally, we may reduce the q-WCG coefficient to a single term when one of the projection numbers m_1, m_2, m takes its maximum or minimum value. The summation in the van der Waerden form (3.53) clearly reduces to a single term when either $m_1 = j_1$ or $m_2 = -j_2$ and, by use of the symmetries proved below in Theorem 3.62, this extends to the other boundary values of m_1, m_2, m. We give the explicit formula for ${}_qC^{j_1\ j_2\ j}_{m_1\ m_2\ j}$ in (5.27), p. 168, where we use this boundary value to derive the Racah form of the q-WCG coefficients. A special case is the value of this coefficient for $j = 0$ in (3.27). The summation in the van der Waerden form (3.53) reduces to a single term also for the 'stretched case' $j = j_1 + j_2$ since, as is evident from the denominator under the summation, only the term $n = 0$ then contributes.

Another case in which the summation can be performed is $m_1 = m_2 = m = 0$. For $q = 1$ the WCG coefficient $C^{j_1\ j_2\ j}_{0\ 0\ 0}$ is zero if $j_1 + j_2 - j$ is odd, as follows by symmetry, and otherwise may be evaluated using Dixon's formula for the sum of a well poised ${}_3F_2$ hypergeometric series. For general values of q, however, the coefficient ${}_qC^{j_1\ j_2\ j}_{0\ 0\ 0}$ is nonzero even when $j_1 + j_2 - j$ is odd, since in this case the relevant symmetry (given in (3.63a) or (3.63b) below) merely shows that

$${}_qC^{j_1\ j_2\ j}_{0\ 0\ 0} = (-)^{j_1+j_2-j}\ {}_{q^{-1}}C^{j_1\ j_2\ j}_{0\ 0\ 0}.$$

The summation in (3.53) can be performed for $m_1 = m_2 = m = 0$ by writing the q-WCG coefficient in terms of a ${}_3\phi_2$ function, as in the following section, and then using a q-analog of Dixon's formula. Such formulas are discussed in Gasper and Rahman [64, p. 36,237], and other summation formulas [64, §1.9] are also applicable to the Majumdar form of the q-WCG coefficients expressed as a ${}_3\phi_2$ function. Let us evaluate now the sum (3.53) for the case $j_1 + j_2 - j = 1$:

$${}_qC^{j_1\ j_2\ j}_{0\ 0\ 0} = \left(q^{\frac{1}{2}} - q^{-\frac{1}{2}}\right)\frac{[j+1]!}{[j_1-1]![j_2-1]!}\sqrt{\frac{[2j+1][2j_1-1]![2j_2-1]!}{[2j+2]!}}, \quad (j_1 + j_2 = j+1)$$

which evidently vanishes for $q = 1$. A particularly significant case is

$$ {}_qC^{1\,1\,1}_{0\,0\,0} = \left(q^{\frac{1}{2}} - q^{-\frac{1}{2}}\right)\sqrt{\frac{[2]}{[4]}}, $$

which, together with

$$ {}_qC^{1\,1\,1}_{1\,0\,1} = q^{\frac{1}{2}}\sqrt{\frac{[2]}{[4]}}, $$

and those coefficients obtained by use of the symmetries in Theorem 3.62 below, comprises all q-WCG coefficients with $j_1 = j_2 = j = 1$.

We list all spin 1 coefficients ${}_qC^{j\ \ 1\ \ j+\Delta j}_{m\ \pm 1\ m\pm 1}$ in Table 3.57 and all the coefficients ${}_qC^{j\ 1\ j+\Delta j}_{m\ 0\ \ m}$ in Table 3.58. These may be calculated directly from the formula (3.53), and have been given previously by Pasquier [25].

❄	$\Delta m = 1$	$\Delta m = -1$
$\Delta j = 1$	$q^{\frac{1}{2}(j-m)}\sqrt{\frac{[j+m+1][j+m+2]}{[2j+1][2j+2]}}$	$q^{-\frac{1}{2}(j+m)}\sqrt{\frac{[j-m+1][j-m+2]}{[2j+1][2j+2]}}$
$\Delta j = 0$	$-q^{-\frac{1}{2}(m+1)}\sqrt{\frac{[2][j-m][j+m+1]}{[2j][2j+2]}}$	$q^{-\frac{1}{2}(m-1)}\sqrt{\frac{[2][j+m][j-m+1]}{[2j][2j+2]}}$
$\Delta j = -1$	$q^{-\frac{1}{2}(j+m+1)}\sqrt{\frac{[j-m-1][j-m]}{[2j][2j+1]}}$	$q^{\frac{1}{2}(j-m+1)}\sqrt{\frac{[j+m-1][j+m]}{[2j][2j+1]}}$

Table 3.57: Spin-1 q-WCG coefficients ${}_qC^{j\ \ 1\ \ \ j+\Delta j}_{m\ \Delta m\ m+\Delta m}$ for $\mathcal{U}_q(\mathfrak{su}(2))$

$\Delta j = 1$	$q^{-\frac{m}{2}}\sqrt{\frac{[2][j-m+1][j+m+1]}{[2j+1][2j+2]}}$
$\Delta j = 0$	$\frac{q^{-\frac{m}{2}}}{\sqrt{[2j][2j+2]}}\left(q^{\frac{1}{2}(j+1)}[j+m] - q^{-\frac{1}{2}(j+1)}[j-m]\right)$
$\Delta j = -1$	$-q^{-\frac{m}{2}}\sqrt{\frac{[2][j-m][j+m]}{[2j][2j+1]}}$

Table 3.58: Spin-1 q-WCG coefficients ${}_qC^{j\ 1\ j+\Delta j}_{m\ 0\ \ m}$ for $\mathcal{U}_q(\mathfrak{su}(2))$

3.5.2 Symmetries of q-Wigner-Clebsch-Gordan Coefficients

In order to derive symmetry properties of ${}_qC^{j_1\ j_2\ j}_{m_1\ m_2\ m}$ we express these coefficients in terms of basic hypergeometric functions ${}_3\phi_2$, which have been described in §2.8.3 (p. 67). We can write ${}_qC^{j_1\ j_2\ j}_{m_1\ m_2\ m}$ in terms of a ${}_3\phi_2$ function by using the identities

$$[m-n]! = \frac{(-1)^n[m]!}{([-m])_n}, \quad [m+n]! = [m]!([m+1])_n, \qquad n,m \in \mathbb{N},$$

and we find

$$\begin{aligned}
{}_qC^{j_1\ j_2\ j}_{m_1\ m_2\ m} &= \delta_{m,m_1+m_2} q^{\frac{1}{4}(j_1+j_2-j)(j_1+j_2+j+1)+\frac{1}{2}(j_1m_2-j_2m_1)} \\
&\times \frac{([j-j_2+m_1+1])_{j_1-m_1}([j-j_1-m_2+1])_{j_2+m_2}}{\left([j_1+j_2+j+1]![j_1+j_2-j]![j-j_1+j_2]![j+j_1-j_2]!\right)^{\frac{1}{2}}} \\
&\times \left(\frac{[2j+1][j_1+m_1]![j_2-m_2]![j+m]![j-m]!}{[j_2+m_2]![j_1-m_1]!}\right)^{\frac{1}{2}} \\
&\times {}_3\phi_2\left(\begin{matrix} q^{-j_1-j_2+j} & q^{-j_1+m_1} & q^{-j_2-m_2} \\ q^{j-j_2+m_1+1} & q^{j-j_1-m_2+1} & \end{matrix} ;q,q\right).
\end{aligned} \tag{3.59}$$

Similarly, the Racah form can be written

$$\begin{aligned}
{}_qC^{j_1\ j_2\ j}_{m_1\ m_2\ m} &= \delta_{m,m_1+m_2}(-)^{j_1-m_1} q^{\frac{1}{4}[j_2(j_2+1)-j_1(j_1+1)-j(j+1)]+\frac{1}{2}m_1(m+1)} \\
&\times \frac{[j_2+j-m_1]!([j_2-j+m_1+1])_{j_1-m_1}}{\left([j_1-j_2+j]![j_2-j_1+j]![j_1+j_2-j]![j_1+j_2+j+1]!\right)^{\frac{1}{2}}} \\
&\times \left(\frac{[j+m]![j_2-m_2]![j_1+m_1]![2j+1]}{[j_2+m_2]![j_1-m_1]![j-m]!}\right)^{\frac{1}{2}} \\
&\times {}_3\phi_2\left(\begin{matrix} q^{j_1+m_1+1} & q^{-j+m} & q^{-j_1+m_1} \\ q^{-j_2-j+m_1} & q^{j_2-j+m_1+1} & \end{matrix} ;q,q\right).
\end{aligned} \tag{3.60}$$

Because the numerator parameters in (3.59) are negative the series terminates and although the denominator parameters $j-j_2+m_1+1$, $j-j_1-m_2+1$ could become negative, they are always larger than the numerator parameters, and so no denominator zeros occur and the ${}_3\phi_2$ function is well defined. A similar remark holds also for (3.60).

The utility of these expressions lies in the fact that properties of the q-WCG coefficients can be derived directly from the well-known properties of basic hypergeometric functions, as we now show (following [140]). For $q=1$, symmetries of the

WCG coefficients have been derived using transformation laws of the ${}_3F_2$ series found originally by Thomae [92, 141] (see for example Biedenharn and Louck [103]). The symmetries form a group of order 72 and in fact we require only one transformation of the ${}_3F_2$ functions together with permutation symmetries in order to derive all 72 symmetries. This transformation can be obtained merely by reversing the order of summation in the terminating ${}_3F_2$ function. The q-analog of this transformation is obtained similarly, by reversing the order of summation, and is sufficient to enable us to derive the q-analog symmetries. The transformation is:

$$\begin{aligned} {}_3\phi_2\left(\begin{matrix} q^{-\alpha} & q^{-\beta} & q^{-n} \\ q^{\delta+1} & q^{\varepsilon+1} & \end{matrix} ; q, z\right) &= q^{-\frac{n}{2}(\alpha+\beta+\delta+\varepsilon+n+3)} \frac{([\alpha-n+1])_n([\beta-n+1])_n}{([\delta+1])_n([\varepsilon+1])_n} \\ &\times(-z)^n {}_3\phi_2\left(\begin{matrix} q^{-\delta-n} & q^{-\varepsilon-n} & q^{-n} \\ q^{1+\alpha-n} & q^{1+\beta-n} & \end{matrix} ; q^{-1}, z^{-1}\right) \end{aligned} \tag{3.61}$$

where n is a positive integer.

THEOREM 3.62 *The q-Wigner-Clebsch-Gordan coefficients satisfy the following symmetry relations:*

$${}_qC^{j_1\ j_2\ j}_{m_1\ m_2\ m} = (-)^{j_1+j_2-j}\, {}_{q^{-1}}C^{j_1\ \ j_2\ \ j}_{-m_1\ -m_2\ -m} \tag{3.63a}$$

$${}_qC^{j_1\ j_2\ j}_{m_1\ m_2\ m} = (-)^{j_1+j_2-j}\, {}_{q^{-1}}C^{j_2\ j_1\ j}_{m_2\ m_1\ m} \tag{3.63b}$$

$${}_qC^{j_1\ j_2\ j}_{m_1\ m_2\ m} = (-)^{j_2+m_2} q^{-\frac{m_2}{2}} \sqrt{\frac{[2j+1]}{[2j_1+1]}}\, {}_{q^{-1}}C^{\ j\ \ j_2\ \ j_1}_{-m\ m_2\ -m_1} \tag{3.63c}$$

$${}_qC^{j_1\ j_2\ j}_{m_1\ m_2\ m_1+m_2} = {}_qC^{\frac{1}{2}(j_1+j_2+m_1+m_2)\ \frac{1}{2}(j_1+j_2-m_1-m_2)\ j}_{\frac{1}{2}(j_1-j_2+m_1-m_2)\ \frac{1}{2}(j_1-j_2-m_1+m_2)\ j_1-j_2} \tag{3.63d}$$

$${}_qC^{j_1\ j_2\ j}_{m_1\ m_2\ m} = (-)^{j_1-m_1} q^{\frac{m_1}{2}} \sqrt{\frac{[2j+1]}{[2j_2+1]}}\, {}_{q^{-1}}C^{j_1\ \ j\ \ j_2}_{m_1\ -m\ -m_2} \tag{3.63e}$$

$${}_qC^{j_1\ j_2\ j}_{m_1\ m_2\ m} = (-)^{j_2+m_2} q^{-\frac{m_2}{2}} \sqrt{\frac{[2j+1]}{[2j_1+1]}}\, {}_{q^{-1}}C^{\ j_2\ \ j\ \ j_1}_{-m_2\ m\ m_1} \tag{3.63f}$$

PROOF: The last two relations can be derived from the first three relations and are included for convenience only. The first three relations are obtained from the transformation (3.61) with a suitable identification of parameters, $n = j_1 + j_2 - j$ for (3.63a,3.63b) and $n = j_2 + m_2$ for (3.63c). The symmetry (3.63d) does not require use of (3.61), but follows simply from the invariance of the basic hypergeometric series under permutations of its numerator or denominator parameters. This also explains why we have $q \to q^{-1}$ in the first three relations but not in (3.63d).□

Special cases of these symmetries are evident in the Tables 3.56-3.58 of coefficients given above; for example, the symmetry ${}_qC^{j\,1\,j'}_{m\,0\,m} = (-)^{j+1-j'}\,{}_{q^{-1}}C^{\,j\ \ 1\ \ j'}_{-m\,0\,-m}$ is apparent in Table 3.58.

By combining (3.63e) with (3.63a) we obtain

$$
{}_qC^{j_1\ j_2\ j}_{m_1\,m_2\,m} = (-)^{j-j_2-m_1} q^{\frac{m_1}{2}} \sqrt{\frac{[2j+1]}{[2j_2+1]}}\;{}_qC^{\ j_1\ \ j\ j_2}_{-m_1\,m\,m_2}\,, \tag{3.64}
$$

a relation which was found, together with (3.63b), by Kirillov and Reshetikhin [65]. These symmetry relations can also be obtained directly from the explicit formula (3.53) (see Nomura [76]).

The formulation of the coefficients in terms of ${}_3\phi_2$ functions is also useful in order to relate the different forms in (3.59) and (3.60). We can obtain the Racah form (3.60) directly from (3.59) with the help of the transformation

$$
{}_3\phi_2\left(\begin{matrix} q^{\alpha_1} & q^{\alpha_2} & q^{-n} \\ & q^{\beta_1} & q^{\beta_2} \end{matrix};q,q\right) = \frac{([\beta_1-\alpha_1])_n q^{\frac{n\alpha_1}{2}}}{([\beta_1])_n}\,{}_3\phi_2\left(\begin{matrix} q^{\alpha_1} & q^{\beta_2-\alpha_2} & q^{-n} \\ & q^{\beta_2} & q^{\alpha_1-\beta_1+1-n} \end{matrix};q,q^{\alpha_2-\beta_1+1}\right) \tag{3.65}
$$

which can be found in Gasper and Rahman [64, (Eq. 3.2.2)], and is a special case of an identity due to Sears [142] relating two ${}_4\phi_3$ functions. In (3.65), choose

$$
\begin{aligned}
\alpha_1 = -j_1 - j_2 + j, \quad \alpha_2 = -j_2 - m_2, \quad n = j_1 - m_1 \\
\beta_1 = j - j_2 + m_1 + 1, \quad \beta_2 = j - j_1 - m_2 + 1
\end{aligned}
$$

and then reverse the order of summation using (3.61), and we obtain in this way the q-analog of the Racah form (3.60).

A third form of the q-WCG coefficients is the q-analog of a form found by Majumdar [143]. By using the transformation (3.65) on the expression (3.59) we obtain this form directly. Specifically, put

$$
\begin{aligned}
n = j_1 + j_2 - j, \quad \alpha_1 = -j_2 - m_2, \quad \alpha_2 = -j_1 + m_1 \\
\beta_1 = j - j_1 - m_2 + 1, \quad \beta_2 = j - j_2 + m_1 + 1
\end{aligned}
$$

in (3.65), then this transformation takes (3.59) into the Majumdar form given in [65].

Further to the symmetries of Theorem 3.62 the q-WCG coefficients satisfy the following relation (see [65] and [76]) which we find to be useful in §5.4:

$$
\sum_{m_1',m_2'} \left(\mathcal{R}^{j_1 j_2}\right)^{m_1',m_2'}_{m_1,m_2}\,{}_qC^{j_1\ j_2\ j}_{m_1'\,m_2'\,m} = (-)^{j_1+j_2-j} q^{\frac{1}{2}\left(j(j+1)-j_1(j_1+1)-j_2(j_2+1)\right)}\,{}_qC^{j_2\ j_1\ j}_{m_2\,m_1\,m}\,, \tag{3.66}
$$

where the matrix elements of $\mathcal{R}$ are given in (2.20). For $q=1$ this symmetry coincides with the $q=1$ case of (3.63b).

3.6 q-$6j$ and q-Racah Coefficients

It is yet another instance of the surprising relationship between mathematics and physics that the q-$6j$ coefficients had been anticipated by Askey and Wilson [61], quite independent of any symmetry that now can be seen to explain their properties. These authors defined these coefficients in terms of q-analogs of hypergeometric functions and this route provides the most expeditious way to the properties we derive. The symmetries of the q-WCG coefficients and the q-$6j$ symbols have been discussed in terms of basic hypergeometric functions also by Rajeswari and Rao [144] and Groza et al. [60].

Consider first the definition of the q-analogs of the Racah and $6j$ symbols. The q-Racah coefficients are obtained by decomposing the tensor product $\mathfrak{V}_{j_1} \otimes \mathfrak{V}_{j_2} \otimes \mathfrak{V}_{j_3}$ of three irreducible representations of $\mathcal{U}_q(\mathfrak{su}(2))$. This can be done in two possible ways (by decomposing first either $\mathfrak{V}_{j_1} \otimes \mathfrak{V}_{j_2}$ or $\mathfrak{V}_{j_2} \otimes \mathfrak{V}_{j_3}$) and the two sets of orthogonal bases obtained thereby are linearly dependent. Specifically, we decompose first

$$\mathfrak{V}_{j_1} \otimes \mathfrak{V}_{j_2} = \sum_{j_{12}} \oplus\, \mathfrak{V}_{j_{12}},$$

where $\mathfrak{V}_{j_{12}}$ is spanned by

$$|(j_1j_2)j_{12}m_{12}\rangle = \sum_{m_1,m_2} {}_qC^{j_1\ j_2\ j_{12}}_{m_1\ m_2\ m_{12}}\ |j_1m_1\rangle \otimes |j_2m_2\rangle,$$

and then decompose

$$\mathfrak{V}_{j_{12}} \otimes \mathfrak{V}_{j_3} = \sum_{j_{123}} \oplus\, \mathfrak{V}_{j_{123}},$$

with basis vectors for $\mathfrak{V}_{j_{123}}$:

$$\begin{aligned} |(j_{12}j_1j_2j_3)j_{123}m_{123}\rangle &= \sum_{m_3,m_{12}} {}_qC^{j_{12}\ j_3\ j_{123}}_{m_{12}\ m_3\ m_{123}}\ |(j_1j_2)j_{12}m_{12}\rangle \otimes |j_3m_3\rangle \\ &= \sum_{\substack{m_1,m_2\\ m_3,m_{12}}} {}_qC^{j_{12}\ j_3\ j_{123}}_{m_{12}\ m_3\ m_{123}}\ {}_qC^{j_1\ j_2\ j_{12}}_{m_1\ m_2\ m_{12}}\ |j_1m_1\rangle \otimes |j_2m_2\rangle \otimes |j_3m_3\rangle. \end{aligned} \tag{3.67}$$

There is a similar expression for the basis vectors determined by decomposing first $\mathfrak{V}_{j_2} \otimes \mathfrak{V}_{j_3}$. The matrix connecting these linearly dependendent orthonormal bases defines the q-Racah coefficients ${}_qW(abcd;ef)$, from which we define the q-$6j$ symbols as in the classical case:

$$\begin{Bmatrix} a & b & e \\ d & c & f \end{Bmatrix}_q \stackrel{\text{def}}{=} (-1)^{a+b+c+d}\ {}_qW(abcd;ef).$$

The expression (3.67) shows that the q-Racah coefficients relate products of two q-WCG coefficients and, by using the orthogonality properties of the q-WCG coefficients, this relation can be obtained in the form:

$$\delta_{cc'}\Big([2e+1][2f+1]\Big)^{\frac{1}{2}} \begin{Bmatrix} a & b & e \\ d & c & f \end{Bmatrix}_q (-)^{a+b+c+d} = \tag{3.68}$$

$$\sum_{\beta,\delta,\gamma} {}_qC^{b\ d\ f}_{\beta\ \delta\ \beta+\delta}\ {}_qC^{e\ d\ c}_{\gamma-\delta\ \delta\ \gamma}\ {}_qC^{a\ b\ e}_{\gamma-\beta-\delta\ \beta\ \gamma-\delta}\ {}_qC^{a\ f\ c'}_{\gamma-\beta-\delta\ \beta+\delta\ \gamma}\ .$$

This provides an expression for the q-$6j$ symbols which can, however, be greatly simplified to a sum over one index. This manipulation has been performed in [65], see also Kachurik and Klimyk [145], and leads to the following formula for the q-$6j$ symbols:

$$\begin{Bmatrix} a & b & e \\ d & c & f \end{Bmatrix}_q = \Delta(abe)\,\Delta(cde)\,\Delta(acf)\,\Delta(bdf)$$
$$\times \sum_n \frac{(-)^n[n+1]!}{[n-a-b-e]![n-a-c-f]![n-b-d-f]![n-d-c-e]!}$$
$$\times \frac{1}{[a+b+c+d-n]![a+d+e+f-n]![b+c+e+f-n]!} \tag{3.69}$$

where the summation is over integers n satisfying:

$$\max(a{+}b{+}e,\ a{+}c{+}f,\ b{+}d{+}f,\ d{+}c{+}e) \leqslant n \leqslant \min(a{+}b{+}c{+}d,\ a{+}d{+}e{+}f,\ b{+}c{+}e{+}f)$$

as is determined from the denominator factors in (3.69). We note that, unlike the q-WCG coefficient, (3.69) is invariant under $q \leftrightarrow q^{-1}$.

The q-analog of the Racah coefficients can be expressed in terms of basic hypergeometric functions, like the q-WCG coefficients, as noted also in [140, 65, 145]. For $q = 1$ the expression of Racah coefficients in terms of ${}_4F_3$ functions with unit argument has been given by Rose [146], Erdélyi [147], and also Jahn and Howell [148], although the identification was incomplete in that the formulas were not valid over the full domain of definition of the quantum numbers occurring in the Racah coefficient. It was shown by Biedenharn and Louck [103] and also Rao and Venkatesh [149, 150, 151] that three different ${}_4F_3$ functions are necessary in order to represent the Racah coefficients over the full range of parameters. The generalization of these formulas to the q-analog case is immediate, and so we can represent the q-Racah coefficients as one of three ${}_4\phi_3$ functions.

For this identification it is necessary to change the summation index n in (3.69) so as to accord with the definition (2.123), p. 68, of basic hypergeometric functions for $p = 3$, that is, we put $m = n - \min\,(a+b+c+d,\ c+d+e+f,\ b+c+e+f)$, and the three possibilities for m give the three expressions of the coefficient in terms of ${}_4\phi_3$. The identity which enables us to make this identification is

$$\sum_n \frac{(-)^n[n+1]!}{[n-\alpha_1]![n-\alpha_2]![n-\alpha_3]![n-\alpha_4]![\beta_1-n]![\beta_2-n]![\beta_3-n]!}$$
$$= \frac{(-)^{\beta_1}[\beta_1+1]!}{[\beta_2-\beta_1]![\beta_3-\beta_1]![\beta_1-\alpha_1]![\beta_1-\alpha_2]![\beta_1-\alpha_3]![\beta_1-\alpha_4]!}$$
$$\times {}_4\phi_3\left(\begin{matrix} q^{\alpha_1-\beta_1} & q^{\alpha_2-\beta_1} & q^{\alpha_3-\beta_1} & q^{\alpha_4-\beta_1} \\ q^{-\beta_1-1} & q^{\beta_2-\beta_1+1} & q^{\beta_3-\beta_1+1} & \end{matrix} ;q,q\right), \qquad (3.70)$$

where

$$\begin{gathered}\beta_1 \geqslant \max(\alpha_1,\alpha_2,\alpha_3,\alpha_4) \\ \beta_1 \geqslant 0, \quad \beta_2 \geqslant \beta_1, \quad \beta_3 \geqslant \beta_1 \\ \alpha_1+\alpha_2+\alpha_3+\alpha_4 = \beta_1+\beta_2+\beta_3, \end{gathered} \qquad (3.71)$$

which can be obtained directly from the definition (2.123). Now we choose

$$\beta_1 = \min(a+b+c+d,\ a+d+e+f,\ b+c+e+f)$$

with β_2, β_3 being the remaining two parameters and $(\alpha_1,\alpha_2,\alpha_3,\alpha_4)$ equal to any permutation of $(a+b+c,\ c+d+e,\ a+c+f,\ b+d+f)$. These parameters satisfy the constraints (3.71), in particular the constraint which ensures that the ${}_4\phi_3$ in (3.70) is balanced.

The q-$6j$ symbols satisfy all the same 144 symmetry relations as for $q=1$. These symmetries all follow by termwise rearrangement of factors in the series (3.69) and are also directly visible from the expression (3.70). All 144 symmetries are equivalent to permutations of the numerator or denominator parameters amongst themselves, exactly as for the $q=1$ case. We refer to [65, 145] for a detailed discussion of these symmetries, including a graphical interpretation.

The q-$6j$ symbols also satisfy the same orthogonality relations as for $q=1$ as was anticipated by Askey and Wilson [61], who obtained q-analogs of the orthogonal Racah polynomials, as mentioned earlier. Other properties have been derived in [65, 145] including recurrence relations for the q-$6j$ symbols, and the q-analogs of the Racah identity and the Biedenharn-Elliot identity, which in terms of the q-Racah coefficients reads:

$${}_qW(a'ab'b;c'e)\,{}_qW(a'ed'd;b'c) = \sum_f [2f+1]\,{}_qW(abcd;ef)\,{}_qW(c'bd'd;b'f)\,{}_qW(a'ad'f;c'c).$$

This fundamental identity has the same significance as for $q=1$, being equivalent to the associative law of multiplication for unit tensor operators. This identity is termed the pentagon identity by Majid [14] and appears as the associativity property of a product functor on a monoidal category.

3.6.1 Asymptotic limit of the q-6j symbol

It was known very early (Biedenharn [152], 1953) that the WCG coefficients can be regarded as a certain limit of Racah coefficients. This fact has numerous consequences and in particular shows how properties of WCG coefficients can be derived from those of Racah coefficients. In order to obtain the q-analog result of this limit relation let us first introduce the following notation for the q-Racah coefficient (following [66] and Biedenharn and Lohe [140]):

Definition 3.72 *Define the symbol* ${}_qW^{a\ b\ c}_{\rho\ \sigma\ \tau}(j)$, *if* $\rho+\sigma=\tau$, *and if each of the triples* $(j-\tau,a,j-\sigma),(j-\sigma,b,j),(j-\tau,c,j)$ *consists of nonnegative integers and half-integers that satisfy the triangle conditions, by:*

$$ {}_qW^{a\ b\ c}_{\rho\ \sigma\ \tau}(j) \stackrel{\text{def}}{=} ([2c+1][2j-2\sigma+1])^{\frac{1}{2}}\,(-)^{a+b+\tau}\begin{Bmatrix} j-\tau & a & j-\sigma \\ b & j & c \end{Bmatrix}_q . $$

Otherwise, define ${}_qW^{a\ b\ c}_{\rho\ \sigma\ \tau}(j)\equiv 0$.

The triangle conditions are stated in (3.16). Since we have

$$ ([2e+1][2f+1])^{\frac{1}{2}}\begin{Bmatrix} a & b & e \\ d & c & f \end{Bmatrix}_q = (-)^{a+b+c+d}\,{}_qW^{\ b\ \ \ \ d\ \ \ \ f}_{e-a\ c-e\ c-a}(c) $$

where $b\geqslant|e-a|$, $d\geqslant|c-e|$, and $f\geqslant|c-a|$, each Racah coefficient is expressible in this notation.

Now we wish to show that the limit $j\to\infty$ of ${}_qW^{b\ d\ f}_{\alpha\ \beta\ \gamma}(j+\gamma)$ is the q-WCG coefficient ${}_qC^{b\ d\ f}_{\alpha\ \beta\ \gamma}$. In order to do this we first put $a=j$, $b=b$, $c=j+\gamma$, $d=d$, $e=j+\alpha$ and $f=f$ in the above equations and write $\beta_i=2j+\delta_i$ where

$$ \delta_1 = \min\left(b+\alpha+d+\beta,\ d-\beta+f+\gamma,\ b+\alpha+f+\gamma\right) $$

and δ_2,δ_3 are the pair of integers remaining in the triple

$$ (b+\alpha+d+\beta,\ d-\beta+f+\gamma,\ b+\alpha+f+\gamma) $$

after removing δ_1. Similarly we may write $\alpha_i=2j+\varepsilon_i$ where $(\varepsilon_1,\ \varepsilon_2,\ \varepsilon_3)$ is any permutation of

$$ (b+\alpha,\ d+\alpha+\gamma,\ f+\gamma) $$

and $\alpha_4=b+d+f$. Using these results, we obtain

$$ {}_qW^{b\ d\ f}_{\alpha\ \beta\ \gamma}(j+\gamma) = \delta_{\alpha+\beta,\gamma}(-)^{b+d+\gamma+\delta_1}\frac{\Delta(bdf)([2j+\delta_1-b-d-f+1])_{b+d+f+1}}{[\delta_2-\delta_1]!\,[\delta_3-\delta_1]!\,[\delta_1-\varepsilon_1]!\,[\delta_1-\varepsilon_2]!\,[\delta_1-\varepsilon_3]!} $$

$$\times \left(\frac{[b-\alpha]!\,[b+\alpha]!\,[d+\beta]!\,[d-\beta]!\,[f+\gamma]!\,[f-\gamma]![2j+2\alpha+1][2f+1]}{([2j+\alpha-b+1])_{2b+1}([2j+\alpha+\gamma-d+1])_{2d+1}([2j+\gamma-f+1])_{2f+1}} \right)^{\frac{1}{2}}$$

$$\times {}_4\phi_3 \left(\begin{matrix} q^{\varepsilon_1-\delta_1} & q^{\varepsilon_2-\delta_1} & q^{\varepsilon_3-\delta_1} & q^{b+d+f-2j-\delta_1} \\ q^{-2j-\delta_1-1} & q^{\delta_2-\delta_1+1} & q^{\delta_3-\delta_1+1} & \end{matrix} ; q, q \right) . \qquad (3.73)$$

In letting $j \to \infty$ we use the limit

$$\frac{([a-2j])_n}{([b-2j])_n} \to q^{\frac{n}{2}(b-a)}$$

in which we assume $q > 1$; the case $q < 1$ is obtained by interchanging q and q^{-1}. The limit of the ${}_4\phi_3$ function is then

$${}_3\phi_2 \left(\begin{matrix} q^{\varepsilon_1-\delta_1} & q^{\varepsilon_2-\delta_1} & q^{\varepsilon_3-\delta_1} \\ q^{\delta_2-\delta_1+1} & q^{\delta_3-\delta_1+1} & \end{matrix} ; q, q \right) .$$

In order to find the limit of the remaining j-dependent factors in (3.73) we use

$$([a+2j])_n \to \frac{q^{\frac{n}{4}(2a+4j+n-1)}}{(q^{\frac{1}{2}} - q^{-\frac{1}{2}})^n}$$

for $q > 1$, which from (3.73) gives the q-factor q^μ, where

$$\mu = \tfrac{1}{4}(b+d+f+1)(2\delta_1 - b - d - f) - \tfrac{1}{2}\alpha(b+d) - \tfrac{1}{2}\gamma(d+f+1) .$$

Taking the case $\delta_1 = b + \alpha + \beta + d$, we find that the limit of (3.73) is *precisely* the q-WCG coefficient (3.59):

LEMMA **3.74**

$$\lim_{j\to\infty} {}_qW^{b\,d\,f}_{\alpha\,\beta\,\gamma}(j+\gamma) = \begin{cases} {}_qC^{b\,d\,f}_{\alpha\,\beta\,\gamma} & \text{if } q > 1 \,; \\ {}_{q^{-1}}C^{b\,d\,f}_{\alpha\,\beta\,\gamma} & \text{if } q < 1 \,. \end{cases}$$

In order to obtain the last equation we used the invariance of the q-Racah coefficient under $q \leftrightarrow q^{-1}$ and so extended the limit to the case $q < 1$. This result on the asymptotic limit has been established also by Nomura [76] and Kachurik and Klimyk [145].

Many properties of the q-WCG coefficients follow from those of the q-Racah coefficients using this limit. For example, as noted by Nomura [76], the invariance of the q-$6j$ symbol under exchange of the first and third columns leads directly to the symmetry shown in (3.64). Another example is that of recurrence relations which can be derived for the q-WCG coefficients from those of the q-$6j$ symbols using this limit [145].

3.7 The Pattern Calculus and Elementary Tensor Operators

For the unitary groups a pattern calculus has been developed by Biedenharn and Louck [126] which allows one to write down matrix elements of elementary tensor operators in $U(n)$. The pattern calculus is actually three things:

(a) an explicit evaluation of all matrix elements of all elementary tensor operators between all unitary irreps in $U(n)$;

(b) a diagrammatic procedure whereby the diagram implies the explicit algebraic matrix element associated by the pattern calculus rules to the diagram;

(c) a procedure for the construction of all tensor operators. The elementary tensor operators are given directly by the pattern calculus and, by extending the calculus so that patterns act on patterns, one obtains an algebraic basis for the construction of general tensor operators.

It is the merit of the pattern calculus that it 'explains' and makes accessible structurally the otherwise difficult-to-comprehend explicit matrix elements, which can be of arbitrarily large complexity and length. The pattern calculus was first constructed (in [126] for $U(n)$) by exploiting the Jordan map. These results were later verified using an algebraic version of the Borel-Weil construction by LeBlanc and Hecht [153] and, using a more conceptual approach, by Gould [154]. These results were extended to the quantum groups $\mathcal{U}_q(\mathfrak{u}(n))$ by Gould and Biedenharn [74], in which matrix elements of q-tensor operators were determined and pattern calculus rules formulated.

Properties of tensor operators for $U(n)$ and $\mathcal{U}_q(\mathfrak{u}(n))$ are described in §3.2 and §3.3, in particular unit q-tensor operators are defined in Corollary 3.24. It is a consequence of the definition of a unit tensor operator for both classical and quantum groups that the matrix elements of the operator:

$$\left\langle \begin{matrix} (\Gamma) \\ [M] \\ (M) \end{matrix} \right\rangle \to \left\langle \begin{matrix} [m^{\text{final}}] \\ (m^{\text{f}}) \end{matrix} \middle| \left\langle \begin{matrix} (\Gamma) \\ [M] \\ (M) \end{matrix} \right\rangle \middle| \begin{matrix} [m^{\text{initial}}] \\ (m^{\text{i}}) \end{matrix} \right\rangle$$

are *exactly* the same as the matrix elements of the matrix which brings the tensor product of the irreps $[M] \otimes [m^{\text{initial}}] \to [m^{\text{final}}]$ to fully reduced (block-diagonal) form, ordered by the standard ordering on irrep labels $[M]$ (Theorem 3.18). In other words the construction of explicit unit tensor operators is equivalent to the complete resolution of the Clebsch-Gordan tensor product, including the resolution of all multiplicities.

A particularly simple class of unit tensor operators is the class of such operators for which the shift labels (Γ) are a permutation of the operator irrep labels $[M]$. For this class, there is no multiplicity. This class includes all **elementary unit tensor operators**, where an elementary unit tensor operator (distinct from the unit operator) is defined to be an irreducible unit tensor operator with irrep labels of the form: $[M] = [\underbrace{1\ldots1}_{k}\underbrace{0\ldots0}_{n-k}]$, $\quad (k = 1, 2, \ldots, n-1)$.

The operator $\langle [M] \rangle \stackrel{\text{def}}{=} \left\langle \begin{smallmatrix} (\Gamma) \\ [M] \\ (M) \end{smallmatrix} \right\rangle$, considered as an explicitly determined algebraic object, is enormously complicated. What is needed is an understanding of the structure implicit in such an object. For $\mathcal{U}_q(\mathrm{u}(n))$ this desired structure is implied by the chain: $\mathcal{U}_q(\mathrm{u}(n)) \supset \mathcal{U}_q(\mathrm{u}(n-1)) \supset \ldots \supset \mathcal{U}(\mathrm{u}(1))$. For this chain we have the decomposition:

$$\left\langle \begin{smallmatrix} [m^{\mathrm{f}}] \\ (m^{\mathrm{f}}) \end{smallmatrix} \middle| \left\langle \begin{smallmatrix} (\Gamma) \\ [M] \\ (M) \end{smallmatrix} \right\rangle \middle| \begin{smallmatrix} [m^{\mathrm{i}}] \\ (m^{\mathrm{i}}) \end{smallmatrix} \right\rangle = \sum_{(\Gamma')} \left\langle \begin{smallmatrix} [m^{\mathrm{f}}]_n \\ [m^{\mathrm{f}}]_{n-1} \end{smallmatrix} \middle\| \begin{bmatrix} (\Gamma) \\ [M] \\ (\Gamma') \end{bmatrix} \middle\| \begin{smallmatrix} [m^{\mathrm{i}}]_n \\ [m^{\mathrm{i}}]_{n-1} \end{smallmatrix} \right\rangle$$
$$\times \left\langle \begin{smallmatrix} [m^{\mathrm{f}}]_{n-1} \\ (m^{\mathrm{f}}) \end{smallmatrix} \middle| \left\langle \begin{smallmatrix} (\Gamma') \\ (M)_{n-1} \end{smallmatrix} \right\rangle \middle| \begin{smallmatrix} [m^{\mathrm{i}}]_{n-1} \\ (m^{\mathrm{i}}) \end{smallmatrix} \right\rangle .$$

This result, stated in words, asserts that a matrix element for a unit tensor operator belonging to $\mathcal{U}_q(\mathrm{u}(n))$ can be expressed as a sum of products of two factors:

(a) a **reduced operator coefficient**:

$$\left\langle \begin{smallmatrix} [m^{\mathrm{f}}]_n \\ [m^{\mathrm{f}}]_{n-1} \end{smallmatrix} \middle\| \begin{bmatrix} (\Gamma) \\ [M] \\ (\Gamma') \end{bmatrix} \middle\| \begin{smallmatrix} [m^{\mathrm{i}}]_n \\ [m^{\mathrm{i}}]_{n-1} \end{smallmatrix} \right\rangle$$

which, by construction, is an invariant under the $\mathcal{U}_q(\mathrm{u}(n-1))$ subgroup; and

(b) an irreducible unit tensor operator matrix element:

$$\left\langle \begin{smallmatrix} [m^{\mathrm{f}}]_{n-1} \\ (m^{\mathrm{f}}) \end{smallmatrix} \middle| \left\langle \begin{smallmatrix} (\Gamma') \\ (M)_{n-1} \end{smallmatrix} \right\rangle \middle| \begin{smallmatrix} [m^{\mathrm{i}}]_{n-1} \\ (m^{\mathrm{i}}) \end{smallmatrix} \right\rangle$$

which belongs to a unit tensor operator in the quantum subgroup $\mathcal{U}_q(\mathrm{u}(n-1))$, where the irrep labels are contained in the patterns (Γ') and $(M)_{n-1}$.

Thus, one has a recursive approach to the explicit determination of tensor operators in which one need only obtain generic matrix elements for the reduced operators denoted $\begin{bmatrix} (\Gamma) \\ [M] \\ (\Gamma') \end{bmatrix}$, by determining the operator from the matrix elements above. (This determination is unique since the labels of the reduced operator determine completely the final state, given the initial state.) It is important to observe that the reduced operators act in a $\mathcal{U}_q(\mathrm{u}(n)) : \mathcal{U}_q(\mathrm{u}(n-1))$ subspace, labelled by a truncated (two-rowed) Gel'fand-Weyl pattern.

3.7.1 The Pattern Calculus Rules for Elementary q-Tensor Operators

Matrix elements for reduced operators have been obtained explicitly by Gould and Biedenharn [74] using projection operator techniques in tensor product vector spaces. Let us demonstrate now that these matrix elements (for elementary operators) are characterized by a very simple structure which can be directly related to the shifts themselves. These relations constitute the **pattern calculus rules.**

The pattern calculus rules determine the explicit reduced matrix element in the form:

$$\left\langle \begin{matrix} [m^{\rm f}]_n \\ [m^{\rm f}]_{n-1} \end{matrix} \middle\| \begin{bmatrix} (\Gamma) \\ [M] \\ (\Gamma_0) \end{bmatrix} \middle\| \begin{matrix} [m^{\rm i}]_n \\ [m^{\rm i}]_{n-1} \end{matrix} \right\rangle = (\pm) q^{\alpha} \left| \frac{N}{D} \right|,$$

directly from the initial states, $[m^{\rm i}]_n$ and $[m^{\rm i}]_{n-1}$, using the operator patterns (Γ) and (Γ_0) and the irrep label $[M]$. The final states are uniquely determined from this information: $[m^{\rm f}]_n = [m^{\rm i} + \Delta(\Gamma)]_n$, $[m^{\rm f}]_{n-1} = [m^{\rm i} + \Delta(\Gamma_0)]_{n-1}$.

The pattern calculus rules apply to extremal operator patterns, that is, operator patterns such that the shift Δ implies the pattern Γ. For such patterns, the shifts Δ are permutations of the maximal weight of the irrep characterizing the operator.

Let us arrange the two shifts Δ and Δ_0 corresponding to the patterns Γ and Γ_0 in the form of a two-rowed pattern, similar to a (truncated) Gel'fand-Weyl pattern, but without the betweenness constraints. We call this the **shift pattern.**

The pattern calculus rules for the elementary tensor operators are as follows:

RULE 1. Write out two rows of dots with n dots in row n and $n-1$ dots in row $n-1$, in the manner displayed below:

RULE 2. In row n, assign to the $i^{\rm th}$ dot the $\mathcal{U}_q(\mathrm{u}(n))$ partial hook defined by $p_{i,n} \stackrel{\rm def}{=} m_{i,n} + n - i$; do the same in row $n-1$ using the $\mathcal{U}_q(\mathrm{u}(n-1))$ partial hooks, $p_{i,n-1} = m_{i,n-1} + n - 1 - i$.

RULE 3. Draw arrows between dots as follows: Select a dot i in row n and a dot j in row $n-1$. Consider the shifts (weights) $\Delta_{in}(\Gamma)$ of the $\mathcal{U}_q(\mathrm{u}(n))$ operator pattern Γ and the shifts $\Delta_{j,n-1}(\Gamma_0)$ of the $\mathcal{U}_q(\mathrm{u}(n-1))$ operator pattern Γ_0. If $\Delta_{in} > \Delta_{j,n-1}$ draw an arrow from dot i to dot j; if $\Delta_{in} < \Delta_{j,n-1}$ draw an arrow from dot j to dot i. If $\Delta_{in} = \Delta_{j,n-1}$ go to another pair. Carry out this procedure for all dots in rows n and $n-1$. This yields a numerator arrow pattern with arrows going between rows.

Carry out this same procedure for dots within row n and dots within row $n-1$. This yields a denominator arrow pattern with arrows going within rows.

RULE 4. Assign to each arrow the q-integer factor

$$[p(\text{tail}) - p(\text{head}) + e(\text{tail})]$$

where p(tail/head) is the partial hook associated with the dot(tail/head), and

$$e(\text{tail}) \stackrel{\text{def}}{=} \begin{cases} 1 & \text{if the tail of the arrow is on row } n-1, \\ 0 & \text{if the tail of the arrow is on row } n. \end{cases}$$

RULE 5. Write out the products:

N^2 = product of all q-integer factors for the numerator arrow pattern,
D^2 = product of all q-integer factors for the denominator arrow pattern.

RULE 6. (sign convention) The $(\pm)$ sign for the matrix element is obtained in this way:

Take the shift Δ — which consists of k ones and $(n-k)$ zeros in some order — and assign the k integers $i_1 < i_2 < \ldots < i_k$ which denote the k places in the shift pattern Δ where the ones occur. For example, $\Delta = 01100 \Rightarrow (i_1 i_2) = (23)$. Similarly assign $j_1 < j_2 < \ldots j_k$ for the shift Δ_0. (Note that we have formally extended the shift Δ_0 to n places by embedding in $\mathcal{U}_q(\mathrm{u}(n))$.)

The sign of the matrix element is then defined to be

$$(-1)^{\frac{1}{2}k(k-1)} \prod_{\ell,s=1}^{k} S(j_\ell - i_s), \quad (i_1 < i_2 < \ldots i_k, \quad j_1 < j_2 < \ldots < j_k),$$

where

$$S(j-i) \stackrel{\text{def}}{=} \begin{cases} +1 & j \geqslant i \\ -1 & j < i \end{cases}.$$

RULE 7. (q-phase factor q^α) There are two cases to consider:

(a) The shift patterns Δ and Δ_0 each have k ones, to which we assign (following Rule 6) the sequences $(i_1, i_2 \ldots i_k)$ and $(j_1, j_2 \ldots j_k)$ respectively.

The q-phase factor is then given by q^α with

$$\alpha = \tfrac{1}{4} \sum_{s=1}^{k} (p_{i_s,n} - p_{j_s,n-1} - 1).$$

(b) The shift pattern Δ contains k ones, and the shift pattern Δ_0 $k-1$ ones. For this case we label the $n-k$ zeros appearing in Δ and Δ_0 by the labels $\bar{\imath}_1 < \bar{\imath}_2 < \ldots < \bar{\imath}_{n-k}$ for Δ, and by $\bar{\jmath}_1 < \bar{\jmath}_2 < \ldots < \bar{\jmath}_{n-k}$ for Δ_0 exactly as we labelled the ones in Rule 6 by $(i_1 \ldots i_k)$ and $(j_1 \ldots j_k)$.

The q-phase factor is then given by q^α with

$$\alpha = \tfrac{1}{4} \sum_{s=1}^{n-k} (p_{\bar{\imath}_s,n} - p_{\bar{\jmath}_s,n-1}).$$

RULE 8. The desired matrix element of the $\mathcal{U}_q(\mathrm{u}(n))$ elementary tensor operator $\langle[\underbrace{1\ldots1}_{k}\underbrace{0\ldots0}_{n-k}]\rangle$ is then given by:

$$\left\langle \begin{matrix} [m_{i,n}+\Delta_n(\Gamma)] \\ [m_{i,n-1}+\Delta_{n-1}(\Gamma_0)] \end{matrix} \middle| \left[\begin{matrix} (\Gamma) \\ [\dot{1}_k \dot{0}_{n-k}] \\ (\Gamma_0) \end{matrix} \right] \middle| \begin{matrix} [m_{i,n}] \\ [m_{i,n-1} \end{matrix} \right\rangle = (\pm 1) q^{\alpha} \left(\frac{N^2}{D^2}\right)^{\frac{1}{2}}, \tag{3.75}$$

where each of the terms has been given by rules above.

In summary, we have the following result, and refer to Gould and Biedenharn [74] for the proof:

LEMMA **3.76** *The pattern calculus rules (1-8) determine explicitly all reduced matrix elements of all elementary tensor operators acting on arbitrary extended weight vectors in the* $\mathcal{U}_q(\mathrm{u}(n)) : \mathcal{U}_q(\mathrm{u}(n-1))$ *subspace. In the limit* $q \to 1$, *these rules become exactly the pattern calculus rules for the unitary groups* $U(n)$ *given in [126].*

Now let us apply the pattern calculus in two examples:

EXAMPLE **3.77** **(the q-tensor operator $\langle 1\ 0\rangle$ in $\mathcal{U}_q(\mathrm{u}(2))$)** Choose the $\mathcal{U}_q(\mathrm{u}(2))$ operator pattern $\Gamma = \begin{smallmatrix} 1 & & 0 \\ & 0 & \end{smallmatrix}$, for which the weight vector is $\Delta(\Gamma) = 0\,1$. Also, choose the $\mathrm{u}(1)$ operator pattern $\Gamma_0 = 1$, for which $\Delta(\Gamma_0) = 1$. The nonzero matrix elements of the Wigner operator, which are also the reduced matrix elements, are:

$$\left\langle \begin{smallmatrix} m_{12} & & m_{22}+1 \\ & m_{11}+1 & \end{smallmatrix} \middle| \left\langle \begin{smallmatrix} & 0 & \\ 1 & & 0 \\ & 1 & \end{smallmatrix} \right\rangle \middle| \begin{smallmatrix} m_{12} & & m_{22} \\ & m_{11} & \end{smallmatrix} \right\rangle = (\pm) q^{\alpha} \left(\frac{N^2}{D^2}\right)^{\frac{1}{2}}.$$

The numerator arrow pattern is given by:

$$\Delta = \begin{matrix} 0 & & 1 \\ & 1 & \end{matrix} \quad \Longrightarrow$$

and we obtain $N^2 = [p_{11} + 1 - p_{12}]$. The denominator pattern is given by:

$$\Delta = \begin{matrix} 0 & & 1 \\ & 1 & \end{matrix} \quad \Longrightarrow$$

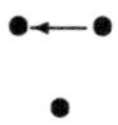

and we obtain $D^2 = [p_{22} - p_{12}]$. In determining the sign, we find $i_1 = 2, j_1 = 1$ and so the sign is negative. The q-factor q^{α} is determined from $\alpha = \frac{1}{4}(p_{i_1 2} - p_{j_1 1} - 1) = \frac{1}{4}(p_{22} - p_{11} - 1)$. Hence we find the matrix element

$$\left\langle \begin{smallmatrix} m_{12} & & m_{22}+1 \\ & m_{11}+1 & \end{smallmatrix} \middle| \left\langle \begin{smallmatrix} & 0 & \\ 1 & & 0 \\ & 1 & \end{smallmatrix} \right\rangle \middle| \begin{smallmatrix} m_{12} & & m_{22} \\ & m_{11} & \end{smallmatrix} \right\rangle = -q^{\frac{1}{4}(m_{22}-m_{11}-1)} \left(\frac{[m_{12}-m_{11}]}{[m_{12}+1-m_{22}]}\right)^{\frac{1}{2}},$$

in agreement with the entry $\Delta m = \frac{1}{2}, \Delta j = -\frac{1}{2}$ given in Table 3.56, (p. 97).

Example **3.78** **(the q-tensor operator $\langle 1\,0\,0\rangle$ in $\mathcal{U}_q(\mathrm{u}(3))$)** Choose the $\mathcal{U}_q(\mathrm{u}(3))$ operator pattern $\Gamma = \begin{smallmatrix} 1 & & 0 & & 0 \\ & 1 & & 0 & \\ & & 0 & & \end{smallmatrix}$. For this pattern the shifts $\Delta(\Gamma)$, which are the weights of the Γ pattern, are $\Delta(\Gamma) = 0\,1\,0$. Hence the operator $\left\langle \begin{smallmatrix} & \Gamma & \\ 1 & 0 & 0 \end{smallmatrix} \right\rangle$ acts on vectors in the irrep spaces according to: $[m_{13}m_{23}m_{33}] \longrightarrow [m_{13}+0\; m_{23}+1\; m_{33}+0]$.

Similarly, choose the $\mathcal{U}_q(\mathrm{u}(2))$ operator pattern $\Gamma_0 = \begin{smallmatrix} 1 & & 0 \\ & 1 & \end{smallmatrix}$ for which $\Delta(\Gamma_0) = 1\,0$. In this case the subgroup irrep is shifted by: $[m_{12}m_{22}] \longrightarrow [m_{12}+1\; m_{22}+0]$.

We write the numerator arrow pattern as follows:

$$\Delta = \begin{matrix} 0 & & 1 & & 0 \\ & 1 & & 0 & \end{matrix} \quad \Longrightarrow$$

Rules 4 and 5 determine that

$$N^2 = [p_{12} + \underset{\uparrow}{1} - p_{13}] \cdot [p_{12} + \underset{\uparrow}{1} - p_{33}] \cdot [p_{23} - p_{22}] \; .$$

To find the denominator arrow pattern consider:

$$\Delta = \begin{matrix} 0 & & 1 & & 0 \\ & 1 & & 0 & \end{matrix} \quad \Longrightarrow$$

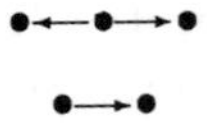

Again from Rules 4 and 5 this implies:

$$D^2 = [p_{23} - p_{13}] \cdot [p_{23} - p_{33}] \cdot [p_{12} + \underset{\uparrow}{1} - p_{22}] \; .$$

Observe also:

(i) The arrows under the 1's in N^2, D^2 indicate the implementation of the all-important 'tail-rule.'
(ii) The sign rule, which is invariant to deformation, yields (-1).
(iii) The q-phase rule yields $q^{\frac{1}{4}(p_{23}-p_{12}-1)}$.

Collecting all factors, we obtain for the reduced matrix element:

$$\left\langle \begin{smallmatrix} m_{13} & & m_{23}+1 & & m_{33} \\ & m_{12}+1 & & m_{22} & \end{smallmatrix} \right| \left[\begin{smallmatrix} & & 0 & & \\ & 1 & & 0 & \\ 1 & & 0 & & 0 \\ & 1 & & 0 & \\ & & 1 & & \end{smallmatrix} \right] \left| \begin{smallmatrix} m_{13} & & m_{23} & & m_{33} \\ & m_{12} & & m_{22} & \end{smallmatrix} \right\rangle = (-1) q^{\frac{1}{4}(p_{23}-p_{13}-1)} \left(\frac{N^2}{D^2} \right)^{\frac{1}{2}},$$

where N^2 and D^2 are given above.

3.7.2 A Conceptual Derivation of the Pattern Calculus Rules

The explicit results for the quantum group pattern calculus rules given in §3.7.1 are structurally similar to the pattern calculus results (Biedenharn and Louck [126]) for the classical Lie group $U(n)$, the most significant difference being the presence of explicit q-factors of the form q^α. This close structural similarity strongly suggests that there should be a direct route to these results for the quantum group pattern calculus. We develop briefly here such a direct, conceptual, approach.

For generic values of q, the Lusztig-Rosso theorem [69, 70] (that quantum group irreps are deformations of the classical group irreps) imposes strong constraints on the representation theory of compact quantum groups and, as we have noted above, the labelling of individual vectors in a $\mathcal{U}_q(\mathrm{u}(n))$ irrep is forced by this theorem to be identical to a Gel'fand-Weyl pattern of integers.

Now let us recall [66] that the zeros of the elementary operators of $U(n)$, which are all due to lexicality constraints, are definitive in determining the pattern calculus, to within ($\pm$) signs which are convention dependent. Knowledge of these zeros therefore determines the linear factors of the numerator in the $U(n)$ case and, from the constraints of the Rosso-Lusztig theorem plus continuity in q, the q-integer factors in the quantum group case. The denominator factors in both cases can be uniquely determined from this information by choosing special cases for which the reduced matrix element is known to be unity. By using permutational symmetry (the generalized Weyl symmetry) of the partial hooks, the general case can then be deduced from these special cases.

These arguments (Gould and Biedenharn [74]) suffice to prove that the explicit, monomial, reduced matrix elements for $\mathcal{U}_q(\mathrm{u}(n))$ are determined by abstract structural arguments (namely lexicality, symmetry, and continuity in q) to within a multiplicative q-phase factor q^α. The pattern calculus rules for $U(n)$ determine the sign of the matrix elements for $\mathcal{U}_q(\mathrm{u}(n))$ and the linear factors for $U(n)$ become q-integer factors for $\mathcal{U}_q(\mathrm{u}(n))$.

This result leaves only the q-phase factor to be determined by an *a priori* argument. To develop this argument, let us consider the q-$6j$ coefficients of $\mathcal{U}_q(\mathrm{u}(n))$, known also as q-Racah coefficients or 'tetrahedral invariant operators'. The q-$6j$ coefficients are most directly understood from the tensor operator viewpoint. The q-tensor operator algebra has as an algebraic basis the unit tensor operators $\langle[M]\rangle$, classified by Gel'fand-Weyl and operator patterns, with invariant operators as scalars. The q-$3j$ coefficients define a multiplication in this algebra, denoted Ⓦ in §3.4, such that the q-$3j$ product of two unit tensor operators yields an irreducible tensor operator: $(\langle[M']\rangle \circledW \langle[M]\rangle) = \mathcal{I}\langle[M'']\rangle$, where $\mathcal{I}$ is an invariant operator. (The product operator is *not* a unit tensor operator but an invariant multiple of such an operator.) This invariant operator $\mathcal{I}$ is determined by none other than the q-$6j$ coefficients which, as stated in Corollary 3.24, may be defined as the matrix elements of the following q-$6j$ invariant operator:

$$\langle[M'']\rangle^\dagger \bullet \langle[M']\rangle \circledW \langle[M]\rangle.$$

The action of this q-$6j$ invariant operator is on the irreps $[m]$ comprising the model space $\mathfrak{M}$. Hence,

$$((\langle[M'']\rangle^{\dagger} \bullet \langle[M']\rangle) \circledw \langle[M]\rangle)[m]$$

is a numerical q-$6j$ coefficient.

Each of these couplings requires an operator pattern assignment (Γ) in order to be uniquely defined. Thus we see that the structure of the q-$6j$ coefficient is determined by *four* triples of irreps (each triple uniquely defined by an associated operator pattern Γ) and by *six* irreps. We may accordingly associate this structure with a tetrahedron, by associating each triple and operator pattern to a face of the tetrahedron, and the six irreps to the lines. (The symmetry of the tetrahedron suggests corresponding symmetries of the q-$6j$ invariant which, however, have not as yet been fully proved for $\mathcal{U}_q(\mathfrak{u}(n)), n \geqslant 3$. Hence one should regard the tetrahedron as merely a convenient mnemonic for the actual coupling relationships.)

It is important to recall, in order to derive the quantum group properties of this tetrahedral coefficient, that each of the four couplings defines an ordering on its triple and, moreover, that the opposite ordering is obtained by the replacement $q \to q^{-1}$. If we now try to impose a consistent ordering on the four faces of the tetrahedron, we see that no consistent ordering is possible and, in fact, the only satisfactory ordering of the four faces requires that each edge be ordered oppositely for each of the two faces to which the edge belongs. Thus each irrep is associated to *two* couplings in which the orderings are *opposite.* This requirement is equivalent to the result that the q-$6j$ coefficient is invariant to the substitution $q \to q^{-1}$. This is a remarkable result which we have already noted for $\mathcal{U}_q(\mathfrak{u}(2))$ in §3.6. From it we deduce that a monomial q-$6j$ coefficient has no explicit q-phase factor.

Monomial $6j$ coefficients are known [103] from $U(n)$ to be uniquely characterized (to within $\pm$ signs) by their zeros and symmetries, which in turn leads to a pattern calculus for these coefficients. By application of the same arguments used above, we conclude that the monomial q-$6j$ coefficients are determined by this same pattern calculus, but with linear integer factors replaced by q-integer factors. Hence there is no undetermined q-phase factor, and the ($\pm$) signs are undeformed.

How does this information help in our problem of determining the q-phase factors for reduced matrix elements? The answer lies in another fundamental structure relation [103, 152] for the tensor operator algebra, that the q-$3j$ coefficients are limits of the q-$6j$ coefficients in the limit that $m_{nn} \to -\infty$, as proved in Lemma 3.74 for $\mathcal{U}_q(\mathfrak{u}(2))$. In particular, monomial reduced matrix elements for elementary tensor operators are the limits of monomial q-$6j$ coefficients. This limit operation defines the q-phase factor, which is found to be precisely the q-phase given by Rule 8 of the pattern calculus rules.

Determination by abstract properties of the q-phase factor completes our aim of determining the pattern calculus for monomial q-$3j$ matrix elements by conceptual arguments. We conclude that the pattern calculus for $\mathcal{U}_q(\mathfrak{u}(n))$ can be determined completely, using abstract structural arguments, from the pattern calculus for $U(n)$.

Chapter 4

The Dual Algebra and the Factor Group

In previous chapters we have developed properties of the q-deformation $\mathcal{U}_q(\mathrm{u}(n))$ of the universal enveloping algebra of the unitary groups, but have not considered the q-analog of the Lie group itself. There is a formulation of quantum groups, due to the Leningrad school, which proceeds by defining quantum matrices with noncommuting entries and which for $q = 1$ reduce to the classical general linear (and hence unitary) matrices. This set of quantum matrices is not a group under matrix multiplication but forms a Hopf algebra which is noncommutative and non-co-commutative. The dual of this Hopf algebra is exactly the q-deformation $\mathcal{U}_q(\mathrm{u}(n))$ of the universal enveloping algebra of the classical group which we considered in previous chapters, and provides the q-analog of the viewpoint that generators of the Lie algebra are the tangent vectors at the identity on the group manifold.

In this chapter we develop properties of the matrix quantum groups, and particularly develop the interpretation that the noncommuting elements of the quantum matrix are components of elementary tensor operators with respect to a direct product algebra which is actually a factor algebra. Hence, by using only the properties of q-tensor operators as developed in Chapter 3 we are able to derive, not postulate, the properties of the matrix quantum groups.

4.1 Introduction

The novel feature of quantum groups, as we have noted before, is the fact that co-multiplication is noncommutative, which is related to the existence of quantum coordinates which do not commute. These ideas appeared first in work by Faddeev, Sklyanin and Takhtajan (see [11], in particular the 1982 Les Houches lectures) in connection with the quantum inverse scattering method, which was applied to a variety of models in which matrices (quantum transport matrices) with noncommuting elements were introduced. This work was later extended by Manin [46] who discussed the concepts of noncommuting coordinates and noncommutative geometry.

The set of $n \times n$ quantum matrices, denoted $\mathfrak{M}_q(n)$, is the q-analog of the general linear matrices $GL(n, \mathbb{C})$, or $SL(n, \mathbb{C})$ if we set the commuting q-determinant equal to 1. Taking for example $n = 2$, these matrices act by matrix multiplication (actually co-multiplication) on complex-valued coordinates $\binom{x}{y}$ which for $q = 1$ comprise a two-component spinor transforming covariantly under $SL(2, \mathbb{C})$. In our context, where we use boson operators as the fundamental constructs, the "coordinates" become commuting operators and, as we have seen, are in fact boson operators themselves. When we generalize these spinor operators to the quantum case their components no longer commute, but the quantum matrices nevertheless act on the q-spinor by matrix multiplication which preserves the q-commutator relations.

We may extend these considerations to a *matrix* of spinor components, on which the group matrix (classical or quantum) acts by right or left matrix multiplication. In the classical ($q = 1$) case for $n = 2$ the matrix of spinor components can be regarded as a 2×2 matrix of complex variables or, in operator form, as a matrix A of four boson creation operators. This matrix is nothing more than the fundamental spin $\frac{1}{2}$ representation matrix $\mathfrak{D}^{[1,0]}(A)$ of $GL(2, \mathbb{C})$ and can be identified with an element of $GL(2, \mathbb{C})$ itself. (The fact that boson operators are components of spin $\frac{1}{2}$ tensor operators in $U(2)$ is easily verified by checking the derivative action of the generators on the elements of the matrix boson.)

In the quantum case, the elements of the matrix $T = (t_i^j)$ of spinor operators no longer commute. However this matrix T, which we find will be equal to the representation matrix ${}^q\mathfrak{D}^{[1,0]}(T)$, may still be identified as an element of the set $\mathfrak{M}_q(2)$ of quantum matrices and consists, by definition, of spin $\frac{1}{2}$ q-tensor operators with respect to two $\mathcal{U}_q(\mathfrak{u}(2))$ groups. These two (left and right) $\mathcal{U}_q(\mathfrak{u}(2))$ groups are determined by the left and right actions of the quantum group on the quantum matrix, and are constructed by co-multiplication from the quantum group. The two quantum groups commute but admit only those representations carrying identical irrep labels and hence we call this direct product of quantum groups a factor group.

In summary, we will identify elements T of $\mathfrak{M}_q(2)$ with the matrix of q-spinor operators according to

$$T = \begin{pmatrix} a & b \\ c & d \end{pmatrix} = {}^q\mathfrak{D}^{[1,0]}(T) = \begin{pmatrix} t_1^1 & t_1^2 \\ t_2^1 & t_2^2 \end{pmatrix}. \tag{4.1}$$

Each of the pairs (t_1^1, t_2^1) and (t_1^2, t_2^2) will be defined to be a spin $\frac{1}{2}$ tensor operator under the left $\mathcal{U}_q(\mathfrak{u}(2))$ algebra, and each of the pairs (t_1^1, t_1^2) and (t_2^1, t_2^2) will be defined to be a spin $\frac{1}{2}$ tensor operator under the right $\mathcal{U}_q(\mathfrak{u}(2))$ algebra. We will verify the identification (4.1) using tensor operator properties and derive the noncommuting properties of elements T of $\mathfrak{M}_q(2)$ which first appeared in the work of Faddeev et al. [11, 24]. We also consider the q-analog of the property that the fundamental $GL(2, \mathbb{C})$ matrix can be regarded as a matrix of boson operators by finding a realization of $T \in \mathfrak{M}_q(2)$ in terms of q-boson operators. (Note that the results of §3.4.2, in particular Example 3.39, show that the fundamental irrep matrix ${}^q\mathfrak{D}^{[1,0]}(T)$ cannot be the matrix of q-boson operators, since q-boson operators do not themselves form q-spinors.)

4.2 Matrix Quantum Groups

The study of specific lattice models, such as the spin $\frac{1}{2}$ XXX model, the Liouville lattice model, and the nonlinear Schrödinger lattice model carried out by the Leningrad school [11, 24] leads to an approach to quantum groups which takes as a starting point the following equations, the "RTT" relations:

$$\mathcal{R}T_1T_2 = T_2T_1\mathcal{R}.$$

Here $\mathcal{R}$ is the $n^2 \times n^2$ R-matrix discussed for $\mathcal{U}_q(\mathfrak{su}(2))$ in §2.1.1 and which satisfies the Yang-Baxter equations (1.10). The $n^2 \times n^2$ matrices T_1, T_2 are given by

$$T_1 = T \otimes I, \quad T_2 = I \otimes T,$$

where T is the $n \times n$ transition matrix, or quantum monodromy matrix, which arises in the quantum inverse scattering method, and I is the $n \times n$ identity matrix. The elements of T, which we denote by $(T)_{ij} = t^j_i$, belong to an associative algebra $\mathcal{A}$ over $\mathbb{C}$ which in general is noncommutative.

For $n = 2$ we may write

$$T = \begin{pmatrix} a & b \\ c & d \end{pmatrix}$$

where $a, b, c, d \in \mathcal{A}$, and we choose for the 4×4 matrix $\mathcal{R}$:

$$\mathcal{R} = \begin{pmatrix} q^{-\frac{1}{2}} & 0 & 0 & 0 \\ 0 & 1 & 0 & 0 \\ 0 & -(q^{\frac{1}{2}} - q^{-\frac{1}{2}}) & 1 & 0 \\ 0 & 0 & 0 & q^{-\frac{1}{2}} \end{pmatrix}.$$

(This form for $\mathcal{R}$ is related to that given in (2.19), denoted $\mathcal{R}_q$, by the relation[1] $\mathcal{R} = q^{-\frac{1}{4}}\widetilde{\mathcal{R}}_{q^{-1}}$; we could equally well choose $\mathcal{R}_q$ for the R-matrix to obtain the same set of relations (4.2)). With this expression for $\mathcal{R}$ the RTT relations may be written equivalently as[2]:

$$ab = q^{-\frac{1}{2}}ba, \quad ac = q^{-\frac{1}{2}}ca, \quad bd = q^{-\frac{1}{2}}db, \quad cd = q^{-\frac{1}{2}}dc,$$
$$bc = cb, \quad ad - da = -(q^{\frac{1}{2}} - q^{-\frac{1}{2}})bc. \tag{4.2}$$

Let us define $\mathfrak{M}_q(2)$ as the set of 2×2 quantum matrices $T = \left(\begin{smallmatrix} a & b \\ c & d \end{smallmatrix}\right)$ with entries satisfying (4.2). $\mathfrak{M}_q(2)$ has the following remarkable properties:

---

[1] $\widetilde{A}$ denotes the transpose of the matrix A.

[2] We have replaced q in [24] and q^{-1} in [46] by $q^{-\frac{1}{2}}$ in accordance with our conventions.

THEOREM 4.3 (*a*) *The set of quantum matrices $\mathfrak{M}_q(2)$ is closed under matrix multiplication, that is, if $T, T' \in \mathfrak{M}_q(2)$ then also $TT' \in \mathfrak{M}_q(2)$. (Here, we assume that all entries in T' commute with all entries in T.)*

(*b*) *The q-determinant of T, defined to be the element*

$$\det_q(T) \overset{\text{def}}{=} ad - q^{-\frac{1}{2}}bc \quad (= da - q^{\frac{1}{2}}bc, \text{ by } (4.2)) \tag{4.4}$$

commutes with each of a, b, c, d, that is, $\det_q(T)$ belongs to the center of $\mathfrak{M}_q(2)$.

(*c*) *If $T, T' \in \mathfrak{M}_q(2)$ then $\det_q(T)\det_q(T') = \det_q(TT')$.*

(*d*) *For each $T \in \mathfrak{M}_q(2)$ such that $\det_q(T) \neq 0$ we define the inverse matrix $T^{-1} \in \mathfrak{M}_{q^{-1}}(2)$ by*

$$T^{-1} = \left(\det_q(T)\right)^{-1} \begin{pmatrix} d & -q^{\frac{1}{2}}b \\ -q^{-\frac{1}{2}}c & a \end{pmatrix},$$

and then $TT^{-1} = T^{-1}T = I$.

(*e*) *The transpose $\tilde{T} \in \mathfrak{M}_q(2)$ for each $T \in \mathfrak{M}_q(2)$.*

(*f*) *The associative algebra $\mathcal{A}$, with generators $\mathbb{1}, a, b, c, d$ satisfying (4.2), is a Hopf algebra with co-multiplication $\Delta : \mathcal{A} \to \mathcal{A} \otimes \mathcal{A}$ defined by:*

$$\Delta(T) = T\dot{\otimes}T, \quad \Delta(I) = I\dot{\otimes}I,$$

where $\dot{\otimes}$ denotes the matrix tensor product of elements of $\mathfrak{M}_q(2)$; and a co-unit $\varepsilon : \mathcal{A} \to \mathbb{C}$ defined by $\varepsilon(T) = I$; and an antipode $\gamma : \mathcal{A} \to \mathcal{A}$ defined by

$$\gamma(T) = \gamma\begin{pmatrix} a & b \\ c & d \end{pmatrix} = \begin{pmatrix} d & -q^{\frac{1}{2}}b \\ -q^{-\frac{1}{2}}c & a \end{pmatrix}.$$

PROOF: We firstly prove that the relations (4.2) are preserved under matrix multiplication, by assuming $ab = q^{-\frac{1}{2}}ba$ and $a'b' = q^{-\frac{1}{2}}b'a'$ for elements T and T' respectively in $\mathfrak{M}_q(2)$, where a, b each commute with a', b'. We find that $ab = q^{-\frac{1}{2}}ba$ is preserved provided also $cd = q^{-\frac{1}{2}}dc$ and $[a, d] = q^{-\frac{1}{2}}bc - q^{\frac{1}{2}}cb$, and likewise for the primed entries. Then we find that these last two relations are also preserved without assuming any further relations. Next, we assume independently the relation $ac = q^{-\frac{1}{2}}ca$ for each of T and T'. This relation is preserved provided also $bd = q^{-\frac{1}{2}}db$ and $[a, d] = q^{-\frac{1}{2}}cb - q^{\frac{1}{2}}bc$ and again these two relations are also preserved without further assumptions. By combining all relations we arrive at (4.2).

Parts (b)-(d) follow by direct calculation, and (e) expresses the symmetry between the elements b and c. The Hopf algebra properties are also verified directly. □

REMARK **4.5** 1. A mnemonic for the first four commutation relations in (4.2) is the diagram

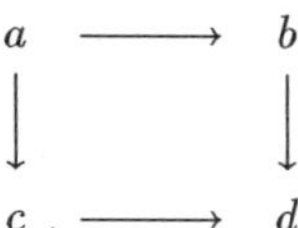

2. The relations (4.2) are self-consistent and so we may derive further relations such as $abc = q^{-1}bca$ regardless of the order in which the commutation relations are applied. Consistency is assured by the existence of an operator realization of the algebraic elements a, b, c, d satisfying (4.2) (see for example Theorem 4.55, or the following Remark 4.56 below).

3. It is significant that since $T^{-1} \in \mathfrak{M}_{q^{-1}}(2)$, the set $\mathfrak{M}_q(2)$ with matrix multiplication does not form a group.

4. It is a curious fact that if we assume two separate sets of relations,

$$ab = q^{-\frac{1}{2}}ba, \quad cd = q^{-\frac{1}{2}}dc, \quad [a, d] = q^{-\frac{1}{2}}bc - q^{\frac{1}{2}}cb,$$

and

$$ac = p^{-\frac{1}{2}}ca, \quad bd = p^{-\frac{1}{2}}db, \quad [a, d] = p^{-\frac{1}{2}}cb - p^{\frac{1}{2}}bc,$$

for parameters $p, q \in \mathbb{C}^{\times}$, matrix multiplication is still preserved, however, we do not obtain a commuting determinant as in part (b) of the theorem, unless $q = p$. The existence of this second deformation can be understood in terms of row and column quantum coordinate pairs in which the coordinates q-commute and p-commute respectively (see Lemma 4.9 below). The 2-parameter deformation determined by these six relations also defines a Hopf algebra (see Petersen [78, §3.6]), and has been discussed by many authors, see for example Schirrmacher et al. [155], Ewen et al. [4, p. 135], but we do not consider it further here. Multiparameter quantum groups are discussed in [4, p. 147].

For $q = 1$ the R-matrix reduces to the identity matrix and so the elements of T commute, as the relations (4.2) show, and consequently $\mathfrak{M}_q(2)$ reduces to the set of all 2×2 matrices with complex entries, which under matrix multiplication forms the group $GL(2, \mathbb{C})$. Hence, we may think of $\mathfrak{M}_q(2)$ as the q-analog of $GL(2, \mathbb{C})$ which we denote $GL_q(2)$. Since $\det_q(T)$ commutes with T, we may form the quotient of $\mathfrak{M}_q(2)$ by the relation $\det_q(T) = \mathbb{1}$, which then defines the quantum group $SL_q(2)$. (Note that by Theorem 4.3(c), the condition $\det_q(T) = \mathbb{1}$ is preserved under matrix multiplication.)

For a precise definition of the quantum groups $GL_q(2)$ and $SL_q(2)$ and more detailed discussion we refer to Manin [46], Faddeev et al. [24] and Kassel [7, Chapter IV] but we point out, to be precise, that $\mathfrak{M}_q(2)$ is defined by Manin to be the coordinate ring of the manifold of the matrices T, and is spanned by monomials of the form

$a^i b^j c^k d^l$, to which we also adjoin the inverse determinant since it is not a polynomial in a, b, c, d. The associative algebra generated by the elements a, b, c, d may be denoted $\mathrm{Fun}_q(GL(2,\mathbb{C}))$ and consists of the functions on the q-deformation of $GL(2,\mathbb{C})$. We will encounter elements of this algebra in the form of the representation matrices ${}^q\mathfrak{D}(T)$ defined on $\mathfrak{M}_q(2)$. The quotient algebra of $\mathrm{Fun}_q(GL(2,\mathbb{C}))$ by the relation $\det_q(T) = \mathbb{1}$ is denoted $\mathrm{Fun}_q(SL(2,\mathbb{C}))$.

In order to define $SU_q(2)$ we note (as observed by Woronowicz [156]) that for $q \in \mathbb{R}^+$ the algebra $\mathrm{Fun}_q(SL(2,\mathbb{C}))$ admits an anti-involution defined by $T^* = T^{-1}$ where

$$T^* = \begin{pmatrix} a^* & c^* \\ b^* & d^* \end{pmatrix}.$$

Hence, we define $SU_q(2)$ as the set of matrices

$$T = \begin{pmatrix} a & b \\ -q^{\frac{1}{2}} b^* & a^* \end{pmatrix}, \tag{4.6}$$

in which the entries satisfy

$$ab = q^{-\frac{1}{2}} ba, \quad ab^* = q^{-\frac{1}{2}} b^* a, \quad bb^* = b^* b, \quad [a,\, a^*] = (q-1) bb^*,$$

together with the conjugate relations

$$b^* a^* = q^{-\frac{1}{2}} a^* b^*, \quad ba^* = q^{-\frac{1}{2}} a^* b,$$

and where the q-determinant $\det_q(T)$ is equal to one:

$$\det_q(T) = aa^* + bb^* = a^* a + q b^* b = \mathbb{1}.$$

The quantum matrix $T \in SU_q(2)$ satisfies $T^* T = T T^* = I$.

4.2.1 The n-Dimensional Matrix Quantum Groups

We may generalize the above considerations from $n = 2$ dimensions to any n given the RTT relations and an explicit expression for the R-matrix. This allows us to determine the explicit relations satisfied by the entries t_i^j of T, first given by Faddeev et al. [157]. The matrix quantum group $GL_q(n)$, which we denote also by $\mathfrak{M}_q(n)$, is the set of $n \times n$ matrices T with entries t_j^i satisfying:

$$\begin{aligned} t_j^i t_l^i &= q^{-\frac{1}{2}} t_l^i t_j^i && (l > j) \\ t_j^i t_j^k &= q^{-\frac{1}{2}} t_j^k t_j^i && (k > i) \\ [t_j^k,\, t_l^i] &= 0 && (k > i,\, l > j) \\ [t_j^i,\, t_l^k] &= -(q^{\frac{1}{2}} - q^{-\frac{1}{2}}) t_l^i t_j^k && (k > i,\, l > j), \end{aligned} \tag{4.7}$$

a set of relations which generalize (4.2). Let us merely state the following properties, the n-dimensional generalizations of Theorem 4.3:

1. $\mathfrak{M}_q(n)$ is closed under matrix multiplication.

2. The q-determinant is defined by

$$\det_q(T) \stackrel{\text{def}}{=} \sum_{\sigma\in S_n} \varepsilon(\sigma) t^1_{\sigma(1)} t^2_{\sigma(2)} \dots t^n_{\sigma(n)},$$

where the sum is over all permutations σ of the symmetric group S_n, and the quantum signature $\varepsilon(\sigma)$ is given for each element of S_n by

$$\varepsilon(\sigma) = \prod_{\substack{j<k \\ \sigma(j)<\sigma(k)}} (-q^{-\frac{1}{2}}) = (-q^{-\frac{1}{2}})^{\ell(\sigma)},$$

where $\ell(\sigma)$ is the length of σ. The q-determinant commutes with t^j_i for each $i, j = 1, \dots, n$.

3. The antipode γ is given by $\gamma(t^j_i) = (-q^{-\frac{1}{2}})^{i-j} A^i_j$ where the cofactor

$$A^j_i \stackrel{\text{def}}{=} \sum_{\sigma\in S_{n-1}} (-q^{-\frac{1}{2}})^{\ell(\sigma)} t^1_{\sigma(1)} \dots \widehat{t^i_{\sigma(i)}} \dots t^n_{\sigma(n)},$$

and where σ is the cyclic permutation $(\sigma_1, \dots \widehat{\sigma_i}, \dots \sigma_n)$. The antipode satisfies $\gamma(T)T = T\gamma(T) = \det_q(T)\, I$.

4. For $q \in \mathbb{R}^+$ the algebra $\mathcal{A}$ of elements t^j_i admits a $*$-anti-involution defined by $\gamma(t^j_i) = (t^*)^i_j$ which enables us to define $SU_q(n)$.

5. The algebra $\mathcal{A}$ generated by $\{\mathbb{1}, t^j_i\}$ is a Hopf algebra [24] with co-multiplication defined similarly to the $n = 2$ case.

The matrix quantum group $SL_q(n)$ and the q-deformation $\mathcal{U}_q(\mathfrak{sl}(n))$ introduced in Chapter 2 can be viewed as algebras which are dual to each other [24], thus providing the q-analog of the viewpoint that the Lie algebra generators are the tangent vectors at the identity on the group manifold. Let $\mathcal{A}' = \mathrm{Hom}(\mathcal{A}, \mathbb{C})$ be the dual space to $\mathcal{A}$ then co-multiplication in $\mathcal{A}$ induces multiplication in $\mathcal{A}'$. Since we shall not use this duality directly we do not give the explicit bilinear form which determines the duality, but refer to Faddeev et al. [24] and Kassel [7, §VII.4] for details, and also to Dobrev and Parashar [158] for a discussion of duality of multiparametric quantum groups.

4.2.2 Noncommuting q-Coordinates and the Quantum Plane

The elements of $\mathfrak{M}_q(2)$ have a natural action (strictly we mean co-action, see Kassel [7, §IV.7]) of matrix multiplication on the coordinates (x, y) or $\binom{x}{y}$ where x, y

are noncommuting elements. It is assumed that x, y each commute with $a, b, c, d \in \mathcal{A}$ and, moreover, that

$$xy = q^{-\frac{1}{2}}yx. \tag{4.8}$$

These elements generate an associative algebra spanned by the basis $x^i y^j$ which, following Manin [46], we call the quantum plane. The properties (4.2) can be viewed as the conditions required to preserve (4.8) under left and right matrix multiplication:

LEMMA **4.9** *The relation $xy = q^{-\frac{1}{2}}yx$ is preserved under left and right matrix multiplication, that is, the elements $\binom{x'}{y'} = \left(\begin{smallmatrix} a & b \\ c & d \end{smallmatrix}\right)\binom{x}{y}$ satisfy $x'y' = q^{-\frac{1}{2}}y'x'$, and the elements $(x''\ y'') = (x\ y)\left(\begin{smallmatrix} a & b \\ c & d \end{smallmatrix}\right)$ satisfy $x''y'' = q^{-\frac{1}{2}}y''x''$, if and only if a, b, c, d satisfy (4.2).*

This result can be regarded as a corollary to Theorem 4.3(a), by considering separately rows and columns of T and T'. Conversely, we can derive all the relations (4.2) by assuming precisely these properties of x and y, for example, $x'y' = q^{-\frac{1}{2}}y'x'$ requires

$$(ax + by)(cx + dy) = q^{-\frac{1}{2}}(cx + dy)(ax + by)$$

which implies

$$ac = q^{-\frac{1}{2}}ca, \quad bd = q^{-\frac{1}{2}}db, \quad [a, d] = q^{-\frac{1}{2}}cb - q^{\frac{1}{2}}bc,$$

and similarly $x''y'' = q^{-\frac{1}{2}}y''x''$ implies

$$ab = q^{-\frac{1}{2}}ba, \quad cd = q^{-\frac{1}{2}}dc, \quad [a, d] = q^{-\frac{1}{2}}bc - q^{\frac{1}{2}}cb,$$

which are the relations (4.2). If we specify that under right matrix multiplication by T on the row (x', y') the relation $x'y' = p^{-\frac{1}{2}}y'x'$ for $p \in \mathbb{R}^+$ be preserved, and that under left multiplication on the column $\binom{x}{y}$ the relation $xy = q^{-\frac{1}{2}}yx$ be preserved, then we obtain the two-parameter Hopf algebra mentioned in Remark 4.5.

As indicated earlier, the pair of noncommuting coordinates x, y can be viewed as components of a spin $\frac{1}{2}$ tensor operator. To verify this we must show that tensor operators in the matrix quantum group satisfy the q-equivariance property, the q-analog of the property stated in Definition 3.3, and to do this requires us to define also the ${}^q\mathfrak{D}$-matrices. This is accomplished in §4.5 and q-equivariance is established in Corollary 4.51, beginning essentially with the concept of a q-tensor operator as defined in Chapter 3 from considerations of $\mathcal{U}_q(\mathfrak{su}(2))$. However let us anticipate this development and point out now several q-spinor properties such as invariant combinations of tensor operator components.

We regard therefore a pair of quantum coordinates (x, y) as comprising a q-spinor if under the action of $SL_q(2)$,

$$(x\ y) \longrightarrow (x\ y)\begin{pmatrix} a & b \\ c & d \end{pmatrix} = (ax + cy,\ bx + dy)$$

where $T \in SL_q(2)$. These spinor components need not necessarily satisfy the relation $xy - q^{-\frac{1}{2}}yx = 0$, however, we can state:

Lemma 4.10 *The combination $q^{\frac{1}{4}}xy' - q^{-\frac{1}{4}}yx'$ of two $SL_q(2)$ spinors $(x\ y)$ and $(x'\ y')$ is an invariant of $SL_q(2)$.*

Proof: From the relations (4.2) we find

$$q^{\frac{1}{4}}(ax+cy)(bx'+dy') - q^{-\frac{1}{4}}(bx+dy)(ax'+cy') = \det_q T\ (q^{\frac{1}{4}}xy' - q^{-\frac{1}{4}}yx')$$

which equals $q^{\frac{1}{4}}xy' - q^{-\frac{1}{4}}yx'$ in $SL_q(2)$. □

This result has already been derived in (3.38), p. 90, by means of a q-WCG coupling of spinors and was deduced there from considerations of $\mathcal{U}_q(\mathfrak{sl}(2))$. If we consider just a single q-spinor $(x,\ y)$, then $q^{\frac{1}{4}}xy - q^{-\frac{1}{4}}yx$ is an invariant but need not be zero, an example of such being the q-symplecton spinor discussed in §8.3.2 (p. 267). However, there are many instances in which $q^{\frac{1}{4}}xy - q^{-\frac{1}{4}}yx$ is zero, such as (3.40). A particularly important example is supplied by the matrix $T \in \mathfrak{M}_q(2)$ itself. As indicated in the introductory remarks to this chapter, the rows and columns of $T \in \mathfrak{M}_q(2)$ may be viewed as q-spinors with respect to two quantum groups, to be specified. The defining relations (4.2) express relations amongst *eight* invariants of these quantum groups. Taking firstly the relation $ab = q^{-\frac{1}{2}}ba$, this states that the invariant $q^{\frac{1}{4}}ab - q^{-\frac{1}{4}}ba$ is zero, and similarly for the first four relations in (4.2). The remaining relations can be expressed as the equality (up to an overall numerical factor) of another four invariants, which of course are each essentially the quantum determinant:

$$q^{\frac{1}{4}}ad - q^{-\frac{1}{4}}bc = q^{\frac{1}{4}}ad - q^{-\frac{1}{4}}cb = -q^{\frac{1}{2}}(q^{\frac{1}{4}}bc - q^{-\frac{1}{4}}da) = -q^{\frac{1}{2}}(q^{\frac{1}{4}}cb - q^{-\frac{1}{4}}da),$$

where the different forms arise from different orders of coupling, and where the factor $-q^{\frac{1}{2}}$ is a constant of proportionality. This is explained more fully in §4.5.

We also define the conjugate q-spinor as a pair of quantum coordinates $\binom{x}{y}$ which transforms according to the antipode $\gamma(T)$, that is, under the action of $SL_q(2)$,

$$\begin{pmatrix} x \\ y \end{pmatrix} \longrightarrow \begin{pmatrix} d & -q^{\frac{1}{2}}b \\ -q^{-\frac{1}{2}}c & a \end{pmatrix} \begin{pmatrix} x \\ y \end{pmatrix} = \begin{pmatrix} dx - q^{\frac{1}{2}}by \\ ay - q^{-\frac{1}{2}}cx \end{pmatrix}.$$

Given a q-spinor $(x\ y)$, the pair $(q^{\frac{1}{4}}y,\ -q^{-\frac{1}{4}}x)$ is a conjugate q-spinor, as may be verified directly. Hence, this definition of the conjugate tensor operator accords precisely with that given in (3.32), p. 88. We can write the transformation between the spinor $(x\ y)$ and its conjugate $\overline{(x\ y)}$ in the form

$$\overline{(x\ y)} = (x\ y)\begin{pmatrix} 0 & -q^{-\frac{1}{4}} \\ q^{\frac{1}{4}} & 0 \end{pmatrix},$$

showing, as noted also in §3.4.1, that the matrix on the right hand side acts as a metric tensor.

We can form invariants of $SL_q(2)$ from not just two spinors, but also from a spinor and its conjugate, as we determined also in (3.34), p. 88. For $SU_q(2)$ we denote the conjugate tensor by $\binom{x^*}{y^*}$, and so under $SU_q(2)$, $\binom{x^*}{y^*} \longrightarrow T^*\binom{x^*}{y^*}$. For any spinor $\boldsymbol{x}' = (x'\ y')$ and a conjugate spinor $\boldsymbol{x}^* = \binom{x^*}{y^*}$ the bilinear $\boldsymbol{x}'\boldsymbol{x}^*$ is an invariant, as is clear because under $SU_q(2)$ $\boldsymbol{x}'\boldsymbol{x}^* \rightarrow \boldsymbol{x}'TT^*\boldsymbol{x}^* = \boldsymbol{x}'\boldsymbol{x}^*$.

4.3 The Classical Unitary Factor Groups

As indicated in the introductory remarks we will derive the relations (4.2), from which we define $\mathfrak{M}_q(2)$ and hence the matrix quantum group $SU_q(2)$, by using only properties of irreps and tensor operators for $\mathcal{U}_q(\mathfrak{su}(2))$ or $\mathcal{U}_q(\mathfrak{u}(2))$. With this purely algebraic approach we explicitly derive all irrep matrices $\mathfrak{D}$ over $\mathfrak{M}_q(2)$ without any direct integration, and are therefore able to provide a complete description of the matrix quantum group $U_q(2)$ taking $\mathcal{U}_q(\mathfrak{u}(2))$ as a starting point.

We need for our development, however, a larger algebra than $\mathcal{U}_q(\mathfrak{u}(2))$ alone and in order to explain how to extend $\mathcal{U}_q(\mathfrak{u}(2))$ to the factor algebra denoted $\mathcal{U}_q\,(\mathfrak{u}(2){*}\mathfrak{u}(2))$ we consider first, for simplicity, the $q = 1$ case and review in this section the salient properties of these factor algebras, and then turn to the quantum extension in §4.4. Let us recall that in §2.4.4 (p. 40) we constructed all $U(2)$ (and by extension also $\mathcal{U}_q(\mathfrak{u}(2))$) irreducible representations, however, we postponed the development of several significant and elegant properties of this construction, which we consider now. We proceed by introducing four complex variables $\{z_i^j,\ i, j = 1, 2\}$ assembled into a matrix $Z = (z_i^j)$, or equivalently, we introduce four boson operators $\{a_i^j\}$ assembled into a matrix boson A, from which we construct *two* commuting left and right $\mathfrak{u}(2)$ algebras. Each of these is a subalgebra of $\mathfrak{u}(4)$ and acts in the space $\mathfrak{P}^4(N)$ of homogeneous polynomials of fixed degree $N = m_{12} + m_{22}$ in four variables. This direct product group acting in $\mathfrak{P}^4(N)$ is denoted $U(2) * U(2)$ because the Casimir invariants of each algebra coincide. Similarly $\mathfrak{u}(2) * \mathfrak{u}(2)$ denotes a factor algebra, meaning that it is the quotient of the direct product algebra by the constraints which impose common invariants. The two $\mathfrak{u}(2)$ algebras in this direct product are sometimes termed complementary (see Moshinsky [111, 159, 160]).

To be more precise[3], we define the factor algebra $\mathcal{U}_q\,(\mathfrak{su}(2){*}\mathfrak{su}(2))$ as the q-deformation of the universal enveloping algebra $\mathcal{U}(\mathfrak{u}(2) \times \mathfrak{u}(2))$ of the direct product group $U(2) \times U(2)$, factored by the relations which equate the invariants of $\mathcal{U}(\mathfrak{u}(2))$ (or $\mathcal{U}(\mathfrak{u}(n))$ in general). Actually, $U(2) * U(2)$ is not a group since tensor products of irreps of $U(2) * U(2)$ contain irreps which do not necessarily have identical irrep labels for the left and right groups, however the algebra $\mathcal{U}_q(\mathfrak{u}(2) * \mathfrak{u}(2))$ is well-defined and this is the object we investigate, and which we refer to as a factor group.

Since the description of factor groups generalizes to arbitrary n without difficulty, let us summarize the main properties for $U(n) * U(n)$. We refer to Louck [108] and

---

[3] **We thank Dan Flath for his helpful comments on this definition.**

the monograph [66, §5.5-5.9] for a more detailed description. Let $\mathfrak{P}(N)$ denote the linear vector space of polynomials $\mathcal{P}(Z)$ in the n^2 complex variables $Z = (z_i^j)$, $i, j = 1, \ldots, n$, where i denotes the row and j the column of the matrix Z, which are homogeneous of degree N, that is

$$\mathfrak{P}(N) = \{ \mathcal{P} \mid \mathcal{P}(\lambda Z) = \lambda^N \mathcal{P}(Z)\}$$

for $\lambda \in \mathbb{R}$ and some $N \in \mathbb{N}$. We may define an inner product on $\mathfrak{P}(N)$ by

$$(\mathcal{P}', \mathcal{P}) = \mathcal{P}'\left(\frac{\partial}{\partial Z}\right) \mathcal{P}(Z)\Big|_{Z=0} .$$

Equivalently, we may introduce n^2 boson operators $A = (a_i^j)$ and their conjugates $\overline{A} = (\overline{a}_i^j)$ which act in the Fock space $\mathfrak{F}^{n^2}$ equipped with the usual boson bra-ket inner product:

$$(\mathcal{P}', \mathcal{P}) = \langle 0| \mathcal{P}'(\overline{A})\, \mathcal{P}(A)|0\rangle.$$

We construct representations of $U(n)$ and $U(n) \times U(n)$ by means of left and right transformations on $\mathfrak{P}(N)$ in the following way:

1. Define a linear operator Θ_g on $\mathfrak{P}(N)$ by

$$\Theta_g \mathcal{P}(Z) = \mathcal{P}(\tilde{g}Z), \quad g \in U(n), \quad \mathcal{P} \in \mathfrak{P}(N),$$

then $g \to \Theta_g$ is a unitary irrep of $U(n)$; hence $U(n)$ acts on $\mathfrak{P}(N)$ by means of a left action.

2. Define a linear operator Θ'_g on $\mathfrak{P}(N)$ by

$$\Theta'_g \mathcal{P}(Z) = \mathcal{P}(Zg), \quad g \in U(n), \quad \mathcal{P} \in \mathfrak{P}(N),$$

then $g \to \Theta'_g$ is a unitary irrep of $U(n)$; in this case $U(n)$ acts on $\mathfrak{P}(N)$ via a right action.

3. Define a linear operator $\Theta_{(g,g')}$ on $\mathfrak{P}(N)$ by

$$\Theta_{(g,g')} \mathcal{P}(Z) = \mathcal{P}(\tilde{g}Zg'), \quad (g, g') \in U(n) \times U(n), \quad \mathcal{P} \in \mathfrak{P}(N), \tag{4.11}$$

then $(g, g') \to \Theta_{(g,g')}$ is a unitary representation of $U(n) \times U(n)$. We note that since $\Theta_{(g,g')} = \Theta_g \Theta'_{g'} = \Theta'_{g'} \Theta_g$ the left and right actions of $U(n)$ commute.

We may express these results in terms of the Lie algebra, in which form the q-generalization is most easily expressed. The generators of the two $U(n)$ groups acting on $\mathfrak{P}(N)$ may be determined by consideration of the n^2 one-parameter subgroups of $U(n)$. Denote the generators of the left and right groups by $\{E_{ij}\}$ and $\{E^{ij}\}$ respectively (which we also refer to as the lower and upper generators, and similarly for the corresponding groups), then we can state:

1. The generators satisfy the standard commutation relations

$$[E_{ij}, E_{kl}] = \delta_{jk}E_{il} - \delta_{il}E_{kj}, \tag{4.12}$$

for $i,j = 1,\ldots,n$, and similarly for the set $\{E^{ij}\}$. By direct calculation we obtain the following differential operator realizations (see Louck [108])

$$E_{ij} = \sum_{k=1}^{n} z_i^k \frac{\partial}{\partial z_j^k}, \qquad E^{ij} = \sum_{k=1}^{n} z_k^i \frac{\partial}{\partial z_k^j},$$

and the commutation relations (4.12) may be verified directly.

2. Since the group actions Θ and Θ' commute we expect, and confirm by direct calculation, that the left and right $\mathfrak{u}(n)$ generators commute:

$$[E_{ij}, E^{kl}] = 0, \quad i,j,k,l = 1,\ldots,n.$$

3. The Gel'fand invariants of the two commuting $U(n)$ groups are identical (Louck [109]). Hence, the only irreps of $U(n) \times U(n)$ which can be carried by $\mathfrak{P}(N)$ are those in which the irrep labels $[M_{1n},\ldots,M_{nn}]$ for each group coincide. We signify this constraint by denoting the direct product of the left and right groups by $U(n) * U(n)$, which is a factor group, and the corresponding factor algebra by $\mathfrak{u}(n) * \mathfrak{u}(n)$.

Since $\mathfrak{P}(N)$ consists of polynomials in n^2 complex variables it also carries representations of $U(n^2)$ which are symmetric, and which are reducible under the subgroup $U(n)*U(n)$ of $U(n^2)$. We may summarize properties of this decomposition as follows:

1. The space $\mathfrak{P}(N)$ carries symmetric irreps $[N\,0\ldots0]$ of $U(n^2)$ which are reducible with respect to $U(n) * U(n) \subset U(n^2)$.

2. The subspaces of $\mathfrak{P}(N)$ which are irreducible with respect to $U(n) * U(n)$ carry the irrep labels $[M] \times [M]$ of $U(n) * U(n)$ and may be denoted $\mathfrak{P}([M] \times [M])$. $[M]$ is a partition of N into nonnegative integers $M_{1n} \geqslant M_{2n} \geqslant \ldots \geqslant M_{nn} \geqslant 0$ such that $\sum_i M_{in} = N$. Each partition occurs exactly once, and we have

$$\mathfrak{P}(N) = \sum_{[M]} \oplus \mathfrak{P}([M] \times [M]).$$

This implies the dimension formula

$$\dim([N\,0\ldots0])_{U(n^2)} = \sum_{\substack{M_{in} \\ \sum_{i=1}^{n} M_{in}=N}} \left(\dim([M_{1n}\ldots M_{nn}])_{U(n)}\right)^2.$$

3. The embedding $U(n^2) \supset U(n) * U(n)$ is not restricted to n^2 but is valid also for nm. One finds $U(nm) \supset U(n) * U(m)$, which implies the restriction that unitary irreps satisfy $m_{in} = 0$ for $n \geqslant i > m$. This embedding is furthermore not restricted to unitary groups but is valid for subgroups as well. We note that since $U(n) \supset S(n)$ (the symmetric group), the embedding shows that $U(n)$ and $S(n)$ commute, which leads to Weyl's theorem on the relation of the $U(n)$ and $S(n)$ groups.

Next, we turn to properties of the basis vectors of $\mathfrak{P}([M] \times [M])$ which play a fundamental role because they may be mapped onto the representation matrix of $U(n)$. Indeed, by calculating the explicit basis vectors we are in effect calculating the elements of the irrep matrix of $U(n)$ using only properties of the Lie algebra $\mathfrak{u}(n)$. This mapping provides therefore a means of passing from the Lie algebra to the Lie group without performing an explicit integration of the generators, and as such will generalize to quantum groups when we introduce tensor operators. The important properties of the basis vectors are:

1. The basis vectors in $\mathfrak{P}([M] \times [M])$ are labelled by two Gel'fand-Weyl patterns carrying a common irrep label $[M]$. We may denote these vectors by a diamond-shaped pattern in which the pattern for the upper group is inverted over the common irrep labels (recall that we used an analogous notation for tensor operators, see Remark 3.7 (p. 78) but in an entirely different context; inverting the patterns is possible because the inequalities of the Gel'fand-Weyl patterns are preserved under inversion). Hence, the orthonormal basis vectors in $\mathfrak{P}([M] \times [M])$ are denoted

$$\left| \begin{matrix} (m') \\ [M] \\ (m) \end{matrix} \right\rangle ,$$

where $(m), (m')$ are the Gel'fand-Weyl patterns for the lower and upper groups respectively.

2. The basis vectors are polynomials in the n^2 variables z_i^j which, upon suppressing labels, we denote also by $\Phi(Z)$. Their explicit form may be calculated by application of the generators to a fixed vector, such as the vector of highest weight with respect to each $U(n)$ group. The generators $\{E_{ij}\}$ act on the lower pattern and $\{E^{ij}\}$ on the upper pattern.

3. The basis vector $\Phi_{\max}(Z)$ of highest weight with respect to each $\mathfrak{u}(n)$ algebra (where highest weight means that the entries in the patterns (m) and (m') take their maximum values) is given by

$$\Phi_{\max}(Z) = \mathcal{N}^{-\frac{1}{2}} \prod_{k=1}^{n} (z_{12\ldots k})^{M_{kn} - M_{k+1,n}}, \tag{4.13}$$

where $z_{12..k}$ is the minor formed from the first k rows and columns of Z, and where the normalization is given by

$$\mathcal{N} = \frac{\prod_{i=1}^{n} P_{in}!}{\prod_{i<j}^{n}(P_{in} - P_{jn})}, \tag{4.14}$$

where $P_{in} \stackrel{\text{def}}{=} M_{in} + n - i$.

As we shall see shortly, it is convenient to define polynomials $\mathcal{P}(Z)$ which differ from the orthonormal Gel'fand-Weyl basis vectors $\Phi(Z)$ in $\mathfrak{P}([M] \times [M])$ only by a normalization. Suppressing labels, we define

$$\mathcal{P}(Z) \stackrel{\text{def}}{=} \mathcal{N}^{\frac{1}{2}}\Phi(Z) \tag{4.15}$$

where $\mathcal{N}$ is given by (4.14). A particular evaluation of $\mathcal{P}(Z)$, which we use below, is for $Z = I_n$, the $n \times n$ unit matrix. In this case we have $\mathcal{P}_{\mathbf{max}}(I_n) = \mathcal{N}^{\frac{1}{2}}\Phi_{\mathbf{max}}(I_n) = 1$, as follows from (4.13).

Let us denote by $\mathfrak{D}^{[M]}(g)$ the representation matrix of $U(n)$ for the irrep $[M]$, that is, the map $g \to \mathfrak{D}^{[M]}(g)$, where $g \in U(n)$, is an irreducible unitary representation of $U(n)$ with labels $[M]$. We call $\mathfrak{D}$ the **representation matrix** of $U(n)$, or irrep matrix, and also for $n = 2$ the **rotation matrix**. The representation matrix has dimension $D([M])$ given by (2.72), p. 47, and the matrix elements are denoted $\mathfrak{D}^{[M]}_{(m)(m')}(g)$ where the row labels are (m) and the column labels (m'). The irrep matrix is a polynomial in the entries g^j_i of the matrix $g \in U(n)$, but may be extended to a polynomial on any set of n^2 complex variables $Z = (z^j_i)$, and hence may be denoted $\mathfrak{D}(Z)$. This extended irrep matrix may be regarded as the representation matrix for a finite dimensional irrep of $GL(n, \mathbb{C})$. A fundamental result is the following (Louck [108]):

LEMMA **4.16** *The orthonormal basis polynomials in* $\mathfrak{P}([M] \times [M])$ *and the irrep matrix* $\mathfrak{D}$, *expressed as functions of* $Z \in GL(n, \mathbb{C})$, *are related by:*

$$\left|\begin{matrix}(m')\\ [M]\\ (m)\end{matrix}\right\rangle(Z) = \mathcal{N}^{-\frac{1}{2}}\,\mathfrak{D}^{[M]}_{(m)(m')}(Z).$$

where $\mathcal{N}$ *is given by (4.14).*

PROOF: Suppressing irrep labels $[M]$, we wish to prove $\mathcal{P}(Z) = \mathfrak{D}(Z)$. We use the property $\widetilde{\mathfrak{D}}(g) = \mathfrak{D}(\widetilde{g})$ of irrep matrices of $U(n)$. Since $\{\mathcal{P}(Z)\}$ carries irreps of both the left and right $U(n)$ groups, we have

$$\mathcal{P}(\widetilde{g}Z) = \widetilde{\mathfrak{D}}(g)\,\mathcal{P}(Z) = \mathfrak{D}(\widetilde{g})\,\mathcal{P}(Z),$$
$$\mathcal{P}(Zg) = \mathcal{P}(Z)\,\mathfrak{D}(g).$$

Replacing $\tilde{g}$ by g in the first equation gives

$$\mathcal{P}(gZ) = \mathfrak{D}(g)\,\mathcal{P}(Z),$$

which for $Z = I_n$ leads to

$$\mathcal{P}(g) = \mathcal{P}(I_n)\mathfrak{D}(g) = \mathfrak{D}(g)\,\mathcal{P}(I_n).$$

By Schur's lemma, $\mathcal{P}(I_n) = \lambda I_d$, where I_d is the unit matrix of dimension $d = D([M])$ and $\lambda \in \mathbb{R}$ depends only on the labels $[M]$. In fact we have $\lambda = 1$, since we have normalized $\mathcal{P}(Z)$ in (4.15) such that a diagonal element of $\mathcal{P}(I_n)$ is equal to unity. The result follows by extending $\mathcal{P}(g) = \mathfrak{D}(g)$ to $\mathcal{P}(Z) = \mathfrak{D}(Z)$ for any $n\times n$ complex matrix Z. □

The irrep matrices satisfy two fundamental product laws, the first being the **group multiplication law**:

$$\mathfrak{D}(g_1)\,\mathfrak{D}(g_2) = \mathfrak{D}(g_1 g_2), \qquad g_1, g_2 \in U(n),$$

which we may extend to elements of $GL(n,\mathbb{C})$; hence we have $\mathfrak{D}(Z)\mathfrak{D}(Z') = \mathfrak{D}(ZZ')$ where Z, Z' are $n\times n$ matrices with complex entries. We may further extend the group multiplication law to operator-valued irrep matrices, in which Z is replaced by the matrix boson A. The second product law is the Wigner product law, discussed below, see (4.19).

So far, we have described the construction of irreps of $U(n) * U(n)$ in terms of left and right actions on polynomial spaces of n^2 complex variables, but we could equally well use the language of boson operators acting in a Fock space. Since this leads naturally to the concept of operator-valued $\mathfrak{D}$-matrices satisfying the group multiplication law, in which the rows and columns of $\mathfrak{D}$ form tensor operators, let us repeat the previous description of factor algebras using boson operator terminology.

The first step is to generalize the Jordan map $\mathcal{J}$ described in §2.3 (p. 24) for $U(n)$ from n commuting bosons to n^2 commuting bosons $\{a_i^j,\ i,j = 1,\ldots,n\}$ and their conjugates, which we assemble into matrices A and $\overline{A}$ respectively. Let the n^2 generators $\{\mathbf{E}\} = \{E_{ij},\ i,j = 1,\ldots,n\}$ be realized by the real-valued $n\times n$ matrices of the fundamental irrep of $U(n)$, that is, $(E_{ij})_{kl} = \delta_{ik}\delta_{jl}$. As indicated earlier in Lemma 2.22, the operator-valued Jordan map, denoted now by $\mathcal{L}$ and defined by

$$\mathcal{L}(\mathbf{E}) \overset{\text{def}}{=} \operatorname{tr}(\tilde{A}\mathbf{E}\overline{A}),$$

preserves the commutation properties of the matrices $\{\mathbf{E}\}$, so that $\mathcal{L}(\mathbf{E})$ forms a realization of the generators of $\mathfrak{u}(n)$. Hence, the operators

$$E_{ij} \overset{\text{def}}{=} \mathcal{L}((\mathbf{E})_{ij}) = \sum_{k=1}^{n} a_i^k \overline{a}_j^k, \quad (i,j = 1,\ldots,n)$$

obey the commutation rules of $\mathfrak{u}(n)$.

This generalized Jordan map allows, however, a *second* mapping from the $n \times n$ matrices $\mathbf{E}$, given by

$$\mathcal{R}(\mathbf{E}) \stackrel{\text{def}}{=} \operatorname{tr}(\widetilde{\mathbf{E}}\widetilde{A}\overline{A}),$$

which also preserves the commutation properties of the matrices $\{\mathbf{E}\}$. Thus we have *two* distinct boson operator realizations of $\mathfrak{u}(n)$ over the same set of bosons, in which the second realization has the explicit form:

$$E^{ij} \stackrel{\text{def}}{=} \mathcal{R}((\mathbf{E})_{ij}) = \sum_{k=1}^{n} a_k^i \overline{a}_k^j, \quad (i,j = 1,\ldots,n).$$

The two mappings correspond to left and right translations, as indicated above, and the two sets of operators commute:

$$[\mathcal{L}(\mathbf{E}),\, \mathcal{R}(\mathbf{E}')] = 0$$

for all matrices $\mathbf{E}, \mathbf{E}'$ in $\mathfrak{u}(n)$. It follows that $\{E^{ij}\}$ and $\{E_{ij}\}$ generate the group $U(n) \times U(n)$ and, since all integer irreps as determined by the Jordan maps $\mathcal{L}$ and $\mathcal{R}$ have the same irrep labels for each of the two $U(n)$ irreps, we have realized the factor algebra $\mathfrak{u}(n) * \mathfrak{u}(n)$.

The carrier space for the irreps of $U(n) * U(n)$ is the Fock space $\mathfrak{F}^{n^2}$ over the n^2 bosons $\{a_i^j\}$ acting on the vacuum ket $|0\rangle$. Denote by $\mathcal{P}(A)$ any polynomial in the matrix boson A (that is, a polynomial in the n^2 boson operators), then $\mathfrak{F}^{n^2}$ is spanned by all such vectors $\mathcal{P}(A)|0\rangle$. For $g \in U(n)$ we may write $g = \exp(-iE)$ where E is Hermitian, and we define L_g as the exponentiated Jordan map of E:

$$L_g = \exp(-i\mathcal{L}(E)). \tag{4.17}$$

By allowing this operator to act on $\mathcal{P}(A)|0\rangle$ we obtain

$$L_g\,\mathcal{P}(A)|0\rangle \;=\; \mathcal{P}(\widetilde{g}A)|0\rangle.$$

This may be proved using $L_g|0\rangle = |0\rangle$, which implies

$$L_g\,\mathcal{P}(A)|0\rangle = L_g\,\mathcal{P}(A)(L_g)^{-1}|0\rangle = \mathcal{P}(L_g A L_g^{-1})|0\rangle,$$

and the result then follows from the Baker-Campbell-Hausdorff formula ([66, §5.6]). This result is an operator formulation of the action of the Lie derivative on the argument of a function, and not on the function itself.

Similarly, if we define $R_g = \exp(-i\mathcal{R}(E))$, we obtain

$$R_g\,\mathcal{P}(A)|0\rangle \;=\; \mathcal{P}(Ag)|0\rangle,$$

which establishes the desired interpretation of the Jordan maps $\mathcal{L}$ and $\mathcal{R}$ as implementing, via exponentiation, left and right translations on the matrix boson A. These left and right actions are, of course, simply restatements of (4.11).

As mentioned above, we can extend the irrep matrices to operator-valued polynomials which are homogeneous of degree $N = \sum_{i=1}^{n} M_{in}$ in the n^2 boson operators. Let us denote by $\widehat{\mathfrak{P}}(N)$ the space spanned by these polynomials. When acting on $|0\rangle$ these polynomials are, by Lemma 4.16, identical to the orthonormal basis polynomials, however, let us now consider their properties as operators. The group multiplication law for the operator-valued irrep matrices may be written:

$$\mathfrak{D}(A)\,\mathfrak{D}(B) = \mathfrak{D}(AB),$$

where A, B are independent $n \times n$ matrix boson operators, AB is the $n \times n$ matrix product, and $\mathfrak{D}(A)\,\mathfrak{D}(B)$ denotes the matrix product in the representative (carrier) space. This equation implies that the rows and columns of $\mathfrak{D}(A)$ form irreducible tensor operators with respect to $U(n) * U(n)$. Specifically, we have:

LEMMA **4.18** *The columns (rows) of $\mathfrak{D}^{[M]}(A)$ form irreducible tensor operators carrying the irrep $[M]$ with respect to the left (right) $U(n)$ group, respectively.*

PROOF: Consider the elements $\mathfrak{D}^{[M]}_{(m)(m')}(A)$ for fixed (m'). Under the left $U(n)$ group this column transforms according to

$$\begin{aligned} \mathfrak{D}^{[M]}_{(m)(m')}(A) \longrightarrow \mathfrak{D}^{[M]}_{(m)(m')}(A') &= \mathfrak{D}^{[M]}_{(m)(m')}(\tilde{g}A) \\ &= \sum_{(m'')} \mathfrak{D}^{[M]}_{(m'')(m)}(g)\,\mathfrak{D}^{[M]}_{(m'')(m')}(A) \end{aligned}$$

which, according to the equivariance condition (Definition 3.3, p. 74) defines an irreducible tensor operator. (Here we used the property $\mathfrak{D}(\tilde{g}) = \widetilde{\mathfrak{D}}(g)$). A similar argument holds for the rows of $\mathfrak{D}(A)$. □

An important special case of $\mathfrak{D}^{[M]}(A)$ is for the irrep $[M] = [1\,0\ldots0]$, which is the defining $n \times n$ irrep, when we have:

$$\mathfrak{D}^{[1\,0\ldots0]}(A) = A.$$

Hence, the rows and columns of A are elementary tensor operators with respect to the right and left transformations of $U(n)$.

As described in §3.2, there exists an algebra of tensor operators, meaning that two tensor operators may be coupled with a WCG coupling coefficient to form a third tensor operator, and this leads to the second product law for the operator-valued $\mathfrak{D}$-matrices, which we call the **Wigner product law**. If we denote the WCG coupling by Ⓦ, which effects the coupling $[M'] \times [M] \to [M'']$, we may write the product law symbolically as

$$\left[\mathfrak{D}^{[M]}(A) \underset{Ⓦ}{\overset{Ⓦ}{}} \mathfrak{D}^{[M']}(A)\right]^{[M'']} = \mathfrak{D}^{[M'']}(A). \tag{4.19}$$

We will write this equation explicitly for $n = 2$ in the following section, see Eqn. (4.27). The importance of the Wigner product law is that it enables us to construct, in principle, the general $\mathfrak{D}$-matrix recursively beginning with only the elementary irrep matrix $\mathfrak{D}^{[1\,0\ldots 0]}(A) = A$, by coupling irrep matrices to form another irrep matrix of higher dimension.

4.3.1 The $U(2)$ Factor Group and the Rotation Matrices

As an example of factor groups let us now specialize to $n = 2$, for which we give explicit results for the irrep matrix $\mathfrak{D}$ and which we generalize to the quantum case. The left (lower) generators

$$E_{ij} \stackrel{\text{def}}{=} a_i^1 \overline{a}_j^1 + a_i^2 \overline{a}_j^2, \quad (i,j = 1,2)$$

obey the commutation rules of $\mathfrak{u}(2)$, as do the upper generators

$$E^{ij} \stackrel{\text{def}}{=} a_1^i \overline{a}_1^j + a_2^i \overline{a}_2^j, \quad (i,j = 1,2).$$

These two sets of operators commute and generate $\mathfrak{u}(2) * \mathfrak{u}(2)$.

The basis vectors are labelled by two $U(2)$ Gel'fand-Weyl patterns:

$$\begin{pmatrix} m_{12} & & m_{22} \\ & m_{11} & \end{pmatrix}, \qquad \begin{pmatrix} m_{12} & & m_{22} \\ & m'_{11} & \end{pmatrix}$$

where, as shown, the $U(2)$ irrep labels $[m_{12}, m_{22}]$ are identical for each $U(2)$ group. By inverting the second pattern we introduce the following concise notation for the eigenvectors in $U(2) * U(2)$:

$$\left| \begin{smallmatrix} & m'_{11} & \\ m_{12} & & m_{22} \\ & m_{11} & \end{smallmatrix} \right\rangle . \tag{4.20}$$

These eigenvectors span the carrier space (the Fock space $\mathfrak{F}^4$) and are homogeneous polynomials in A of degree $m_{12} + m_{22}$ and act on $|0\rangle$. Since by Lemma 4.16 they are also essentially the irrep matrices of $U(2)$, we may calculate the explicit irrep matrices using only the properties of $\mathfrak{u}(2) * \mathfrak{u}(2)$, by determining the state of highest weight in $\mathfrak{F}^4$ with respect to both $\mathfrak{u}(2)$ algebras, and then using the lowering generators to calculate the full set of basis vectors. The $U(2)$ representation matrix is related to the $SU(2)$ matrix $\mathfrak{D}^j(Z)$, and the basis elements take the form [66, p. 222]:

$$\left| \begin{smallmatrix} & m'_{11} & \\ m_{12} & & m_{22} \\ & m_{11} & \end{smallmatrix} \right\rangle = \left(\frac{m_{22}!(m_{12}+1)!}{(m_{12} - m_{22} + 1)} \right)^{-\frac{1}{2}} (\det A)^{m_{22}} \, \mathfrak{D}^{\frac{1}{2}(m_{12}-m_{22})}_{m,m'}(A)\, |0\rangle \tag{4.21}$$

where

$$m = m_{11} - \tfrac{1}{2}(m_{12} + m_{22}), \quad m' = m'_{11} - \tfrac{1}{2}(m_{12} + m_{22}). \tag{4.22}$$

The matrix $\mathfrak{D}$ in (4.21) is the $SU(2)$ irrep matrix $\mathfrak{D}^j(g)$, the rotation matrix for the irrep $j = \frac{1}{2}(m_{12} - m_{22})$, except that the matrix boson A replaces the 2×2 unitary matrix $g \in SU(2)$, and is given explicitly below by (4.24). (In [66, p. 222] the notation

$2k = m_{12}+m_{22}$ is used, together with j, m, m' defined as above, and the basis vectors are denoted $|k, j; m, m'\rangle$).

Using (4.21) we may now define the operator-valued irrep matrices $\mathfrak{D}^{[m_{12},m_{22}]}$ for $U(2)$ as follows, in which we distinguish the $SU(2)$ and $U(2)$ $\mathfrak{D}$-matrices by the form of the superscript:

$$\mathfrak{D}^{[m_{12},m_{22}]}_{m_{11},m'_{11}}(A) \stackrel{\text{def}}{=} (\det A)^{m_{22}}\, \mathfrak{D}^{\frac{1}{2}(m_{12}-m_{22})}_{m,m'}(A), \tag{4.23}$$

where m and m' are defined in (4.22). Hence, the expression (4.21) is equivalent to the $n = 2$ case of Lemma 4.16, that is, the eigenvectors of $U(2)*U(2)$ define operator-valued polynomial functions which are, to within normalization, the operator-valued irrep matrices. For the subgroup $SU(2) * SU(2)$ these matrices are operator-valued equivalents of the irrep eigenvectors $|k, j;\ m, m'\rangle$. Suitably normalized, they are precisely the representation matrix elements $\mathfrak{D}^{j}_{mm'}(Z)$ expressed, not as functions of the group element, but as functions of Z, the coordinates on which $SU(2) * SU(2)$ acts.

The operator-valued matrices $\mathfrak{D}(A)$ can be expressed in the following elegant form:

$$\mathfrak{D}^{j}_{m',m}(A) = \Big((j+m)!(j-m)!(j+m')!(j-m')!\Big)^{\frac{1}{2}} \sum_{\boxed{\alpha}} \frac{(a^1_1)^{\alpha^1_1}\,(a^1_2)^{\alpha^1_2}\,(a^2_1)^{\alpha^2_1}\,(a^2_2)^{\alpha^2_2}}{(\alpha^1_1)!\,(\alpha^1_2)!\,(\alpha^2_1)!\,(\alpha^2_2)!}, \tag{4.24}$$

where $\boxed{\alpha}$ denotes the square array of nonnegative integers (α^j_i) satisfying

$$\begin{aligned} \alpha^1_1 + \alpha^2_1 = j + m', \quad \alpha^1_2 + \alpha^2_2 = j - m', \\ \alpha^1_1 + \alpha^1_2 = j + m, \quad \alpha^2_1 + \alpha^2_2 = j - m. \end{aligned} \tag{4.25}$$

These constraints effectively reduce the summation to one over a single index.

The rows and columns of $\mathfrak{D}(A)$ are tensor operators with respect to the left and right $U(2)$ groups. It can be seen explicitly from the expression (4.24) that the fundamental 2×2 irrep operator matrix is precisely the matrix boson A itself: $\mathfrak{D}^{\frac{1}{2}}(A) = A$. The rows and columns of A transform as spinor ($j = \frac{1}{2}$) operators with respect to the right and left $U(2)$ groups respectively.

These boson operator matrices necessarily obey, as operator-valued irreps, the group multiplication law for the group $U(2)$. That is, the fact that the 2×2 matrix product $g_1 g_2 \in U(2)$ for every $g_1, g_2 \in U(2)$ implies

$$\mathfrak{D}(g_1)\,\mathfrak{D}(g_2) = \mathfrak{D}(g_1 g_2),$$

which in turn implies

$$\mathfrak{D}(A)\,\mathfrak{D}(B) = \mathfrak{D}(AB), \tag{4.26}$$

where A, B are independent 2×2 matrix boson operators, AB is the 2×2 matrix product, and $\mathfrak{D}(A)\,\mathfrak{D}(B)$ denotes the matrix product in the representative (carrier) space.

The second "product law" for the boson operator irrep matrices is the Wigner product law, and is most easily stated for $SU(2)$ operator matrices, when $m_{22} = 0$ and for which therefore there is no determinantal factor. In the more concise j, m notation, obtained by putting $m_{12} = 2j$, $m_{11} = j + m$, we have

$$\mathfrak{D}^{j_3}_{m_3,m_3'}(A) = \sum_{m_1,m_1',m_2,m_2'} C^{j_1\ j_2\ j_3}_{m_1\ m_2\ m_3} C^{j_1\ j_2\ j_3}_{m_1'\ m_2'\ m_3'} \mathfrak{D}^{j_1}_{m_1,m_1'}(A)\, \mathfrak{D}^{j_2}_{m_2,m_2'}(A). \qquad (4.27)$$

This equation also expresses, of course, the reduction of the direct product representation into its irreducible constituents by means of the WCG coefficients.

This determination of the $U(2)$ operator-valued irrep matrices, which comprise a complete set of functions on the group, and their two fundamental properties, the product laws (4.26) and (4.27), are indeed the claimed results for a purely algebraic program, but we have not yet shown that their derivation is algebraic, since we used exponentiation in (4.17) to integrate the infinitesimal generators. We can, however, sidestep this integration by observing that the Wigner product law (4.27) allows one to construct — using the WCG coefficients — the general operator matrix (4.24) from the fundamental 2×2 irrep, the 2×2 boson operator matrix A itself. Thus, knowledge of the WCG coefficients actually suffices to determine the final result (4.24) in a purely algebraic fashion.

The matrix boson A in this algebraic approach plays the role of operator-valued coordinates which, however, were assumed from the beginning to commute (for $q = 1$). Nevertheless, the algebraic build-up process using the WCG coefficients is meaningful even when the components of the matrix boson are noncommuting. Accordingly, this algebraic method can extend directly to a construction for noncommuting quantum coordinates using q-WCG coefficients but, as we will show below, the matrix boson itself generalizes, not to a "matrix q-boson", but to a matrix of fundamental 2×2 q-tensor operators.

4.4 Extension to the Quantum Factor Algebra

The approach which we have outlined in the previous sections for the classical groups generalizes directly to $\mathcal{U}_q(\mathfrak{u}(2))$ and will yield not only all irrep matrices over the quantum plane but the commutation relations for the coordinates themselves. Beginning with $\mathcal{U}_q(\mathfrak{u}(2))$ and a full knowledge of its irreps and tensor operators, we will define quantum coordinates as components of elementary q-tensor operators, that is, as q-spinors with respect to left and right $\mathcal{U}_q(\mathfrak{u}(2))$ algebras to be defined. Hence, we identify the rows and columns of an element $T \in \mathfrak{M}_q(2)$ as elementary q-tensor operators and the relations (4.2), which define $\mathfrak{M}_q(2)$, will be *derived* using only the properties of q-tensor operators in $\mathcal{U}_q(\mathfrak{u}(2))$ as discussed in Chapter 3.

Let us firstly extend to $\mathcal{U}_q(\mathfrak{u}(2))$ the concept of a factor algebra. We begin by introducing four independent, commuting, q-boson operators a^i_j, $(i, j = 1, 2)$ and their conjugates together with their associated number operators N^i_j obeying the q-boson commutation rules of (2.27) and (2.33). Just as in the preceding section, but using now co-multiplication, we may construct two distinct sets of generators of $\mathcal{U}_q(\mathfrak{u}(2))$ denoted[4] $\{E_{ij}\}$ and $\{E^{ij}\}$:

$$\begin{aligned}
E_{12} &= q^{\frac{1}{4}(N_1^2-N_2^2)}a_1^1\overline{a}_2^1 + q^{\frac{1}{4}(N_2^1-N_1^1)}a_1^2\overline{a}_2^2 \\
E_{21} &= q^{\frac{1}{4}(N_1^2-N_2^2)}a_2^1\overline{a}_1^1 + q^{\frac{1}{4}(N_2^1-N_1^1)}a_2^2\overline{a}_1^2 \\
E_{11} &= N_1^1 + N_1^2 \\
E_{22} &= N_2^1 + N_2^2 \\
E^{12} &= q^{\frac{1}{4}(N_2^1-N_2^2)}a_1^1\overline{a}_1^2 + q^{\frac{1}{4}(N_1^2-N_1^1)}a_2^1\overline{a}_2^2 \\
E^{21} &= q^{\frac{1}{4}(N_2^1-N_2^2)}a_1^2\overline{a}_1^1 + q^{\frac{1}{4}(N_1^2-N_1^1)}a_2^2\overline{a}_2^1 \\
E^{11} &= N_1^1 + N_2^1 \\
E^{22} &= N_1^2 + N_2^2.
\end{aligned} \tag{4.28}$$

The generators E_{12} and E_{21} were denoted J_+ and J_- in (2.59) in §2.4.4. It follows from the properties of co-multiplication that each of these two sets of generators realizes the commutation relations of $\mathcal{U}_q(\mathfrak{u}(2))$.

Before determining further properties of these generators let us define the important operator a_{12}^q, which also appeared in §2.4.4, and then state several useful symmetry relations:

DEFINITION **4.29**

$$a_{12}^q \stackrel{\text{def}}{=} q^{\frac{1}{4}(N_2^1+N_1^2+1)}a_1^1a_2^2 - q^{-\frac{1}{4}(N_1^1+N_2^2+1)}a_2^1a_1^2.$$

We may refer to a_{12}^q as a q-determinant, as will be justified in Theorem 4.55 below. As noted in §2.4.4, a_{12}^q does not commute with the q-boson operators a_j^i, however, it does commute with the tensor operator components given in (3.40) and those in Theorem 4.55.

LEMMA **4.30** (a) *Under the interchange $a_1^i \leftrightarrow a_2^i$ for $i = 1,2$ and $q \leftrightarrow q^{-1}$ (from which follows $N_1^i \leftrightarrow N_2^i$ for $i = 1,2$), we have*

$$E_{12} \longleftrightarrow E_{21}, \quad E_{11} \longleftrightarrow E_{22}, \quad a_{12}^q \longleftrightarrow -a_{12}^q,$$

and the generators E^{ij} remain invariant.

(b) *Under the interchange $a_i^1 \leftrightarrow a_i^2$ for $i = 1,2$ and $q \leftrightarrow q^{-1}$ (from which follows $N_i^1 \leftrightarrow N_i^2$ for $i = 1,2$), we have*

$$E^{12} \longleftrightarrow E^{21}, \quad E^{11} \longleftrightarrow E^{22}, \quad a_{12}^q \longleftrightarrow -a_{12}^q,$$

and the generators E_{ij} remain invariant.

[4]We drop the q-affix in this Chapter as convenient.

(c) *Under the interchange* $a_2^1 \leftrightarrow a_1^2$, *from which follows* $N_2^1 \leftrightarrow N_1^2$, *we have*

$$E_{ij} \longleftrightarrow E^{ij}, \quad i,j=1,2,$$

and a_{12}^q *remains invariant.*

(d) *Under the interchange* $a_1^1 \leftrightarrow a_2^2$, *from which follows* $N_1^1 \leftrightarrow N_2^2$, *we have*

$$E_{ij} \longleftrightarrow E^{ji}, \quad i,j=1,2,$$

and a_{12}^q *remains invariant.*

We will find these symmetries useful in evaluating commutators in the following series of lemmas.

Now we may show that the two sets of generators defined by (4.28) commute:

LEMMA **4.31** *The operators* $\{E_{ij}\}$ *and* $\{E^{ij}\}$ *each generate* $\mathcal{U}_q(\mathrm{u}(2))$ *algebras (called the lower and upper, or left and right, algebras respectively), and the two sets of generators mutually commute:*

$$[E_{ij}, E^{kl}] = 0, \quad i,j,k,l = 1,2.$$

PROOF: The commutation relations involving the diagonal generators follow immediately. The relation $[E_{12}, E^{12}] = 0$ is established by direct calculation; we move all q-boson operators in the product $E_{12}E^{12} - E^{12}E_{12}$ to the right, and then use the q-boson relation (2.26), p. 26, to cancel all terms. (We need to use (2.26) also with $q \leftrightarrow q^{-1}$, which is permissable since q-boson operators are invariant under the interchange $q \leftrightarrow q^{-1}$ when defined by (2.33).)

Now we may complete the proof by taking advantage of symmetries of the generators E_{ij} and E^{ij} under interchange of q-boson operators, as shown in Lemma 4.30; for example, by using Lemma 4.30(a) we obtain $[E_{21}, E^{12}] = 0$, and similarly for the remaining relations. □

Next, we observe that the Casimir (quadratic) invariants for the $\mathcal{U}_q(\mathrm{u}(2))$ algebras coincide, so that the eigenvalues with respect to the Fock space eigenvectors are also identical:

LEMMA **4.32** *Let the Casimir invariants (2.11) for the left (right)* $\mathcal{U}_q(\mathrm{u}(2))$ *algebras be denoted* C_L (C_R) *respectively, then* $C_L = C_R$. *The linear invariant operators for each* $\mathcal{U}_q(\mathrm{u}(2))$ *algebra also coincide:*

$$E_{11} + E_{22} = E^{11} + E^{22}.$$

PROOF: We postpone the proof of $C_L = C_R$ to Lemma 4.39 below. The invariance of the linear operator is clear; this operator is simply the total number operator for the q-quanta. □

Thus, the algebra generated by $\{E_{ij}, E^{kl}\}$ is a factor algebra, which we denote $\mathcal{U}_q(\mathfrak{u}(2)*\mathfrak{u}(2))$, and the algebra generated by the two commuting sets of $\mathcal{U}_q(\mathfrak{su}(2))$ generators:

$$J_+ = E_{12}, \quad J_- = E_{21}, \quad J_z = \tfrac{1}{2}(E_{11} - E_{22}),$$
$$K_+ = E^{12}, \quad K_- = E^{21}, \quad K_z = \tfrac{1}{2}(E^{11} - E^{22}),$$

is also a factor algebra, denoted $\mathcal{U}_q(\mathfrak{su}(2)*\mathfrak{su}(2))$.

Now let us further explore properties of a^q_{12}. Firstly:

LEMMA **4.33** *The q-determinant a^q_{12} is an invariant of $\mathcal{U}_q(\mathfrak{su}(2)*\mathfrak{su}(2))$.*

PROOF: Let

$$a^+_{12} \stackrel{\text{def}}{=} q^{\frac{1}{4}(N^1_2+N^2_1+1)} a^1_1 a^2_2, \qquad a^-_{12} \stackrel{\text{def}}{=} q^{-\frac{1}{4}(N^1_1+N^2_2+1)} a^1_2 a^2_1, \tag{4.34}$$

and hence $a^q_{12} = a^+_{12} - a^-_{12}$. We calculate the following commutator:

$$[E_{12}, a^+_{12}] = q^{\frac{1}{4}(2N^1_2-2N^2_2+N^2_1-N^1_1)} a^1_1 a^2_1.$$

Under the symmetry transformation in Lemma 4.30(b), the right hand side of this equation remains invariant, as does E_{12}, but $a^+_{12} \leftrightarrow a^-_{12}$. Hence, we obtain

$$[E_{12}, a^+_{12}] = [E_{12}, a^-_{12}],$$

and so $[E_{12}, a^q_{12}] = 0$. By further use of the symmetries given in Lemma 4.30 we deduce $[E_{ij}, a^q_{12}] = 0 = [E^{ij}, a^q_{12}]$ for $i \neq j$. The commutators

$$[E_{11} - E_{22}, a^q_{12}] = 0 = [E^{11} - E^{22}, a^q_{12}]$$

are derived directly. □

In his development of angular momentum properties using techniques of boson operators in 1953 Schwinger observed that (for $q = 1$) the determinantal operator $a_{12} = a^1_1 a^2_2 - a^1_2 a^2_1$ and its Hermitean conjugate generate the Lie algebra $\mathfrak{su}(1,1)$ (see [71, p. 241] and [103, Chapter 4, §5, p. 124]), an observation which is useful in the discussion of unit Racah operators. We may generalize this result using the q-analog a^q_{12} of a_{12} but, firstly, we must consider the quantum group $\mathcal{U}_q(\mathfrak{su}(1,1))$ corresponding to $\mathfrak{su}(1,1)$. The generators $H_\pm, H_z$ of $\mathcal{U}_q(\mathfrak{su}(1,1))$ satisfy the defining relations

$$[H_z, H_\pm] = \pm H_\pm, \quad [H_+, H_-] = -[2H_z], \tag{4.35}$$

(noting the minus sign in the last relation). The Casimir invariant C_H of this algebra takes the form

$$C_H \stackrel{\text{def}}{=} -H_- H_+ + [H_z][H_z + 1],$$

as may be verified directly in the same way as for $\mathcal{U}_q(\mathfrak{su}(2))$ in Lemma 2.13. The generalization of Schwinger's result is:

LEMMA 4.36 *The operators*

$$H_+ = a^q_{12}, \quad H_- = \overline{a}^q_{12}, \quad H_z = \tfrac{1}{2}(N^1_1 + N^2_1 + N^1_2 + N^2_2 + 2),$$

satisfy the commutation relations (4.35) of $\mathcal{U}_q(\mathfrak{su}(1,1))$.

PROOF: It is convenient to define the Hermitean conjugates of the operators a^+_{12}, a^-_{12} defined in (4.34):

$$\overline{a}^+_{12} \stackrel{\text{def}}{=} q^{\frac{1}{4}(N^1_2+N^2_1+1)}\overline{a}^1_1\overline{a}^2_2, \qquad \overline{a}^-_{12} \stackrel{\text{def}}{=} q^{-\frac{1}{4}(N^1_1+N^2_2+1)}\overline{a}^1_2\overline{a}^2_1. \tag{4.37}$$

These operators satisfy

$$[a^+_{12}, \overline{a}^-_{12}] = 0 = [\overline{a}^+_{12}, a^-_{12}],$$

and so

$$[H_+, H_-] = [a^+_{12}, \overline{a}^+_{12}] + [a^-_{12}, \overline{a}^-_{12}].$$

Next, we calculate the commutator

$$[a^+_{12}, \overline{a}^+_{12}] = -q^{\frac{1}{2}(N^1_2+N^2_1+1)}[N^1_1 + N^2_2 + 1],$$

where we used the identity (2.12) with $c = 1$. There is a similar equation for $[a^-_{12}, \overline{a}^-_{12}]$ obtained with the help of Lemma 4.30(b), and upon using the identity

$$[n + n'] = q^{-\frac{n}{2}}[n'] + q^{\frac{n'}{2}}[n], \tag{4.38}$$

these equations lead to

$$[H_+, H_-] = -[N^1_1 + N^2_2 + 1 + N^1_2 + N^2_1 + 1],$$

as required. The remaining commutation relations are easily derived. □

The previous Lemmas 4.31,4.33 may now be combined into the result that we have *three* commuting algebras:

LEMMA 4.39 *The generators of the left $\mathcal{U}_q(\mathfrak{su}(2))$ algebra $\{J_\pm, J_z\}$, of the right $\mathcal{U}_q(\mathfrak{su}(2))$ algebra $\{K_\pm, K_z\}$, and of the $\mathcal{U}_q(\mathfrak{su}(1,1))$ algebra $\{H_\pm, H_z\}$, mutually commute. Furthermore, the Casimir invariants of these algebras are identical:*

$$C_L = C_R = C_H.$$

PROOF: The fact that the generators of $\mathcal{U}_q(\mathfrak{su}(1,1))$ commute with those of the factor algebra $\mathcal{U}_q\,(\mathfrak{su}(2)*\mathfrak{su}(2))$ follows from Lemma 4.33. To prove the equality of the invariants consider first $C_L - C_H$. We expand the term J_-J_+ in C_L by moving q-boson operators to the right, and expressing each factor in terms of number operators to the extent possible. There are two terms which do not involve number operators, but contain factors $a_1^1 a_2^2 \overline{a}_2^1 \overline{a}_1^2$ and $a_2^1 a_1^2 \overline{a}_1^1 \overline{a}_2^2$. The term $H_- H_+$ may be expanded using the operators $a_{12}^{\pm}$ and their conjugates defined in (4.34) and (4.37), and we obtain

$$H_- H_+ = \overline{a}_{12}^q a_{12}^q = \overline{a}_{12}^+ a_{12}^+ + \overline{a}_{12}^- a_{12}^- - \overline{a}_{12}^- a_{12}^+ - \overline{a}_{12}^+ a_{12}^-.$$

The last two terms here contain the same factors $a_1^1 a_2^2 \overline{a}_2^1 \overline{a}_1^2$ and $a_2^1 a_1^2 \overline{a}_1^1 \overline{a}_2^2$ arising in the expansion of $J_- J_+$, and cancel them. The remaining terms in $H_- H_+$ can be expressed as functions of number operators, which may be combined with the other terms in $C_L - C_H$, all of which may be expressed in terms of number operators. Now we use the familiar identity (4.38) and finally (2.12) with

$$a = H_z, \quad b = J_z, \quad c = a + b + 1,$$

to show that all terms cancel, that is, $C_L = C_H$. By using the transformation (c) of Lemma 4.30, which leaves invariant the generators of $\mathcal{U}_q(\mathfrak{su}(1,1))$, we also deduce that $C_R = C_H$. □

Properties of the three commuting quantum groups have also been considered by Smirnov and Kibler [2, p. 691], who have applied these commuting groups to establishing recurrence relations for q-WCG coefficients using Schwinger's approach. A generalization of the above construction of complementary quantum groups has been carried out by Quesne [161].

4.4.1 Basis Polynomials in an Irrep of the Quantum Factor Algebra

Having established the fundamental properties of $\mathcal{U}_q(\mathfrak{u}(2)*\mathfrak{u}(2))$ we may now proceed as in the $q = 1$ case and construct irreps in the space $\mathfrak{F}^4$ of homogeneous polynomials of degree $m_{12} + m_{22}$ in the four q-boson operators, acting on $|0\rangle$. The eigenvectors for the general (integer) irrep are uniquely labelled, exactly as in §4.3.1 for $U(2)*U(2)$, by the extended Gel'fand-Weyl patterns shown in (4.20). We calculate the explicit form of these orthonormal vectors by applying lowering operators E_{21} and E^{21} to the vector of highest weight with respect to both $\mathcal{U}_q(\mathfrak{u}(2))$ algebras. As we will show in §4.6 (Lemma 4.58), the orthonormal vectors are equal to the operator-valued irrep matrix ${}^q\mathfrak{D}$ for $\mathcal{U}_q(\mathfrak{u}(2))$ acting on $|0\rangle$ where the operator entries are realized in terms of q-boson operators; this is the q-analog extension of Lemma 4.16. However, this result is not sufficient to enable us to determine the required operator expression for ${}^q\mathfrak{D}$ from which we wish to deduce properties of noncommuting coordinates. Instead, we will construct ${}^q\mathfrak{D}$ in §4.5 as a polynomial in $\widehat{\mathfrak{P}}^4$ over four noncommuting variables using tensor operator concepts.

We have already calculated in §2.4.4 the vectors which are of highest weight with respect to the upper (right) $\mathcal{U}_q(\mathfrak{u}(2))$ algebra; they are given by (from (2.62)):

$$\left| \begin{matrix} & m_{12} & \\ m_{12} & & m_{22} \\ & m_{11} & \end{matrix} \right\rangle = \mathcal{N}_{\max}^{-\frac{1}{2}} (a_{12}^q)^{m_{22}} (a_1^1)^{m_{11}-m_{22}} (a_2^1)^{m_{12}-m_{11}} |0\rangle,$$

where

$$\mathcal{N}_{\max} = \frac{[m_{12}+1]![m_{12}-m_{11}]![m_{11}-m_{22}]![m_{22}]!}{[m_{12}-m_{22}+1]!}.$$

We recall that since a^q_{12} does not commute with the q-boson operators a^j_i the order of the operators in these basis vectors must be maintained. We now apply $K_- = E^{21}$ to this maximal state $n = m_{12} - m'_{11}$ times to obtain the general eigenstate, remembering that K_- commutes with a^q_{12}, and using the known matrix elements of $(K_-)^n$:

$$(K_-)^n \left| \begin{smallmatrix} & m_{12} & \\ m_{12} & & m_{22} \\ & m_{11} & \end{smallmatrix} \right\rangle = \sqrt{\frac{[n]![m_{12}-m_{22}]!}{[m_{12}-m_{22}-n]!}} \left| \begin{smallmatrix} & m_{12}-n & \\ m_{12} & & m_{22} \\ & m_{11} & \end{smallmatrix} \right\rangle .$$

Hence we obtain the following explicit expression for the orthonormal eigenvectors in $\mathfrak{F}^4$:

$$\begin{aligned} \left| \begin{smallmatrix} & m'_{11} & \\ m_{12} & & m_{22} \\ & m_{11} & \end{smallmatrix} \right\rangle &= \mathcal{N}^{-\frac{1}{2}} q^{-\frac{1}{4}(m_{12}-m'_{11})(m_{11}-m_{22})} (a^q_{12})^{m_{22}} \\ &\times \sum_n \frac{q^{\frac{n}{4}(m_{12}-m_{22})} (a^1_1)^{m_{11}-m_{22}-n} (a^1_2)^{m'_{11}-m_{11}+n} (a^2_1)^n (a^2_2)^{m_{12}-m'_{11}-n}}{[n]![m_{11}-m_{22}-n]![m'_{11}-m_{11}+n]![m_{12}-m'_{11}-n]!} |0\rangle, \end{aligned} \tag{4.40}$$

where the summation is over integers n such that

$$\max(0,\, m_{11}-m'_{11}) \leqslant n \leqslant \min(m_{11}-m_{22},\, m_{12}-m'_{11}),$$

and where the normalization is given by

$$\mathcal{N} = \frac{[m_{12}+1]![m_{22}]!}{[m_{12}-m_{22}+1][m'_{11}-m_{22}]![m_{11}-m_{22}]![m_{12}-m_{11}]![m_{12}-m'_{11}]!} . \tag{4.41}$$

We may verify this expression for the eigenvectors by induction on $n = m_{12} - m'_{11}$.

It is interesting to note that the polynomial in (4.40) is in fact a terminating basic hypergeometric function ${}_2\phi_1$. This identification can be made using the formulas of §2.8.3 (p. 67) and we find, choosing the case $m'_{11} \geqslant m_{11}$,

$$\begin{aligned} \left| \begin{smallmatrix} & m'_{11} & \\ m_{12} & & m_{22} \\ & m_{11} & \end{smallmatrix} \right\rangle &= \mathcal{M}^{-\frac{1}{2}} (a^q_{12})^{m_{22}} (a^1_1)^{m_{11}-m_{22}} (a^1_2)^{m'_{11}-m_{11}} (a^2_2)^{m_{12}-m'_{11}} \\ &\times {}_2\phi_1 \left(\begin{matrix} q^{m_{22}-m_{11}} \quad q^{m'_{11}-m_{12}} \\ q^{m'_{11}-m_{11}+1} \end{matrix} ;\, q, \frac{a^1_2 a^2_1}{a^1_1 a^2_2} q^{\mu} \right) |0\rangle, \end{aligned} \tag{4.42}$$

where $\mu = 1 + \frac{3}{4}(m_{12} - m_{22})$, and

$$\mathcal{M} = q^{\frac{1}{2}(m_{12}-m'_{11})(m_{11}-m_{22})} \frac{[m_{12}+1]![m_{22}]![m_{11}-m_{22}]![m'_{11}-m_{11}]!^2[m_{12}-m'_{11}]!}{[m_{12}-m_{22}+1][m'_{11}-m_{22}]![m_{12}-m_{11}]!}.$$

This basic hypergeometric function terminates because the numerator parameters $m_{22}-m_{11}$ and $m'_{11}-m_{12}$ are each negative. We have expressed the argument formally using inverse q-boson operators but, of course, the eigenfunctions are polynomials in these q-boson operators and may equivalently be replaced by their realizations in terms of complex variables.

Properties of representations can be translated into properties of the ${}_2\phi_1$ function in the usual way; for example the action of the raising and lowering generators $J_\pm$ and $K_\pm$ implies the existence of first order finite difference relations amongst ${}_2\phi_1$ functions, when we realize q-boson annihilation operators as finite difference operators (see §2.4.1). The eigenfunction equation satisfied by the basis polynomials is equivalent to a second order finite difference equation for the basic hypergeometric function, and orthonormality and completeness, which follow from the q-analog of the Peter-Weyl theorem [156, 162], are equivalent to finite sum integral and completeness relations for these same functions.

We note again that the expression (4.40) for these basis functions, as operator-valued polynomials acting on $|0\rangle$ does not lead us directly to our goal of determining the operator-valued ${}^q\mathfrak{D}$-matrices as polynomials in q-spinor components although, conversely, we will be able to derive (4.40) from the operator-valued ${}^q\mathfrak{D}$-matrix.

4.4.2 Derivation of q-WCG Coefficients

The properties of the set $\mathfrak{F}^4$ of polynomials provide a convenient means to calculate an explicit form of the q-WCG coefficients which we discussed in detail in §3.5. As mentioned there, several methods have been used to determine these coefficients and the following is a method which gives these coefficients in the form (3.53), which is the q-analog of the van der Waerden form.

The method follows that in [66, §5.7, p. 223] and uses the properties of the two sets of q-boson operators and corresponding Fock spaces which we introduced in §4.4. We may form two realizations of $\mathcal{U}_q(\mathfrak{su}(2))$ from the q-bosons $\mathbf{a}^i = (a_1^i, a_2^i)$ (for $i = 1, 2$), distinct from the factor algebra realization, according to

$$J_+(i) = a_1^i\overline{a}_2^i, \quad J_-(i) = a_2^i\overline{a}_1^i, \quad J_z(i) = \tfrac{1}{2}(N_1^i - N_2^i), \quad i = 1, 2$$

(no summation over i). Clearly, the two sets of operators mutually commute:

$$[\mathbf{J}(1), \mathbf{J}(2)] = 0,$$

and have distinct Casimir invariants: $C_1 \neq C_2$. Let us denote the corresponding irrep labels by j_1, j_2, that is, the Casimir invariants C_i have eigenvalues $[j_i][j_i+1]$ for $i = 1, 2$. As discussed in §2.4.2, these labels are eigenvalues of the Euler operators $N_1^i + N_2^i$. The magnetic quantum numbers m_1, m_2 are eigenvalues of the diagonal generators $J_z(1), J_z(2)$ respectively. The basis vectors in the irrep labelled j_i are

$\mathcal{P}_{j_i m_i}(\mathbf{a}^i)|0\rangle$ where, as shown in §2.4.2,

$$\mathcal{P}_{j_i m_i}(\mathbf{a}^i) = \frac{(a_1^i)^{j_i+m_i}(a_2^i)^{j_i-m_i}}{\sqrt{[j_i+m_i]![j_i-m_i]!}} \qquad (i=1,2).$$

The space $\mathfrak{F}^4$ is spanned therefore by the vectors

$$|j_1m_1;j_2m_2\rangle \stackrel{\text{def}}{=} \mathcal{P}_{j_1m_1}(\mathbf{a}^1)\,\mathcal{P}_{j_2m_2}(\mathbf{a}^2)|0\rangle,$$

and is of dimension $(2j_1+1)(2j_2+1)$.

On the other hand, we have already found a set of q-boson polynomials which span $\mathfrak{F}^4$, those given by (4.40), and which are therefore related to the vectors $|j_1m_1;j_2m_2\rangle$ by a linear transformation. Let us denote the vectors (4.40) by

$$|k,j;m,m'\rangle, \qquad j=\tfrac{1}{2}(m_{12}-m_{22}),\ k=\tfrac{1}{2}(m_{12}+m_{22}),$$

and determine the relationship between $|j_1m_1;j_2m_2\rangle$ and $|k,j;m,m'\rangle$. By comparing the generators $\mathbf{J}(i)$ with E^{ii} and E_{ii} given by (4.28) we find

$$J_z(1)+J_z(2) = \tfrac{1}{2}(N_1^1+N_1^2-N_2^1-N_2^2) = \tfrac{1}{2}(E_{11}-E_{22})$$

and hence $m_1+m_2=m$, (keeping in mind the identification (4.22)). Similarly, by noting that

$$K_z = \tfrac{1}{2}(E^{11}-E^{22}) = \tfrac{1}{2}(N_1^1+N_2^1-N_1^2-N_2^2),$$

we find

$$m' = j_1 - j_2.$$

The polynomial degree, which is the total number of quanta, is given by

$$2k = 2j_1 + 2j_2,$$

and so we may express the labels k,j,m,m' in terms of j_1,j_2,j,m. Consequently, the orthonormal vectors (4.40) may also be expressed in terms of the labels j_1,j_2,j,m, where $-j \leqslant m \leqslant j$. The dimension of $\mathfrak{F}^4$ is $\sum_j(2j+1) = (2j_1+1)(2j_2+1)$.

Upon substituting

$$m_{12}=j+j_1+j_2, \quad m_{22}=j_1+j_2-j, \quad m_{11}=j_1+j_2+m, \quad m'_{11}=2j_1$$

into (4.40), we find the following explicit expression for the vector $|j_1,j_2;jm\rangle$:

$$|j_1,j_2;jm\rangle = \mathcal{N}^{-\frac{1}{2}}q^{-\frac{1}{4}(j-j_1+j_2)(j+m)}(a_{12}^q)^{j_1+j_2-j}$$
$$\times\sum_n \frac{q^{\frac{1}{2}nj}(a_1^1)^{j+m-n}(a_2^1)^{j_1-j_2-m+n}(a_1^2)^n(a_2^2)^{j-j_1+j_2-n}}{[n]![j+m-n]![j_1-j_2-m+n]![j-j_1+j_2-n]!}\,|0\rangle, \quad (4.43)$$

where

$$\mathcal{N} = \frac{[j+j_1+j_2+1]![j_1+j_2-j]!}{[2j+1][j_1-j_2+j]![j+m]![j-m]![j-j_1+j_2]!}.$$

The linear transformation linking the two vectors, $|j_1,j_2;jm\rangle$ and $|j_1m_1;j_2m_2\rangle$, is in fact the matrix of q-WCG coefficients which couples the two commuting angular momenta $\mathbf{J}(1)$ and $\mathbf{J}(2)$ to the left $\mathcal{U}_q(\mathfrak{su}(2))$ algebra obtained by co-multiplication as above, and generated by J_+, J_-, J_z. Hence we have, as shown in (3.17),

$$|j_1,j_2;jm\rangle = \sum_{m_1,m_2} {}_qC^{j_1\ j_2\ j}_{m_1\ m_2\ m}|j_1m_1;j_2m_2\rangle$$

and so we may calculate the q-WCG coefficients by means of the formula

$${}_qC^{j_1\ j_2\ j}_{m_1\ m_2\ m} = \langle j_1m_1;j_2m_2|j_1,j_2;jm\rangle. \tag{4.44}$$

Let us now perform this calculation. Powers of $a^q_{12} = a^+_{12} - a^-_{12}$ may be expanded by noting that a^+_{12} and a^-_{12}, defined in (4.34), q-commute:

$$a^+_{12}a^-_{12} = q\, a^-_{12}a^+_{12},$$

and so $(a^q_{12})^r = (a^+_{12} - a^-_{12})^r$ may be expanded using the q-binomial theorem 2.102 (p. 60). Hence, we derive the following operator equations:

$$\begin{aligned}(a^q_{12})^r &= \sum_{s=0}^{r}\frac{[r]!}{[s]![r-s]!}(-)^s q^{\frac{s}{2}(r-s)}(a^-_{12})^s(a^+_{12})^{r-s} \qquad (4.45)\\ &= \sum_{s=0}^{r}\frac{[r]!}{[s]![r-s]!}(-)^s q^{\frac{s}{2}(r-s)}\\ &\quad\times q^{-\frac{s}{4}(N^1_1+N^2_2+1)}q^{\frac{1}{4}(r-s)(N^1_2+N^2_1+1-2s)}(a^1_1)^{r-s}(a^1_2)^s(a^2_1)^s(a^2_2)^{r-s}.\end{aligned}$$

We use this expansion with $r = m_{22} = j_1 + j_2 - j$.

Next, we substitute the expression (4.43) for $|j_1,j_2;jm\rangle$ into the formula (4.44), where the vector $\langle j_1m_1;j_2m_2|$ is given by

$$\langle j_1m_1;j_2m_2| = \langle 0|\frac{(\overline{a}^1_1)^{j_1+m_1}(\overline{a}^1_2)^{j_1-m_1}(\overline{a}^2_1)^{j_2+m_2}(\overline{a}^2_2)^{j_2-m_2}}{\left([j_1+m_1]![j_1-m_1]![j_2+m_2]![j_2-m_2]!\right)^{\frac{1}{2}}}.$$

The bra-ket inner product is evaluated using orthogonality of the Fock space polynomials, shown in (2.36), p. 29; we deduce that the q-WCG coefficients are zero unless four conditions are met, one of which is $s + n = j_2 + m_2$. Hence, only one term in the sum over s contributes, and the three remaining conditions then reduce to $m = m_1 + m_2$. In order to compare the resulting expression directly with the explicit q-WCG coefficients shown in (3.53), p. 95, we relabel the summation index n and sum over $n' = j_1 - j - m_2 + n$. Upon collecting all terms we obtain the required expression (3.53), but with m, m_1, m_2 replaced by their negatives, q replaced by q^{-1}, together with a sign change; specifically, we obtain the q-WCG coefficient in the form

$$(-)^{j_1+j_2-j}\,{}_{q^{-1}}C^{\,j_1\;\;\,j_2\;\;\,j}_{-m_1\,-m_2\,-m}\,.$$

According to the symmetry relation (3.63a) this is in fact equal to precisely ${}_qC^{\,j_1\;j_2\;j}_{m_1\,m_2\,m}$, as required.

The explicit algebraic form for the q-WCG coefficients may be expressed in terms of the ${}^q\mathfrak{D}$-matrix for $\mathcal{U}_q(\mathfrak{su}(2))$ using Lemma 4.58 below, which expresses the vectors $|j_1, j_2; jm\rangle$ in terms of the ${}^q\mathfrak{D}$-matrix, and leads also to an interpretation of the q-WCG coefficients as "discretized" irrep matrices; this is described for $SU(2)$ in [66, §5.8] but we do not develop this idea further here.

4.5 Commutation Rules for Elements of the Quantum Matrix

We now turn to the derivation of properties of operators which transform as q-spinors with respect to the factor algebra $\mathcal{U}_q\,(\mathfrak{u}(2)*\mathfrak{u}(2))$. The problem we consider is this: do the matrix elements t^j_i of the fundamental operator-valued irrep matrix ${}^q\mathfrak{D}$ satisfy the required commutation relations (4.2)? Let us denote the elements of T by (t^j_i) and denote by $\widehat{\mathfrak{P}}^4$ the set of polynomials in t^j_i. The irrep matrix ${}^q\mathfrak{D}$ is an element of $\widehat{\mathfrak{P}}^4$ and we will find that the fundamental irrep matrix is T itself. We seek to identify elements of $\mathfrak{M}_q(2)$ with the matrix of q-spinors according to

$$T = \begin{pmatrix} a & b \\ c & d \end{pmatrix} = {}^q\mathfrak{D}^{\frac{1}{2}}(T) = \begin{pmatrix} t^1_1 & t^2_1 \\ t^1_2 & t^2_2 \end{pmatrix}.$$

We begin therefore by defining the elements of T as q-spinors:

DEFINITION **4.46** *Each of the pairs* $(t^1_1,\, t^1_2)$ *and* $(t^2_1,\, t^2_2)$ *is defined to be a* q*-spinor under the left* $\mathcal{U}_q(\mathfrak{u}(2))$ *algebra, and each of the pairs* $(t^1_1,\, t^2_1)$ *and* $(t^1_2,\, t^2_2)$ *is defined to be a* q*-spinor under the right* $\mathcal{U}_q(\mathfrak{u}(2))$ *algebra.*

We now prove

LEMMA **4.47** *The elements* t^j_i *satisfy the relations (4.2).*

We will need an extension of the equality of irreducible representation labels, as expressed in Lemma 4.32, to a certain class of q-tensor operators. This extension is:

LEMMA **4.48** *The* q*-tensor operators* t^j_i*, which act in* $\widehat{\mathfrak{P}}^4$*, transform as fundamental* q*-tensor operators with respect to each* $\mathcal{U}_q(\mathfrak{u}(2))$ *with identical operator pattern labels.*

PROOF: The elements of $\widehat{\mathfrak{P}}^4$ may equivalently be written as the operator elements of the ${}^q\mathfrak{D}$-matrix acting to the right on the vacuum. Thus the action of T on $\widehat{\mathfrak{P}}^4$ is equivalent to the product of ${}^q\mathfrak{D}$-operators acting on the vacuum. This product of ${}^q\mathfrak{D}$-operators is exactly the q-analog of the Wigner definition of the q-WCG operators. By the extension of Wigner's theorem [138] the operators have the operator pattern labels restriction as stated. □

Now let us prove Lemma 4.47 by considering the possible couplings of the spinor components to form invariants. Tensor coupling in $\mathcal{U}_q(\mathfrak{u}(2)*\mathfrak{u}(2))$ involves q-WCG coefficients in both upper and lower $\mathcal{U}_q(\mathfrak{u}(2))$ spaces; due to the constraint on the two irrep labels it suffices to couple in either space, but if the omitted coupling is not possible the resulting q-tensor operator vanishes. Consider any of the four spinors comprising T, namely the two rows and the two columns of T. From these four spinors we may form six invariants (with respect to one of the $\mathcal{U}_q(\mathfrak{u}(2))$ algebras), using the formula (3.38). For example, we may couple t_i^1 and t_j^1 to give a scalar operator, with irrep labels [1 1] in the lower $\mathcal{U}_q(\mathfrak{u}(2))$ algebra generated by $\{E_{ij}\}$. Hence, using the q-WCG coefficients, we find

$$q^{\frac{1}{4}}t_1^1t_2^1 - q^{-\frac{1}{4}}t_2^1t_1^1 \tag{4.49}$$

is a tensor operator in $\mathcal{U}_q(\mathfrak{u}(2)*\mathfrak{u}(2))$ with, by construction, lower $\mathcal{U}_q(\mathfrak{u}(2))$ labels of [1 1]. However, in the upper $\mathcal{U}_q(\mathfrak{u}(2))$ the tensor product of a vector $\left|\begin{smallmatrix}1\,0\\1\end{smallmatrix}\right\rangle$ with a vector $\left|\begin{smallmatrix}1\,0\\1\end{smallmatrix}\right\rangle$ can lead only to a vector $\left|\begin{smallmatrix}2\,0\\2\end{smallmatrix}\right\rangle$. This is because maximal states "add" under the q-WCG tensor product. Thus the constructed operator is labelled [2 0] with respect to the upper algebra. This is a contradiction unless the resultant tensor product (4.49) is identically zero, since the only allowed irreps of $\mathcal{U}_q(\mathfrak{u}(2)*\mathfrak{u}(2))$, as stated in Lemma 4.32 and extended to operators in Lemma 4.48, are those sharing the same labels. Thus (4.49) is identically zero. If we now identify $a = t_1^1$ and $c = t_2^1$ we have proved that the commutation relation $ac = q^{-\frac{1}{2}}ca$ in (4.2) must be satisfied.

The same coupling scheme is valid for the remaining relations in the first line of (4.2) involving all four components of T, and hence we have established these relations as well.

Next, let us consider the equations which relate the two forms of the determinant (4.4). We can view these relations as originating from the construction of a quadratic invariant operator in the q-tensor algebra of $\mathcal{U}_q(\mathfrak{u}(2)*\mathfrak{u}(2))$. There are two invariants — which differ at most by a constant — due to the two possibilities of coupling order.

Let us carry out the construction of the two quadratic invariants by firstly coupling t_i^1 and t_j^2 to give an invariant I_1, using again the formula (3.38), to obtain

$$I_1 = q^{\frac{1}{4}}t_1^1t_2^2 - q^{-\frac{1}{4}}t_2^1t_1^2.$$

Secondly, we couple t_j^2 and t_i^1 to give another invariant I_2:

$$I_2 = q^{\frac{1}{4}}t_1^2t_2^1 - q^{-\frac{1}{4}}t_2^2t_1^1.$$

In this coupling process we could equally well have coupled on the *upper labels* rather than the lower labels to form I_1 and I_2. The q-WCG coefficients are unchanged, but the orders of two of the terms in each invariant is reversed. Thus we obtain

$$\begin{aligned} I_1 &= q^{\frac{1}{4}} t_1^1 t_2^2 - q^{-\frac{1}{4}} t_1^2 t_2^1 \\ I_2 &= q^{\frac{1}{4}} t_2^1 t_1^2 - q^{-\frac{1}{4}} t_2^2 t_1^1, \end{aligned}$$

which by comparison with the previous expressions show that we must have the structural relation $t_2^1 t_1^2 - t_1^2 t_2^1 = 0$ or $bc = cb$, another of the relations (4.2). Furthermore, we may also assert that the two invariants I_1, I_2 are proportional and after evaluating the constant of proportionality by letting the invariants act on irrep vectors, we are lead to the final relation in (4.2) or (4.4).

Accordingly we have verified directly from the q-tensor calculus of $\mathcal{U}_q(\mathfrak{u}(2)*\mathfrak{u}(2))$ that the matrix elements of the fundamental tensor operator matrix T satisfy the commutation relations and structural identities of $\mathfrak{M}_q(2)$. In §4.6 we confirm this by explicit calculation.

As noted in Theorem 4.3, the algebraic relations (4.2) suffice to prove that

$${}^q\mathfrak{D}^{[1,0]}(T)\,{}^q\mathfrak{D}^{[1,0]}(T') = {}^q\mathfrak{D}^{[1,0]}(TT')$$

with T, T' and TT' each elements of $\mathfrak{M}_q(2)$. This result extends to general irreps $[m_{12}\, m_{22}]$ of $\mathcal{U}_q(\mathfrak{u}(2))$ since we may construct the ${}^q\mathfrak{D}$ matrices using the Wigner product law:

$${}^q\mathfrak{D}^{j_3}_{m_3,m_3'}(T) = \sum_{m_1,m_1',m_2,m_2'} {}_qC^{j_1\ j_2\ j_3}_{m_1\ m_2\ m_3}\ {}_qC^{j_1\ j_2\ j_3}_{m_1'\ m_2'\ m_3'}\ {}^q\mathfrak{D}^{j_1}_{m_1,m_1'}(T)\ {}^q\mathfrak{D}^{j_2}_{m_2,m_2'}(T).$$

Thus, the two fundamental product laws for operator-valued $\mathcal{U}_q(\mathfrak{u}(2))$ matrix irreps are valid.

Because the rows and columns for ${}^q\mathfrak{D}$ may be formed by coupling with q-WCG coefficients, we deduce the q-analog to Lemma 4.18:

LEMMA **4.50** *The columns (rows) of ${}^q\mathfrak{D}^j(T)$ form irreducible tensor operators carrying the irrep j with respect to the left (right) $\mathcal{U}_q(\mathfrak{su}(2))$ algebras, respectively.*

As a corollary, we deduce from the group multiplication law ${}^q\mathfrak{D}(T)\,{}^q\mathfrak{D}(T') = {}^q\mathfrak{D}(TT')$ that q-tensor operators, which we defined in Definition 3.14 (p. 82) in terms of $\mathcal{U}_q(\mathfrak{su}(2))$ and satisfy the derivative property given in Theorem 3.30 (p. 88), also satisfy the q-equivariance property, the q-analog of Definition 3.3:

COROLLARY **4.51** *Tensor operators $\mathbf{T}_{jm}$ in $SU_q(2)$ satisfy equivariance, that is, under the action of $SU_q(2)$:*

$$\mathbf{T}_{jm} \longrightarrow \mathbf{T}'_{jm} = \sum_{m'} \mathbf{T}_{jm'}\,{}^q\mathfrak{D}^j_{m'm}(g), \quad g \in SU_q(2).$$

We have already discussed a special case of this result in §4.2.2, where we considered q-spinors $(x\ y)$ which transformed according to $(x\ y) \longrightarrow (x\ y)T$, where $T \in SU_q(2)$; this follows from the above corollary by noting that ${}^q\mathfrak{D}^{\frac{1}{2}}(T) = T$.

The representation matrices ${}^q\mathfrak{D}(T)$ are elements of $\widehat{\mathfrak{P}}^4$, that is, polynomials in the four spinor components t^j_i. The basis vectors in $\mathfrak{F}^4$ of $\mathcal{U}_q\,(\mathfrak{u}(2){*}\mathfrak{u}(2))$ may be obtained from these matrices, when T is realized as in Theorem 4.55 below, by allowing ${}^q\mathfrak{D}$ to act on the vacuum (see Lemma 4.58 below). As mentioned, we can in principle determine the explicit ${}^q\mathfrak{D}$-matrices as functions of the elementary tensor operator components by means of the Wigner product law, but in practice we find it simpler to derive these matrices by letting the quantum matrix act on noncommuting coordinates (that is, on q-spinors). This calculation is carried out in §4.7 where we obtain:

LEMMA **4.52** *The irrep matrices of $\mathcal{U}_q(\mathfrak{su}(2))$, which are elements of $\widehat{\mathfrak{P}}^4$, have the explicit form:*

$$\begin{aligned} {}^q\mathfrak{D}^j_{m',m}(T) &= \Big([j+m]![j-m]![j+m']![j-m']!\Big)^{\frac{1}{2}} \\ &\quad \times \sum_{\boxed{\alpha}} q^{\frac{1}{4}(\alpha^1_1+\alpha^2_2)(\alpha^1_2+\alpha^2_1)+\frac{1}{2}(\alpha^1_2\alpha^2_1)} \frac{(t^1_1)^{\alpha^1_1}\,(t^1_2)^{\alpha^1_2}\,(t^2_1)^{\alpha^2_1}\,(t^2_2)^{\alpha^2_2}}{[\alpha^1_1]!\,[\alpha^1_2]!\,[\alpha^2_1]!\,[\alpha^2_2]!}, \end{aligned} \tag{4.53}$$

where the sum is over the square array $\boxed{\alpha}$ of nonnegative integers shown in (4.25).

We have expressed these matrices as functions over $\mathfrak{M}_q(2)$, but they reduce to functions on $SU_q(2)$ using the expression (4.6) for T. For $q \to 1$, (4.53) reduces to the operator-valued representation matrix given in (4.24) in terms of boson operators. Forms similar to (4.53) have been given by other authors (Masuda et al. [162] and Nomura [163]) in different contexts, but without a tensor operator interpretation. These polynomials are the q-analogs of the Jacobi polynomials, well-known for $SU(2)$ when the rotation matrices are expressed in terms of Euler angles. They are related [162] to the little q-Jacobi polynomials introduced by Hahn [164] in 1949.

Properties of the ${}^q\mathfrak{D}$-matrices are developed in Chapter 5, including the proof of the Wigner product law, but let us note immediately two significant properties:

$${}^q\mathfrak{D}^{\frac{1}{2}}(T) = T, \qquad {}^q\widetilde{\mathfrak{D}}(T) = {}^q\mathfrak{D}(\widetilde{T}).$$

As is the case for $q = 1$, we may define irrep matrices for $\mathcal{U}_q(\mathfrak{u}(2))$ from those of $\mathcal{U}_q(\mathfrak{su}(2))$ by a formula similar to (4.23):

$${}^q\mathfrak{D}^{[m_{12},m_{22}]}_{m_{11},m'_{11}}(T) \stackrel{\text{def}}{=} (\det_q(T))^{m_{22}}\ {}^q\mathfrak{D}^{\frac{1}{2}(m_{12}-m_{22})}_{m,m'}(T), \tag{4.54}$$

where m, m' are given by (4.22) and $j = \frac{1}{2}(m_{12} - m_{22})$.

4.5.1 Generalization to the Quantum Hyperplane

The construction given in §4.4 for $\mathcal{U}_q(\mathrm{u}(2))$ generalizes directly to $\mathcal{U}_q(\mathrm{u}(n))$ including the construction of the factor algebra $\mathcal{U}_q(\mathrm{u}(n)*\mathrm{u}(n))$. Lemmas 4.31 and 4.48 generalize as well. $\mathfrak{M}_q(n)$ is defined in §4.2.1 and has properties similar to $\mathfrak{M}_q(2)$. Matrices in $\mathfrak{M}_q(n)$ act on q-spinors with n noncommuting components x_i which comprise the quantum hyperplane and satisfy $x_i x_j = q^{-\frac{1}{2}} x_j x_i$ for $j > i$.

Now let us consider properties of q-spinors with respect to $\mathcal{U}_q(\mathrm{u}(n)*\mathrm{u}(n))$ and show that the q-spinor components t^i_j comprising the $n \times n$ quantum matrix T satisfy the commutation relations of elements of $\mathfrak{M}_q(n)$, following exactly the methods used in §4.5, as we now sketch abstractly. Consider two q-tensor components t^i_j and t^i_l, $j < l$, which are chosen to lie in the same column (here i). The product of these two elements necessarily belongs to the carrier space $\mathfrak{V}$ of representations of $\mathcal{U}_q(\mathrm{u}(n)*\mathrm{u}(n))$ with Gel'fand-Weyl labels $[2\,0\ldots0]\times[2\,0\ldots0]$ — since this is enforced by the condition that the two indices i are the same — and to a unique vector in this space. The product of these two elements *in opposite order* leads to exactly this same unique vector but with a different numerical coefficient. Using tables of the fundamental q-WCG coefficients, given by the pattern calculus rules stated in §3.7, one finds the following explicit result for the commutation relation: $t^i_j t^i_l = q^{-\frac{1}{2}} t^i_l t^i_j$, $j < l$, that is, the first set of relations in (4.7). Similarly, by exchanging columns for rows exactly the same abstract relations hold, and we derive (again using q-WCG tables) the second set of relations in (4.7).

Consider also the (generic) products $t^k_j t^i_l, (i < k, j < l)$, and $t^i_j t^k_l$, $(i < k, j < l)$, each taken in the two possible orders. The resulting *four* products, when acting in $\mathfrak{V}$, are each numerical multiples of two abstract vectors in $\mathfrak{V}$, one vector in $[2\,0\ldots0]\times[2\,0\ldots0]$, the second vector in $[1\,1\,0\ldots0]\times[1\,1\,0\ldots0]$. Thus there exist *two* linear relations between the *four* products, as determined by the q-WCG coefficients. These relations are:

$$t^i_l t^k_j = t^k_j t^i_l$$

and

$$t^k_l t^i_j - t^i_j t^k_l = \left(q^{-\frac{1}{2}} - q^{\frac{1}{2}}\right) t^i_l t^k_j,$$

which comprise the remaining relations in (4.7).

This abstract argument shows that the basic results of the q-tensor operator approach are generic and demonstrate the importance of q-tensor operators in understanding the quantum hyperplane, and hence also the relationship of $U_q(n)$ to $\mathcal{U}_q(\mathrm{u}(n))$.

4.6 A q-Boson Realization of Noncommuting Elements

In the previous section we developed abstract arguments showing that q-spinors in the factor algebra satisfy the defining relations of $\mathfrak{M}_q(2)$. However we also require explicit expressions for the q-spinors in order to derive a q-analog of the equality between the matrix boson A and the fundamental irrep on which the group acts, that is, we wish to find an explicit representation of $T \in \mathfrak{M}_q(2)$ in terms of q-boson operators with entries a, b, c, d satisfying (4.2).

We cannot identify T with the q-boson matrix $A^q = ((a^q)_i^j)$, because q-boson operators do not form the components of a spinor with respect to the factor algebra $\mathcal{U}_q(\mathfrak{u}(2)*\mathfrak{u}(2))$ generated by $\{E_{ij}, E^{kl}\}$ in (4.28). Moreover, q-boson operators themselves clearly do not satisfy (4.2). Neither is there any apparently simple way in which we can construct the matrix of spinor components T from spinors such as $(a_1^1 q^{-\frac{1}{4}N_2^1}, a_2^1 q^{\frac{1}{4}N_1^1})$ considered in (3.40). One approach would be to construct T from its matrix elements using the q-WCG coefficients to determine invariant factors, as in [66, Chapter 5].

However, we will proceed in the following way: firstly, we note that it is sufficient to determine t_1^1 explicitly in terms of q-boson operators since, given this element, the remaining elements t_2^1, t_1^2, t_2^2 may be determined using the commutation relations (3.37) satisfied by a spinor. That is, t_2^1 can be determined from t_1^1 using the action of E_{21}, and t_1^2 can be determined using E^{21}. Then t_2^2 can be determined in either of two ways, by letting E_{21} act on t_1^2, or by letting E^{21} act on t_2^1.

In order to determine t_1^1 we note, secondly, that $a = t_1^1$ commutes with the q-determinant $\det_q(T) = ad - q^{-\frac{1}{2}}bc$ which, up to an overall scalar multiple, we have in fact already determined. For $q = 1$ the determinant is given by $\det(A) = a_{12} = a_1^1 a_2^2 - a_2^1 a_1^2$ and is an invariant of $SU(2) * SU(2)$. The q-analog of a_{12}, denoted a_{12}^q, is given in Definition 4.29 and is an invariant of $\mathcal{U}_q(\mathfrak{su}(2)*\mathfrak{su}(2))$ as shown in Lemma 4.33. We anticipate therefore that a_{12}^q will be proportional to $\det_q(T)$ and so will commute with all the entries of $T \in \mathfrak{M}_q(2)$ and in particular with t_1^1. Hence, we may construct t_1^1 from operators which commute with a_{12}^q, namely $a_1 q^{-\frac{1}{4}N_2^1}$, $q^{N_2^1 - N_1^2}$ and $q^{N_1^1 - N_2^2}$ and so we assume the following form for t_1^1 (which, however, is not the most general form for the component of highest weight):

$$t_1^1 = a_1 q^{-\frac{1}{4}N_2^1 + \alpha(N_2^1 - N_1^2) + \beta(N_1^1 - N_2^2)}$$

where α and β are constants to be determined.

Because t_1^1 is the first component of a spinor with respect to both the left and right $\mathcal{U}_q(\mathfrak{su}(2))$ algebras it must satisfy, according to (3.37), $E_{12}t_1^1 = q^{-\frac{1}{4}} t_1^1 E_{12}$ and $E^{12}t_1^1 = q^{-\frac{1}{4}} t_1^1 E^{12}$. These two equations serve to determine $\alpha = \beta = \frac{1}{8}$. Therefore we obtain the required explicit representation for t_1^1 and hence (after a long calculation) we obtain also all the elements t_i^j comprising the matrix T. Of particular note is the form for t_2^2, which involves the conjugate q-boson $\overline{a}_1^1$ and which is necessarily present in order to realize the full algebra given in (4.2). We may summarize the results as follows:

THEOREM **4.55** *Let*

$$\begin{aligned} t^1_1 &= a^1_1 q^{\frac{1}{8}(N^1_1-N^1_2-N^2_1-N^2_2)} \\ t^2_1 &= a^2_1 q^{\frac{1}{8}(3N^1_1-N^1_2+N^2_1-N^2_2)} \\ t^1_2 &= a^1_2 q^{\frac{1}{8}(3N^1_1+N^1_2-N^2_1-N^2_2)} \\ t^2_2 &= a^2_2 q^{\frac{1}{8}(-N^1_1+3N^1_2+3N^2_1+N^2_2)} + (q^{\frac{1}{2}} - q^{-\frac{1}{2}}) a^1_2 a^2_1 \overline{a}^1_1 q^{\frac{1}{8}(N^1_1+N^1_2+N^2_1-N^2_2-2)}. \end{aligned}$$

Then

(a) *for $q = 1$ the matrix $T = (t^j_i)$ reduces to the matrix boson A.*

(b) *the columns of T each form a spinor with respect to the lower (left) quantum group generated by $\{E_{ij}\}$, and the rows each form a spinor with respect to the upper (right) quantum group generated by $\{E^{ij}\}$.*

(c) *T is an element of $\mathfrak{M}_q(2)$, that is, if we write $T = \left(\begin{smallmatrix} a & b \\ c & d \end{smallmatrix}\right)$, then the entries a, b, c, d satisfy the algebraic relations (4.2).*

(d) *the q-determinant of T, defined in (4.4), is given by*

$$\det_q(T) = q^{-\frac{3}{8}} a^q_{12},$$

(where a^q_{12} is given by Definition 4.29) and commutes with each element t^j_i of T.

PROOF: By direct calculation. Observe that the task is simplified with the help of Lemma 4.30(c). Under this symmetry the elements t^1_1 and t^2_2 remain invariant and t^1_2 is interchanged with t^2_1. □

REMARK **4.56** 1. We note that this operator realization of elements of $\mathfrak{M}_q(2)$ is not unique, for we can obtain another related realization by forming the antipode matrix

$$\gamma(T) = \begin{pmatrix} d & -q^{\frac{1}{2}} b \\ -q^{-\frac{1}{2}} c & a \end{pmatrix},$$

from the elements a, b, c, d of the theorem, then exchanging $q \leftrightarrow q^{-1}$ so that the resulting matrix lies in $\mathfrak{M}_q(2)$, and relabelling the q-boson operators. Alternatively, we may use the symmetries of Lemma 4.30 to map one operator realization to another.

2. We may rescale the entries of T by an arbitrary power of q, which will leave the conclusions of the theorem unaltered except that the constant of proportionality between $\det_q(T)$ and a^q_{12} will change.

3. Other representations of $\mathfrak{M}_q(2)$ in terms of q-boson operators are known, the simplest of which is perhaps the following which uses a single q-boson operator and its conjugate (due to Mo-Lin Ge et al. [165]):

$$T = \begin{pmatrix} a & q^{\frac{N}{2}} \\ -q^{\frac{1}{2}(N+1)} & -(q^{\frac{1}{2}} - q^{-\frac{1}{2}})q^{\frac{1}{2}(N+1)}\bar{a} \end{pmatrix} .$$

The q-determinant for this quantum matrix is 1, but by rescaling can be set to an arbitrary value. Clearly there exist relations between the elements of this matrix, for example $c = -q^{\frac{1}{2}}b$. There is a related representation which may be deduced from the inverse matrix.

Other realizations are due to Damaskinsky and Sokolov [166], but differ from that of Theorem 4.55. A q-tensor operator interpretation for these realizations requires a suitable realization of the corresponding algebra $\mathcal{U}_q(\mathfrak{su}(2))$.

4. We have restricted our discussion in this section to the case $n = 2$, however Theorem 4.55 may be generalized to the n-dimensional case by consideration of irreducible tensor operators in the n-fold tensor product $\mathfrak{F}^{n^2}$ of Fock spaces to obtain a realization of $\mathfrak{M}_q(n)$. The construction of irreducible tensor operators in $\mathfrak{F}^{n^2}$ has been discussed by Quesne [2, p. 609].

Whereas we may construct the quantum matrix T and consequently determine the properties of the matrix quantum group from the q-spinors t_i^j, we may also construct the generators of $\mathcal{U}_q\,(\mathfrak{u}(2)*\mathfrak{u}(2))$ from this realization of the elements of T. We have the following corollary to Theorem 4.55:

COROLLARY **4.57** *The generators of* $\mathcal{U}_q\,(\mathfrak{u}(2)*\mathfrak{u}(2))$ *are related to the elements* t_i^j *by*

$$\begin{aligned} E_{12}q^{\frac{1}{2}E_{11}} &= t_1^1\bar{t}_2^1 + t_1^2\bar{t}_2^2 \\ q^{\frac{1}{2}E_{11}}E_{21} &= t_2^1\bar{t}_1^1 + t_2^2\bar{t}_1^2 \\ E^{12}q^{\frac{1}{2}E^{11}} &= t_1^1\bar{t}_1^2 + t_2^1\bar{t}_2^2 \\ q^{\frac{1}{2}E^{11}}E^{21} &= t_1^2\bar{t}_1^1 + t_2^2\bar{t}_2^1 \,. \end{aligned}$$

PROOF: By direct calculation. It is sufficient to derive one equation and then obtain the others with the help of Lemma 4.30, and by taking Hermitean conjugates. □

We observe that the generators E_{11}, E^{11} appear in a distinguished way in these relations; the symmetry with E_{22} and E^{22} is regained by noting that there is another realization of T obtained by using the symmetries of Lemma 4.30 (as remarked above), which leads to equivalent relations involving E_{22} and E^{22}. These formulas are similar to that found in (3.50), in which a spin 1 tensor operator is constructed from spin $\frac{1}{2}$ components.

Having obtained the explicit form of ${}^q\mathfrak{D}(T)$ as a polynomial in the elements t_i^j together with a q-boson realization of T, we may determine the q-analog of Lemma 4.16 which relates the basis polynomials spanning $\mathfrak{F}^4$ to the operator-valued irrep matrix $\mathfrak{D}(A)$ acting on the vacuum:

LEMMA **4.58** *The orthonormal basis vectors (4.40) in $\mathfrak{F}^4$ are given by*

$$\left| \begin{matrix} & m'_{11} & \\ m_{12} & & m_{22} \\ & m_{11} & \end{matrix} \right\rangle = \mathcal{N}^{-\frac{1}{2}}\, q^{\nu}\; {}^q\mathfrak{D}^{[m_{12},m_{22}]}_{m_{11},m'_{11}}(T)\,|0\rangle,$$

where the elements of T are realized as in Theorem 4.55, the specific form of ${}^q\mathfrak{D}$ is given by Eqns. (4.53, 4.54), the normalization is given by

$$\mathcal{N} = \frac{[m_{22}]![m_{12}+1]!}{[m_{12}-m_{22}+1]},$$

and

$$\nu = -\tfrac{1}{16}(m_{12}-m_{22})(m_{12}-m_{22}-1) + \tfrac{3}{8}m_{22}.$$

PROOF: As the formula for ${}^q\mathfrak{D}$ in Lemma 4.52 shows, we must compute

$$q^{\frac{1}{4}(\alpha_1^1+\alpha_2^2)(\alpha_2^1-\alpha_1^2)+\frac{1}{2}(\alpha_2^1\alpha_1^2)}(t_1^1)^{\alpha_1^1}\,(t_2^1)^{\alpha_2^1}\,(t_1^2)^{\alpha_1^2}\,(t_2^2)^{\alpha_2^2}\,|0\rangle \tag{4.59}$$

with t_i^j given by Theorem 4.55. The following operator identity is useful:

$$\left(aq^{\frac{N}{8}}\right)^n = a^n q^{\frac{n}{16}(2N+n-1)}, \quad n \in \mathbb{N},$$

where a is a q-boson creation operator with a corresponding number operator N. Clearly, the last term of t_2^2 involving the q-boson annihilation operator $\overline{a}_1^1$ gives zero contribution, and so the expression (4.59) is equal to

$$q^{\nu'}(a_1^1)^{\alpha_1^1}\,(a_2^1)^{\alpha_2^1}\,(a_1^2)^{\alpha_1^2}\,(a_2^2)^{\alpha_2^2}\,|0\rangle,$$

where ν' depends only on the indices α_i^j. In order to compare with (4.40) we sum over the array $\boxed{\alpha}$ in (4.53) by substituting

$$\alpha_1^1 = m_{11}-m_{22}-n,\ \alpha_2^1 = m'_{11}-m_{11}+n,\ \alpha_1^2 = n,\ \alpha_2^2 = m_{12}-m'_{11}-n,$$

where $n \in \mathbb{N}$ (note that in the explicit form (4.53) the indices m, m' are reversed compared to the order used in this lemma, and also we use the Gel'fand-Weyl notation given by (4.22)). We find then that $\nu' = \frac{n}{4}(m_{12}-m_{22}) + \rho$ where ρ is independent of n. Upon collecting all q-factors, including those arising by substituting for $\det_q(T) = q^{-\frac{3}{8}}a_{12}^q$, we obtain

$$\rho + \nu - \tfrac{3}{8}m_{22} = -\tfrac{1}{4}(m_{12}-m'_{11})(m_{11}-m_{22}),$$

which gives the explicit q-factor in (4.40). The normalization $\mathcal{N}$ combines with the square root factors in (4.53) to give precisely the normalization in (4.40), and the denominator factors in (4.40) arise from the denominator factors in (4.53). □

The precise value of the explicit factor q^{ν} in Lemma 4.58 is not significant since, as explained in the remark above, we may rescale T by an arbitrary q-factor without altering the fundamental properties of T or ${}^{q}\mathfrak{D}$.

4.7 Irreps of the Matrix Quantum Group

We may now define representations of $SU_q(2)$ analogously to those for $SU(2)$ as described in §4.3, in which the operator Θ acts in the space of homogeneous polynomials in the noncommuting coordinates. Specifically, we can define left and right actions in the space $\widehat{\mathfrak{P}}^2$ consisting of homogeneous polynomials in two noncommuting variables $\boldsymbol{z} = (z_1, z_2)$ where

$$z_1 z_2 = q^{-\frac{1}{2}} z_2 z_1 .$$

A representation Θ_g, for $g \in SU_q(2)$, may be defined by

$$\Theta_g\, \mathcal{P}(\boldsymbol{z}) = \mathcal{P}(\tilde{g}\boldsymbol{z}), \quad \mathcal{P} \in \widehat{\mathfrak{P}}^2,$$

(regarding $\boldsymbol{z}$ as a column vector) and we have $\Theta_{g_1 g_2} = \Theta_{g_1}\Theta_{g_2}$ for all $g_1, g_2 \in SU_q(2)$. We may also define a right action in $\widehat{\mathfrak{P}}^2$:

$$\Theta'_g\, \mathcal{P}(\boldsymbol{z}) = \mathcal{P}(\boldsymbol{z}g), \quad \mathcal{P} \in \widehat{\mathfrak{P}}^2,$$

noting that $\tilde{g} \in SU_q(2)$ for each $g \in SU_q(2)$, and again the mapping $g \to \Theta'_g$ is a representation of $SU_q(2)$.

This may be extended to left and right actions in the space $\widehat{\mathfrak{P}}^4$ of homogeneous polynomials on the quantum coordinates $T = (t^j_i)$ (satisfying (4.2)) by the q-analog of the formula (4.11):

$$\Theta_{(g,g')}\, \mathcal{P}(T) = \mathcal{P}(\tilde{g}Tg'), \quad \mathcal{P} \in \widehat{\mathfrak{P}}^4.$$

where $g, g' \in SU_q(2)$. In this case the mapping $(g, g') \to \Theta_{(g,g')}$ is a representation of $SU_q(2) * SU_q(2)$.

We may realize the noncommuting variables in terms of q-boson operators acting on the vacuum, and hence construct irreps of $SU_q(2)$ over polynomials in complex variables or, equivalently, in a Fock space. For example we may realize T as in Theorem 4.55 to construct irreps of $SU_q(2)*SU_q(2)$ over the Fock space $\mathfrak{F}^4$ constructed from four q-boson operators, spanned by the vectors given in Lemma 4.58.

The quantum coordinates $\boldsymbol{z}$ may be realized as operators in $\mathfrak{F}^2$ by the q-spinor pair given in (3.40), p. 90:

$$z_1 = a_1\, q^{-\frac{N_2}{4}}, \quad z_2 = a_2\, q^{\frac{N_1}{4}}, \tag{4.60}$$

and the irrep Θ' may now be defined by the right action

$$\Theta'_g \mathcal{P}(z)\,|0\rangle = \mathcal{P}(zg)\,|0\rangle, \quad \mathcal{P} \in \mathfrak{F}^2. \tag{4.61}$$

We may use this definition to determine the irrep matrix ${}^q\mathfrak{D}$ for $SU_q(2)$ by allowing Θ' to act on the basis functions $|jm\rangle$, which are given by (2.48) but are expressed in terms of the noncommuting components z in (4.60) by:

$$|jm\rangle = q^{\frac{1}{4}(j-m)(j+m)}([j+m]![j-m]!)^{-\frac{1}{2}}(z_1)^{j+m}(z_2)^{j-m}\,|0\rangle, \tag{4.62}$$

Under the right transformation Θ'_g these vectors are transformed into a linear combination of themselves, and the coefficients determine the ${}^q\mathfrak{D}$-matrices:

$$\begin{aligned} |jm\rangle \longrightarrow |jm\rangle &= \sum_{m'} {}^q\mathfrak{D}^j_{m'm}(g)\,|jm'\rangle \qquad (4.63)\\ &= q^{\frac{1}{4}(j-m)(j+m)}([j+m]![j-m]!)^{-\frac{1}{2}}((zg)_1)^{j+m}((zg)_2)^{j-m}\,|0\rangle. \end{aligned}$$

More generally we can choose g to be an element of $\mathfrak{M}_q(2)$, and the irrep matrices of $SU_q(2)$ can then be deduced using (4.6).

Let us now prove Lemma 4.52, which gives the explicit expression for ${}^q\mathfrak{D}$. We expand the right hand side of (4.63) using the q-binomial theorem 2.102, first noting that:

$$(zg)_1 = az_1 + cz_2, \quad (zg)_2 = bz_1 + dz_2,$$

where

$$g = \begin{pmatrix} a & b \\ c & d \end{pmatrix} = \begin{pmatrix} t^1_1 & t^2_1 \\ t^1_2 & t^2_2 \end{pmatrix} \in \mathfrak{M}_q(2).$$

Since

$$(az_1)(cz_2) = q^{-1}(cz_2)(az_1), \quad (bz_1)(dz_2) = q^{-1}(dz_2)(bz_1),$$

we can apply the q-binomial theorem and then identify ${}^q\mathfrak{D}(g)$ using (4.63). We also need the identity

$$z_2^n z_1^m = q^{\frac{1}{2}nm} z_1^m z_2^n, \quad n, m \in \mathbb{N}.$$

We express the result for ${}^q\mathfrak{D}$ in the form (4.53) by identifying the summation indices α^j_i as the exponents of the elements t^j_i.

As a corollary to this derivation, we deduce from the representation property $\Theta'_{g_1g_2} = \Theta'_{g_1}\Theta'_{g_2}$ that for all $g_1, g_2 \in \mathfrak{M}_q(2)$,

$${}^q\mathfrak{D}(g_1g_2) = {}^q\mathfrak{D}(g_1)\,{}^q\mathfrak{D}(g_2).$$

4.7.1 Fractional Linear Transformations

In §2.4.3 we described a realization of $\mathcal{U}_q(\mathfrak{su}(2))$ in which irreps were constructed in the space of polynomials of degree $2j$ on a projective space $\mathbb{P}_1(\mathbb{C})$. In terms of the group $SU(2)$ we defined an action by means of fractional linear transformations (2.52), p. 36, which we now generalize to $SU_q(2)$. This generalization requires consideration of noncommuting coordinates, and may be derived in a manner similar to that for $SU(2)$.

We begin with the representation (4.61) defined by right transformations:

$$\Theta'_g f(z_1, z_2) = f(az_1 + cz_2,\ bz_1 + dz_2), \tag{4.64}$$

where now $g \in SU_q(2)$ or, more generally we take g to be a quantum matrix in $\mathfrak{M}_q(2)$ with elements a, b, c, d, and z_1, z_2 are now noncommuting coordinates satisfying $z_1 z_2 = q^{-\frac{1}{2}} z_2 z_1$. We define a representation on the space $\widehat{\mathcal{P}}$ of polynomials in one complex variable ζ with coefficients in $\mathcal{A}$, the associative algebra generated by the elements a, b, c, d satisfying (4.2). Let us assume that suitable inverse elements exist so that we may allow inverses of the form $(a + c\zeta)^{-1}$. We let $\zeta = z_2 z_1^{-1} = q^{\frac{1}{2}} z_1^{-1} z_2$ which we identify as the complex variable.

The fact that z_1 q-commutes with ζ leads to explicit q-factors when deriving the fractional linear transformation, beginning with the representation (4.64). We take account of these q-factors by the device of a number operator N for ζ:

$$N = \zeta \frac{\partial}{\partial \zeta},$$

which satisfies $q^{\frac{N}{2}} \zeta q^{-\frac{N}{2}} = q^{\frac{1}{2}} \zeta$. We then derive the following q-commutation rule: Let

$$X = (a + c\zeta) q^{-\frac{N}{2}}, \qquad Y = (b + d\zeta) q^{-\frac{N}{2}},$$

then $XY = q^{-\frac{1}{2}} YX$. In particular, we will encounter the combination

$$\begin{aligned} YX^{-1} &= (b + d\zeta)(a + c\zeta)^{-1} \\ = q^{-\frac{1}{2}} X^{-1} Y &= \left(a + q^{\frac{1}{2}} c\zeta\right)^{-1} \left(q^{-\frac{1}{2}} b + d\zeta\right). \end{aligned}$$

The number operator also provides a convenient means of writing a product of noncommuting factors as a power of X, as follows:

$$\begin{aligned} \left(q^{\frac{N}{2}}\right)^n \left((a + c\zeta) q^{-\frac{N}{2}}\right)^n &= \prod_{r=n}^{1} \left(a + q^{\frac{r}{2}} c\zeta\right) \\ &= \left(a + q^{\frac{n}{2}} c\zeta\right) \left(a + q^{\frac{1}{2}(n-1)} c\zeta\right) \ldots \left(a + q^{\frac{1}{2}} c\zeta\right), \end{aligned}$$

where the order of the noncommuting factors is as shown in the last line.

Now we may define the q-analog of fractional linear transformations:

LEMMA 4.65 *Define the operator Θ_g for each $g = \begin{pmatrix} a & b \\ c & d \end{pmatrix} \in \mathfrak{M}_q(2)$ in $\mathcal{A}[\zeta]$ by:*

$$\Theta_g \varphi(\zeta) = \left(q^{\frac{N}{2}}\right)^{2j} \left((a + c\zeta) q^{-\frac{N}{2}}\right)^{2j} \varphi\left((b + d\zeta)(a + c\zeta)^{-1}\right).$$

Then $g \longrightarrow \Theta_g$ is a representation of $\mathfrak{M}_q(2)$.

PROOF: It is convenient to write the definition of Θ_g in the form

$$\Theta_g\varphi(\zeta) = \left(q^{\frac{N}{2}}\right)^{2j} X^{2j}\varphi\left(YX^{-1}\right).$$

We must prove $\Theta_{g'g} = \Theta_{g'}\Theta_g$ for all $g', g \in \mathfrak{M}_q(2)$. Let

$$g' = \begin{pmatrix} a' & b' \\ c' & d' \end{pmatrix}, \qquad X' = (a' + c'\zeta)q^{-\frac{N}{2}}, \qquad Y' = (b' + d'\zeta)q^{-\frac{N}{2}}.$$

We have

$$\begin{aligned}
\Theta_{g'g}\varphi(\zeta) &= \left(q^{\frac{N}{2}}\right)^{2j}\left((aa' + cb' + (ac' + cd')\zeta)q^{-\frac{N}{2}}\right)^{2j} \\
&\quad \times\varphi\left((ba' + db' + (bc' + dd')\zeta)(aa' + cb' + (ac' + cd')\zeta)^{-1}\right) \\
&= \left(q^{\frac{N}{2}}\right)^{2j}(aX' + cY')^{2j}\varphi\left((bX' + dY')(aX' + cY')^{-1}\right), \qquad (4.66)
\end{aligned}$$

and also,

$$\begin{aligned}
\Theta_{g'}\Theta_g\varphi(\zeta) &= \left(q^{\frac{N}{2}}\right)^{2j}(X')^{2j}\prod_{r=2j}^{1}\left(a + q^{\frac{r}{2}}cY'(X')^{-1}\right) \\
&\quad \times\varphi\left((b + dY'(X')^{-1})(a + cY'(X')^{-1})^{-1}\right). \qquad (4.67)
\end{aligned}$$

The argument of φ in this equation is seen to equal the argument of φ on the right hand side of (4.66). The coefficient of φ in (4.67) is given by

$$\begin{aligned}
&\left(q^{\frac{N}{2}}\right)^{2j}(X')^{2j}\prod_{r=2j}^{1}\left(a + q^{\frac{r}{2}}cY'(X')^{-1}\right) \\
&\qquad = \left(q^{\frac{N}{2}}\right)^{2j}\left(a + cY'(X')^{-1}\right)(X')^{2j}\prod_{r=2j-1}^{1}\left(a + q^{\frac{r}{2}}cY'(X')^{-1}\right) \\
&\qquad = \left(q^{\frac{N}{2}}\right)^{2j}(aX' + cY')(X')^{2j-1}\prod_{r=2j-1}^{1}\left(a + q^{\frac{r}{2}}cY'(X')^{-1}\right) \\
&\qquad = \left(q^{\frac{N}{2}}\right)^{2j}(aX' + cY')^{2j},
\end{aligned}$$

where in the second line we used the property $X^nY = q^{-\frac{n}{2}}YX^n$ (for the primed operators) for $n = 2j$. The last line follows by recursive use of the previous lines, by q-commuting powers of X' as necessary. Hence the right hand sides of (4.66) and (4.67) are equal as required. □

A basis for the space $\widehat{\mathcal{P}}$ is the set of monomials $\{1, \zeta, \zeta^2, \ldots, \zeta^{2j}\}$ and this leads to an alternative derivation of the explicit expression for the irrep matrices ${}^q\mathfrak{D}$ given in Lemma 4.52, using (4.63).

Chapter 5

Quantum Rotation Matrices

An explicit algebraic expression for the quantum rotation (representation) matrices of $\mathcal{U}_q(\mathfrak{su}(2))$ and $\mathcal{U}_q(\mathfrak{u}(2))$ has been developed in Chapter 4, and the existence of the associated irreducible tensor operators has been demonstrated in Lemma 4.50 and Corollary 4.51.

It might appear that an explicit general result would suffice for understanding the quantum rotation functions, but this is far from true and a detailed discussion of some features of these functions is essential. We need not seek far to find the reason for this situation: the associative algebra $\mathcal{A}$, over which the quantum rotation functions are constructed and which we refer to as the quantum plane, consists of noncommuting coordinates and this change in viewpoint leads to profound changes not only in the applicable (noncommutative) algebraic techniques, but — for physics — in the very interpretation and meaning of this structure itself.

This is evident when we consider the form of the quantum rotation matrices. The primary structure is the quantum plane itself, comprising 2×2 matrices T of coordinate operators designating a "point" in the quantum plane. One has on this quantum plane an algebraic structure $\mathfrak{M}_q(2)$ which constitutes a Hopf algebra. A *second* algebraic structure comprising the irrep matrices ${}^q\mathfrak{D}^{[m_{12},m_{22}]}(T)$ of $\mathcal{U}_q(\mathfrak{u}(2))$ is then realized over this quantum plane, which implements the homomorphism: $T \rightarrow {}^q\mathfrak{D}^{[m_{12},m_{22}]}(T)$ such that

$$T_1 T_2 = T_{12} \Longrightarrow {}^q\mathfrak{D}^{[m_{12},m_{22}]}(T_1)\, {}^q\mathfrak{D}^{[m_{12},m_{22}]}(T_2) = {}^q\mathfrak{D}^{[m_{12},m_{22}]}(T_{12}),$$

where the product $T_1 T_2$ is the matrix product (actually co-multiplication) in the quantum plane and the product ${}^q\mathfrak{D}^{[m_{12},m_{22}]}(T_1)\, {}^q\mathfrak{D}^{[m_{12},m_{22}]}(T_2)$ is a $(m_{12}-m_{22}+1) \times (m_{12}-m_{22}+1)$ matrix product in the irrep space of the factor algebra $\mathcal{U}_q\,(\mathfrak{u}(2)*\mathfrak{u}(2))$.

Thus the quantum rotation matrices comprise a relatively complicated structure involving a *pairing* of two very different algebraic entities: (a) the Hilbert space vectors

$$\left| \begin{matrix} & m_{12} & \\ m_{12} & & m_{22} \\ & m_{11} & \end{matrix} \right\rangle$$

of unitary irreps for $\mathcal{U}_q(\mathfrak{u}(2)*\mathfrak{u}(2))$ and (b) the elements T of the algebra $\mathfrak{M}_q(2)$ comprising the quantum plane.

We begin our discussion by listing the definition and fundamental properties of the quantum rotation matrices, including the tensor operator properties of the rows and columns, and the product laws, and then turn to special cases of these matrices, which provide examples of the properties discussed. Then we turn to the definition of the generating function and use it to derive symmetries of the ${}^q\mathfrak{D}$-matrices, and follow this with a discussion of equivariance properties of tensor operators. In Chapter 3 we defined tensor operators via the induced action of the generators of $\mathcal{U}_q(\mathfrak{su}(2))$ and the ${}^q\mathfrak{D}$-matrices now allow us to extend this action to the matrix quantum group to obtain the q-analog of equivariance. Finally we return to the Wigner product law and use it to determine orthogonality relations of the ${}^q\mathfrak{D}$-matrices, recurrence relations, and the ${}^q\mathfrak{D}$-matrix form of the RTT relations.

5.1 Fundamental Properties of the Quantum Rotation Matrices

In §4.7 we constructed irreps of $SU_q(2)$ in the Fock space $\mathfrak{F}^2$ from which we determined an explicit expression for the quantum rotation matrix ${}^q\mathfrak{D}$ as the matrix of this irrep, as stated in Lemma 4.52. Let us adopt this expression now as the definition:

DEFINITION **5.1** *The quantum rotation matrices on $\mathfrak{M}_q(2)$ are defined by*

$$
\begin{aligned}
{}^q\mathfrak{D}^j_{m',m}(T) \;&\stackrel{\text{def}}{=}\; \Big([j+m]![j-m]![j+m']![j-m']!\Big)^{\frac{1}{2}} \\
&\times \sum_{\boxed{\alpha}} q^{\frac{1}{4}(\alpha^1_1+\alpha^2_2)(\alpha^1_2+\alpha^2_1)+\frac{1}{2}(\alpha^1_2\alpha^2_1)} \frac{(t^1_1)^{\alpha^1_1}\,(t^1_2)^{\alpha^1_2}\,(t^2_1)^{\alpha^2_1}\,(t^2_2)^{\alpha^2_2}}{[\alpha^1_1]!\,[\alpha^1_2]!\,[\alpha^2_1]!\,[\alpha^2_2]!},
\end{aligned}
$$

where the sum is over the square array $\boxed{\alpha}$ of nonnegative integers (α^j_i) satisfying

$$
\begin{aligned}
\alpha^1_1+\alpha^2_1 = j+m', \quad \alpha^1_2+\alpha^2_2 = j-m', \\
\alpha^1_1+\alpha^1_2 = j+m, \quad \alpha^2_1+\alpha^2_2 = j-m,
\end{aligned}
$$

and the elements (t^j_i) on which ${}^q\mathfrak{D}$ is defined comprise the quantum matrix T:

$$
T = \begin{pmatrix} t^1_1 & t^2_1 \\ t^1_2 & t^2_2 \end{pmatrix} = \begin{pmatrix} a & b \\ c & d \end{pmatrix},
$$

where

$$
\begin{aligned}
ab = q^{-\frac{1}{2}}ba, \quad ac = q^{-\frac{1}{2}}ca, \quad bd = q^{-\frac{1}{2}}db, \quad cd = q^{-\frac{1}{2}}dc, \\
bc = cb, \quad ad - da = -(q^{\frac{1}{2}} - q^{-\frac{1}{2}})bc.
\end{aligned} \tag{5.2}
$$

Although we have defined the rotation matrix on $\mathfrak{M}_q(2)$, we can specialize to $SL_q(2)$ by imposing $\det_q(T) = ad - q^{-\frac{1}{2}}bc = \mathbb{1} = da - q^{\frac{1}{2}}bc$, and then further specialize to $SU_q(2)$ by introducing the $*$ anti-involution as described in §4.2 (for $q \in \mathbb{R}^+$). Hence, the quantum rotation matrices for $SU_q(2)$ are given by:

$$
\begin{aligned}
{}^q\mathfrak{D}^j_{m',m}(T) \quad &\stackrel{\text{def}}{=} \quad \Big([j+m]![j-m]![j+m']![j-m']!\Big)^{\frac{1}{2}} \\
&\times \sum_{\boxed{\alpha}} q^{\frac{1}{4}(\alpha^1_1+\alpha^2_2)(\alpha^1_2+\alpha^2_1)+\frac{1}{2}\alpha^1_2(\alpha^2_1+1)}(-)^{\alpha^1_2} \frac{(a)^{\alpha^1_1}\,(b^*)^{\alpha^1_2}\,(b)^{\alpha^2_1}\,(a^*)^{\alpha^2_2}}{[\alpha^1_1]!\,[\alpha^1_2]!\,[\alpha^2_1]!\,[\alpha^2_2]!}\,,
\end{aligned}
\tag{5.3}
$$

where

$$
T = \begin{pmatrix} a & b \\ -q^{\frac{1}{2}}b^* & a^* \end{pmatrix} \in SU_q(2).
$$

We define irrep matrices for $\mathcal{U}_q(\mathfrak{u}(2))$ by:

$$
{}^q\mathfrak{D}^{[m_{12},m_{22}]}_{m_{11},m'_{11}}(T) \stackrel{\text{def}}{=} (\det_q(T))^{m_{22}}\ {}^q\mathfrak{D}^{\frac{1}{2}(m_{12}-m_{22})}_{m,m'}(T), \tag{5.4}
$$

where m, m' are given in terms of the Gel'fand-Weyl labels by

$$
\begin{aligned}
m &= m_{11} - \tfrac{1}{2}(m_{12}+m_{22}) \\
m' &= m'_{11} - \tfrac{1}{2}(m_{12}+m_{22}).
\end{aligned}
$$

Let us summarize the properties of ${}^q\mathfrak{D}(T)$, as given in Definition 5.1:

LEMMA **5.5** *The quantum rotation matrices* ${}^q\mathfrak{D}(T)$

1. *are of dimension* $2j+1$ *and each entry is a polynomial in the elements* (t^j_i) *of total degree* $2j$;
2. *satisfy*
$$
{}^q\widetilde{\mathfrak{D}}(T) = {}^q\mathfrak{D}(\tilde{T});
$$
3. *satisfy the group multiplication law*
$$
{}^q\mathfrak{D}(T_1T_2) = {}^q\mathfrak{D}(T_1)\,{}^q\mathfrak{D}(T_2)
$$
for all $T_1, T_2 \in \mathfrak{M}_q(2)$;
4. *satisfy the Wigner product law*
$$
{}^q\mathfrak{D}^{j_3}_{m_3,m'_3}(T) = \sum_{m_1,m'_1,m_2,m'_2} {}_qC^{j_1\ j_2\ j_3}_{m_1\ m_2\ m_3}\ {}_qC^{j_1\ j_2\ j_3}_{m'_1\ m'_2\ m'_3}\ {}^q\mathfrak{D}^{j_1}_{m_1,m'_1}(T)\ {}^q\mathfrak{D}^{j_2}_{m_2,m'_2}(T), \tag{5.6}
$$
where ${}_qC^{\cdots}_{\cdots}$ *denotes the* q*-WCG coefficients.*

In addition: the columns (rows) of ${}^q\mathfrak{D}^j(T)$ form irreducible tensor operators carrying the irrep j with respect to the left (right) $\mathcal{U}_q(\mathfrak{su}(2))$ algebras, respectively.

PROOF: Property 3 follows by construction, as explained in §4.7. Property 4 can be regarded as the defining property of the ${}^q\mathfrak{D}$-matrices, indeed this is how these matrices were defined in §4.5, where we constructed the elements of the fundamental quantum matrix ${}^q\mathfrak{D}^{\frac{1}{2}}(T) = T$ by defining them as spinors with respect to the factor algebra $\mathcal{U}_q\left(\mathfrak{su}(2)*\mathfrak{su}(2)\right)$. The general ${}^q\mathfrak{D}$-matrices were then constructed recursively, in principle, by means of the q-WCG coupling according to the Wigner product law (5.6). However, we found it more convenient there to derive a closed form expression for ${}^q\mathfrak{D}(T)$ by means of the action of the elements of $\mathfrak{M}_q(2)$ on noncommuting coordinates, as described in §4.7. Hence, we have yet to prove that the ${}^q\mathfrak{D}$-matrices satisfy the Wigner product law, and this we do now.

The Wigner product law (5.6) shows us how to construct the ${}^q\mathfrak{D}$-matrix for spin j_3 given ${}^q\mathfrak{D}^{j_1}(T)$ and ${}^q\mathfrak{D}^{j_2}(T)$. Hence, if we are given the fundamental spin $\frac{1}{2}$ matrix ${}^q\mathfrak{D}^{\frac{1}{2}}(T)$ we can perform a q-WCG coupling to construct ${}^q\mathfrak{D}^1(T)$, from which in turn we construct ${}^q\mathfrak{D}^{\frac{3}{2}}(T)$, and so on recursively. It is sufficient, therefore, to prove (5.6) for the case $j_1 = j$, $j_2 = \frac{1}{2}$ and $j_3 = j+\frac{1}{2}$. Then, if ${}^q\mathfrak{D}^j(T)$ is known we may construct ${}^q\mathfrak{D}^{j+\frac{1}{2}}(T)$ using the fundamental q-WCG coefficients given in (3.43), p. 91. We now substitute $j_1 = j, j_2 = \frac{1}{2}, j_3 = j+\frac{1}{2}$ and $m_3 = m'+\frac{1}{2}$, $m_3' = m+\frac{1}{2}$ in (5.6), for which the sum on the right hand side has four terms corresponding to m_2, $m_2' = \pm\frac{1}{2}$ with, of course, $m_1+m_2 = m_3$ and $m_1'+m_2' = m_3'$. Upon substituting for the q-WCG coefficients we obtain:

$$\begin{aligned}
[2j+1]\,{}^q\mathfrak{D}^{j+\frac{1}{2}}_{m'+\frac{1}{2},m+\frac{1}{2}}(T) &= q^{-\frac{1}{4}(-2j+m+m')}\sqrt{[j+m+1][j+m'+1]}\;{}^q\mathfrak{D}^j_{m',m}(T)\,a \\
&+ q^{-\frac{1}{4}(m+m'+1)}\sqrt{[j-m][j+m'+1]}\;{}^q\mathfrak{D}^j_{m',m+1}(T)\,b \\
&+ q^{-\frac{1}{4}(m+m'+1)}\sqrt{[j+m+1][j-m']}\;{}^q\mathfrak{D}^j_{m'+1,m}(T)\,c \\
&+ q^{-\frac{1}{4}(2j+m+m'+2)}\sqrt{[j-m][j-m']}\;{}^q\mathfrak{D}^j_{m'+1,m+1}(T)\,d\,.
\end{aligned} \tag{5.7}$$

These relations constitute **recurrence relations** for the ${}^q\mathfrak{D}$-matrices which, given ${}^q\mathfrak{D}^0(T) = \mathbb{1}$, determine all ${}^q\mathfrak{D}$-matrices uniquely.

We must prove that the explicit expression of Definition 5.1 satisfies this recurrence relation. It is helpful to choose one of the integers (α_i^j) as the summation index, for example $\alpha_1^1 = n$. We can change the positions of the noncommuting elements with the help of the following relations, proved by induction on $n \in \mathbb{N}$:

$$\begin{gathered}
b^n a = q^{\frac{n}{2}} a b^n, \quad c^n a = q^{\frac{n}{2}} a c^n, \quad d^n b = q^{\frac{n}{2}} b d^n, \quad d^n c = q^{\frac{n}{2}} c d^n \\
d^n a = a d^n + \left(q^{\frac{1}{2}} - q^{-\frac{1}{2}}\right) q^{\frac{1}{2}(n-1)}[n] b c d^{n-1}\,.
\end{gathered}$$

By direct calculation, we find that (5.7) is satisfied provided that the following identity is valid:

$$\begin{aligned}[2j+1] &= q^{\frac{1}{2}(n-2m'-m-j-2)}[\alpha] + q^{\frac{1}{2}(n-2m-m'-j-2)}[\beta] \\ &\quad + q^{-\frac{1}{2}(\alpha+\beta+n)}[n-m-m'-1] + q^{\frac{1}{2}(2j+1-n)}[n] \\ &\quad + q^{-\frac{1}{2}(m+m'+1)}\left(q^{\frac{1}{2}} - q^{-\frac{1}{2}}\right)[\alpha][\beta],\end{aligned}$$

where

$$\alpha = j+m-n+1, \quad \beta = j+m'-n+1.$$

The identity may be verified directly to complete the proof of (5.7), and hence of the Wigner product law. □

5.1.1 Special Cases

For $j=0$ the ${}^q\mathfrak{D}$-matrix is 1 dimensional and is equal to the unit $\mathbb{1}$ in $\mathcal{A}$. For $j=\frac{1}{2}$ the ${}^q\mathfrak{D}$-matrix is 2 dimensional and is equal to T itself: ${}^q\mathfrak{D}^{\frac{1}{2}}(T) = T$. For $j=1$ the ${}^q\mathfrak{D}$-matrix is 3 dimensional and is given by Table 5.8. We also give the 4 dimensional matrix in Table 5.9, for $j=\frac{3}{2}$.

*	$m=1$	$m=0$	$m=-1$
$m'=1$	a^2	$\sqrt{[2]}\, q^{\frac{1}{4}} ab$	b^2
$m'=0$	$\sqrt{[2]}\, q^{\frac{1}{4}} ac$	$ad + q^{\frac{1}{2}} bc$	$\sqrt{[2]}\, q^{\frac{1}{4}} bd$
$m'=-1$	c^2	$\sqrt{[2]}\, q^{\frac{1}{4}} cd$	d^2

Table 5.8: The spin-1 quantum rotation matrix ${}^q\mathfrak{D}^1_{m',m}(T)$

In addition, the summation in the Definition 5.1 of ${}^q\mathfrak{D}^j_{m',m}(T)$ reduces to a single term for the boundary values $m=\pm j$ and $m'=\pm j$, leading to the following special cases:

$$\begin{aligned}{}^q\mathfrak{D}^j_{m',j}(T) &= \sqrt{\frac{[2j]!}{[j+m']![j-m']!}}\; q^{\frac{1}{4}(j+m')(j-m')} a^{j+m'} c^{j-m'} \\ {}^q\mathfrak{D}^j_{m',-j}(T) &= \sqrt{\frac{[2j]!}{[j+m']![j-m']!}}\; q^{\frac{1}{4}(j+m')(j-m')} b^{j+m'} d^{j-m'}\end{aligned}$$

$*$	$m=\frac{3}{2}$	$m=\frac{1}{2}$	$m=-\frac{1}{2}$	$m=-\frac{3}{2}$
$m'=\frac{3}{2}$	a^3	$\sqrt{[3]}\; q^{\frac{1}{2}} a^2 b$	$\sqrt{[3]}\; q^{\frac{1}{2}} ab^2$	b^3
$m'=\frac{1}{2}$	$\sqrt{[3]}\; q^{\frac{1}{2}} a^2 c$	$a^2 d + [2]q\; abc$	$qcb^2 + [2]q^{\frac{1}{2}} abd$	$\sqrt{[3]}\; q^{\frac{1}{2}} b^2 d$
$m'=-\frac{1}{2}$	$\sqrt{[3]}\; q^{\frac{1}{2}} ac^2$	$qc^2 b + [2]q^{\frac{1}{2}} acd$	$ad^2 + [2]q\; cbd$	$\sqrt{[3]}\; q^{\frac{1}{2}} bd^2$
$m'=-\frac{3}{2}$	c^3	$\sqrt{[3]}\; q^{\frac{1}{2}} c^2 d$	$\sqrt{[3]}\; q^{\frac{1}{2}} cd^2$	d^3

Table 5.9: The spin-$\frac{3}{2}$ quantum rotation matrix ${}^q\mathfrak{D}^{\frac{3}{2}}_{m',m}(T)$

$$\begin{aligned} {}^q\mathfrak{D}^j_{j,m}(T) &= \sqrt{\frac{[2j]!}{[j+m]![j-m]!}}\; q^{\frac{1}{4}(j+m)(j-m)} a^{j+m} b^{j-m} \\ {}^q\mathfrak{D}^j_{-j,m}(T) &= \sqrt{\frac{[2j]!}{[j+m]![j-m]!}}\; q^{\frac{1}{4}(j+m)(j-m)} c^{j+m} d^{j-m} \end{aligned} \tag{5.10}$$

The special case ${}^q\mathfrak{D}^j_{0,0}(T)$ cannot in general be reduced to a monomial in the entries of T, however it may be simply expressed in terms of a basic hypergeometric function:

$${}^q\mathfrak{D}^j_{0,0}(T) = {}_2\phi_1\left(\begin{matrix} q^{-j} & q^{j+1} \\ & q \end{matrix}\; ;\; q, -q^{\frac{1}{2}} bc\right),$$

which is a q-analogue of the Legendre polynomials. This expression, which is valid for $\det_q(T) = \mathbb{1}$, follows from (5.24) below.

5.2 Generating Function

The explicit expression for ${}^q\mathfrak{D}$ in Definition 5.1 was obtained in §4.7 by a matrix group action which leads to the following generating function for the quantum rotation matrices. Let z_1, z_2 be noncommuting elements satisfying

$$z_1 z_2 = q^{-\frac{1}{2}} z_2 z_1,$$

then the **generating function** for ${}^q\mathfrak{D}(T)$ is given by:

$$q^{\frac{1}{4}(j-m)(j+m)} \frac{(az_1 + cz_2)^{j+m} (bz_1 + dz_2)^{j-m}}{\sqrt{[j-m]![j+m]!}}$$

$$= \sum_{m'} {}^q\mathfrak{D}^j_{m',m}(T)\, q^{\frac{1}{4}(j-m')(j+m')} \frac{(z_1)^{j+m'}(z_2)^{j-m'}}{\sqrt{[j-m']![j+m']!}}. \tag{5.11}$$

This formula could equally well be expressed in terms of one variable $\zeta = z_2 z_1^{-1}$ by using the homogeneity of this expression; effectively one defines the irrep matrices in this way by means of the fractional linear transformations discussed in §4.7.1. Formulas have been given by Nomura [120].

A property which follows from (5.11) is:

Lemma 5.12 *The quantum rotation matrix satisfies:*

$${}^q\mathfrak{D}^j_{m,m'}(T) = (-)^{m'-m} q^{\frac{1}{2}(m'-m)}\, {}^{q^{-1}}\mathfrak{D}^j_{-m',-m}(\gamma(T)),$$

where the antipode $\gamma(T)$ is given by

$$\gamma(T) = \begin{pmatrix} d & -q^{\frac{1}{2}}b \\ -q^{-\frac{1}{2}}c & a \end{pmatrix}.$$

Proof: Replace T by $\gamma(T)$ in (5.11), at the same time exchanging q and q^{-1} (since $\gamma(T) \in \mathfrak{M}_{q^{-1}}(2)$) and also z_1 and z_2. Then put the resulting formula in the form (5.11) by replacing $m, m' \to -m, -m'$ and $z_1 \to -q^{\frac{1}{2}} z_1$ (which preserves the relation $z_1 z_2 = q^{-\frac{1}{2}} z_2 z_1$), to establish the result. □

Next, we define the **conjugate quantum rotation matrix** ${}^q\mathfrak{D}^*$ using the $*$ anti-involution:

$${}^q\mathfrak{D}^{*j}_{m',m}(T) \stackrel{\text{def}}{=} \left({}^q\mathfrak{D}^j(T)^*\right)_{m',m} = \left({}^q\mathfrak{D}^j_{m,m'}(T)\right)^*. \tag{5.13}$$

Observe that we transpose the matrix in applying the $*$-operation, for example ${}^q\mathfrak{D}^{*\frac{1}{2}}(T) = {}^q\mathfrak{D}^{\frac{1}{2}}(T)^* = T^*$ where

$$T = \begin{pmatrix} a & b \\ c & d \end{pmatrix}, \qquad T^* = \begin{pmatrix} a^* & c^* \\ b^* & d^* \end{pmatrix}.$$

Lemma 5.14 *The conjugate quantum rotation matrix satisfies*

$${}^q\mathfrak{D}^{*j}_{m',m}(T) \;=\; {}^{q^{-1}}\mathfrak{D}^j_{m',m}(T^*).$$

Proof: The $*$-operation applied to the generating function (5.11) gives

$$q^{-\frac{1}{4}(j-m)(j+m)} \frac{(a^* z_1^* + c^* z_2^*)^{j+m}(b^* z_1^* + d^* z_2^*)^{j-m}}{\sqrt{[j-m]![j+m]!}}$$

$$= \sum_{m'} \left({}^q\mathfrak{D}^j_{m',m}(T)\right)^* q^{-\frac{1}{4}(j-m')(j+m')} \frac{(z_1^*)^{j+m'}(z_2^*)^{j-m'}}{\sqrt{[j-m']![j+m']!}},$$

where we have q-commuted the factors involving z_1^*, z_2^* (remembering that the action of T preserves the q-commutators), and where z_1^*, z_2^* also q-commute but with $q \to q^{-1}$. By comparison with (5.11) we deduce

$$\left({}^q\mathfrak{D}^j_{m',m}(T)\right)^* = {}^{q^{-1}}\mathfrak{D}^j_{m',m}(\tilde{T}^*) = {}^{q^{-1}}\mathfrak{D}^j_{m,m'}(T^*),$$

as required (upon exchanging m, m'). □

EXAMPLE **5.15** From Table 5.9 we find:

$${}^q\mathfrak{D}^{\frac{3}{2}}_{-\frac{1}{2},\frac{1}{2}}(T) = qc^2b + [2]q^{\frac{1}{2}}acd = q^{-1}c^2b + [2]q^{-\frac{1}{2}}dca = -q^{\frac{1}{2}}\,{}^{q^{-1}}\mathfrak{D}^{\frac{3}{2}}_{-\frac{1}{2},\frac{1}{2}}(\gamma(T)),$$

which is an example of Lemma 5.12. As an example of Lemma 5.14 consider the element ${}^q\mathfrak{D}^{\frac{3}{2}}_{\frac{1}{2},-\frac{1}{2}}(T)$ in Table 5.9, then

$$\begin{aligned}{}^q\mathfrak{D}^{\frac{3}{2}}_{\frac{1}{2},-\frac{1}{2}}(T) &= qcb^2 + [2]q^{\frac{1}{2}}abd = q^{-1}b^2c + [2]q^{-\frac{1}{2}}dba \\ &= \left(q^{-1}(b^*)^2c^* + [2]q^{-\frac{1}{2}}a^*b^*d^*\right)^* = \left({}^{q^{-1}}\mathfrak{D}^{\frac{3}{2}}_{-\frac{1}{2},\frac{1}{2}}(T^*)\right)^*.\end{aligned}$$

5.2.1 Symmetries of the Quantum Rotation Matrix

For $q = 1$ the finite group of transformations of the $\mathfrak{D}$-matrices are well known and are discussed in [66, §5.9]. Let us now generalize these results to the quantum case. We have already noted the following symmetries for arguments which are elements of $\mathfrak{M}_q(2)$ or $\mathfrak{M}_{q^{-1}}(2)$:

$$\begin{aligned}{}^q\mathfrak{D}^j_{m,m'}\begin{pmatrix} a & b \\ c & d\end{pmatrix} &= {}^q\mathfrak{D}^j_{m',m}\begin{pmatrix} a & c \\ b & d\end{pmatrix}, \\ {}^q\mathfrak{D}^j_{m,m'}\begin{pmatrix} a & b \\ c & d\end{pmatrix} &= (-)^{m'-m}q^{\frac{1}{2}(m'-m)}\,{}^{q^{-1}}\mathfrak{D}^j_{-m',-m}\begin{pmatrix} d & -q^{\frac{1}{2}}b \\ -q^{-\frac{1}{2}}c & a\end{pmatrix}.\end{aligned} \tag{5.16}$$

The first of these relations is the transposition symmetry, and the second relates the ${}^q\mathfrak{D}$-matrices of T and its antipode $\gamma(T)$, which we may rewrite in the form:

$${}^q\mathfrak{D}^j_{m,m'}\begin{pmatrix} a & b \\ c & d\end{pmatrix} = {}^{q^{-1}}\mathfrak{D}^j_{-m,-m'}\begin{pmatrix} d & c \\ b & a\end{pmatrix}. \tag{5.17}$$

We now extend the domain of ${}^q\mathfrak{D}$ so that it is defined on the four arguments (t^j_i) as given in the Definition 5.1 but which need not be elements of $\mathfrak{M}_q(2)$. We still reserve, however, the notation a, b, c, d for elements satisfying (5.2). We find directly from the definition ([163]):

$${}^q\mathfrak{D}^j_{m,m'}\begin{pmatrix} a & b \\ c & d\end{pmatrix} = q^{\frac{1}{2}(j-m)(j+m)}\,{}^{q^{-1}}\mathfrak{D}^j_{m,-m'}\begin{pmatrix} b & a \\ d & c\end{pmatrix},$$

which is a symmetry under column exchange. Next, we rewrite this equation by replacing $q \to q^{-1}$ and exchanging a and d and also b and c (which leaves invariant the defining relations (5.2)), and then combine the result with (5.17) to obtain:

$$ {}^{q}\mathfrak{D}^{j}_{m,m'}\begin{pmatrix} a & b \\ c & d \end{pmatrix} = q^{-\frac{1}{2}(j-m)(j+m)}\ {}^{q}\mathfrak{D}^{j}_{-m,m'}\begin{pmatrix} c & d \\ a & b \end{pmatrix}, $$

which is a symmetry of row interchange.

Hence, we have generalized the three symmetries of transposition and row and column exchange discussed for $q = 1$ in [66]. In addition, we have an operation of conjugation for which ${}^{q}\mathfrak{D}(T) \to {}^{q}\mathfrak{D}(\gamma(T))$. Then Lemma 5.12 expresses the symmetry of the ${}^{q}\mathfrak{D}$-matrix under this operation.

In the case of $SU_q(2)$ we can also include the symmetry of the $*$-operation, which is related to conjugation:

$$ {}^{q}\mathfrak{D}^{*j}_{m,m'}\begin{pmatrix} a & b \\ -q^{\frac{1}{2}}b^{*} & a^{*} \end{pmatrix} = {}^{q^{-1}}\mathfrak{D}^{j}_{m,m'}\begin{pmatrix} a^{*} & -q^{\frac{1}{2}}b \\ b^{*} & a \end{pmatrix} = (-)^{m-m'} q^{\frac{1}{2}(m-m')}\, {}^{q}\mathfrak{D}^{j}_{-m',-m}\begin{pmatrix} c & b \\ -q^{\frac{1}{2}}b^{*} & a^{*} \end{pmatrix}, $$

where the right hand side follows from Lemma 5.12.

5.3 Tensor Operator Properties of the Quantum Rotation Matrices

As already noted, the rows and columns of ${}^{q}\mathfrak{D}(T)$ form tensor operators with respect to the right and left quantum groups in the factor algebra $\mathcal{U}_q\,(\mathfrak{su}(2)*\mathfrak{su}(2))$. From the group multiplication law we deduce the equivariance of tensor operators (as stated in Corollary 4.51):

Tensor operators $\mathbf{T}_{jm}$ *in* $SU_q(2)$ *satisfy equivariance, that is, under the action of* $SU_q(2)$:

$$ \mathbf{T}_{jm} \longrightarrow \mathbf{T}'_{jm} = \sum_{m'} \mathbf{T}_{jm'}\ {}^{q}\mathfrak{D}^{j}_{m'm}(T), \quad T \in SU_q(2). \tag{5.18} $$

The group multiplication law, which is the homomorphism of the co-multiplication property in the matrix quantum group, is therefore also a statement that the rows and columns of ${}^{q}\mathfrak{D}$ form tensor operators.

We can extend equivariance to the conjugate tensor operator, defined in (3.32), p. 88, using the conjugate rotation matrix ${}^{q}\mathfrak{D}^{*}$ defined in (5.13). Consider therefore $SU_q(2)$, for which $T^{*}T = I = TT^{*}$, then:

LEMMA **5.19** *The conjugate quantum rotation matrix satisfies*

$$ {}^{q}\mathfrak{D}^{*j}_{m,m'}(T) = (-)^{m-m'} q^{\frac{1}{2}(m-m')}\, {}^{q}\mathfrak{D}^{j}_{-m',-m}(T), \quad T \in SU_q(2). \tag{5.20} $$

The proof uses the definition (5.3), by taking the $*$-conjugate and relabelling the summation index.

An example of this property is:

$$\left({}^q\mathfrak{D}^{\frac{3}{2}}_{\frac{1}{2},-\frac{1}{2}}(T)\right)^* = -q^{\frac{3}{2}}b(b^*)^2 + [2]q^{\frac{1}{2}}ab^*a^* = -q^{-\frac{1}{2}}\ {}^q\mathfrak{D}^{\frac{3}{2}}_{-\frac{1}{2},\frac{1}{2}}(T).$$

By comparison with the definition of conjugate tensor operator in (3.32) we deduce that the columns of ${}^q\mathfrak{D}^*$ are conjugate tensor operators under $SU_q(2)$, and hence:

LEMMA **5.21** *Conjugate tensor operators* $\overline{\mathbf{T}}_{jm}$ *in* $SU_q(2)$ *satisfy equivariance, that is, under the action of* $SU_q(2)$:

$$\overline{\mathbf{T}}_{jm} \longrightarrow \overline{\mathbf{T}}'_{jm} = \sum_{m'} \overline{\mathbf{T}}_{jm'}\ {}^q\mathfrak{D}^{*j}_{m,m'}(T), \quad T \in SU_q(2).$$

The proof follows directly from (5.18) and (5.20).

5.4 The Wigner Product Law

We investigate now further consequences of the Wigner product law, which reads:

$$\delta^{j_3 j_3'}\ {}^q\mathfrak{D}^{j_3}_{m_3,m_3'}(T) = \sum_{m_1,m_1',m_2,m_2'} {}_qC^{j_1\ j_2\ j_3}_{m_1\ m_2\ m_3}\ {}_qC^{j_1\ j_2\ j_3'}_{m_1'\ m_2'\ m_3'}\ {}^q\mathfrak{D}^{j_1}_{m_1,m_1'}(T)\ {}^q\mathfrak{D}^{j_2}_{m_2,m_2'}(T). \tag{5.22}$$

This law expresses the tensor operator coupling of two ${}^q\mathfrak{D}$-matrices to form a third ${}^q\mathfrak{D}$-matrix, and may also be viewed as the decomposition of the co-multiplied irrep matrices into irreducible constituents and so is a consequence of co-multiplication in the quantum universal enveloping algebra $\mathcal{U}_q(\mathfrak{su}(2))$. By means of orthogonality of the q-WCG coefficients (3.54) we can write the Wigner product in the following useful forms:

$$\sum_{m_1,m_2} {}_qC^{j_1\ j_2\ j_3}_{m_1\ m_2\ m_3}\ {}^q\mathfrak{D}^{j_1}_{m_1,m_1'}(T)\ {}^q\mathfrak{D}^{j_2}_{m_2,m_2'}(T) = \sum_{m_3'} {}_qC^{j_1\ j_2\ j_3}_{m_1'\ m_2'\ m_3'}\ {}^q\mathfrak{D}^{j_3}_{m_3,m_3'}(T) \tag{5.23a}$$

$$\sum_{m_1',m_2'} {}_qC^{j_1\ j_2\ j_3}_{m_1'\ m_2'\ m_3'}\ {}^q\mathfrak{D}^{j_1}_{m_1,m_1'}(T)\ {}^q\mathfrak{D}^{j_2}_{m_2,m_2'}(T) = \sum_{m_3} {}_qC^{j_1\ j_2\ j_3}_{m_1\ m_2\ m_3}\ {}^q\mathfrak{D}^{j_3}_{m_3,m_3'}(T) \tag{5.23b}$$

$${}^q\mathfrak{D}^{j_1}_{m_1,m_1'}(T)\ {}^q\mathfrak{D}^{j_2}_{m_2,m_2'}(T) = \sum_{j_3,m_3,m_3'} {}_qC^{j_1\ j_2\ j_3}_{m_1\ m_2\ m_3}\ {}^q\mathfrak{D}^{j_3}_{m_3,m_3'}(T)\ {}_qC^{j_1\ j_2\ j_3}_{m_1'\ m_2'\ m_3'}. \tag{5.23c}$$

By putting $m_3 = 0, j_3 = 0$ and $m_1' + m_2' = 0$ in (5.23a) together with the value of the q-WCG coefficient ${}_qC^{j_1\ j_2\ 0}_{m_1\ m_2\ 0}$ given in (3.27), p. 87, we obtain the following orthogonality relation satisfied by the ${}^q\mathfrak{D}$-matrices:

$$\sum_m (-)^{m_1-m} q^{\frac{1}{2}(m_1-m)}\ {}^q\mathfrak{D}^j_{-m,-m_1}(T)\ {}^q\mathfrak{D}^j_{m,m_2}(T) = \delta_{m_1,m_2}.$$

For $T \in SU_q(2)$ we can use (5.20) to obtain

$$\sum_m {}^q\mathfrak{D}^{*j}_{m,m_1}(T)\ {}^q\mathfrak{D}^j_{m,m_2}(T) = \delta_{m_1,m_2},$$

which of course is the irrep form of $T^*T = I$. Similarly, there is another orthogonality relation corresponding to $TT^* = I$ obtained by putting $m'_3 = 0, j_3 = 0, m_1 + m_2 = 0$ in (5.23b).

The form (5.23c) is useful for deriving recurrence relations for the ${}^q\mathfrak{D}$-matrices which are simpler than (5.7). These relations are obtained by putting $j_2 = \frac{1}{2}$ in (5.23c) for which the sum on the right hand side reduces to two terms (on putting $m_3 = m_1 + m_2$ and $m'_3 = m'_1 + m'_2$) in which $j_3 = j_1 \pm \frac{1}{2}$. There are four such recurrence relations, corresponding to the four possible values of m_2, m'_2. As an example, let us take $m_2 = \frac{1}{2} = m'_2$, then:

$$\begin{aligned}[2j+1]\ {}^q\mathfrak{D}^j_{m,m'}(T)\ a &= q^{\frac{1}{4}(2j-m-m')}\sqrt{[j+m+1][j+m'+1]}\ {}^q\mathfrak{D}^{j+\frac{1}{2}}_{m+\frac{1}{2},m'+\frac{1}{2}}(T) \\ &\quad + q^{-\frac{1}{4}(2j+m+m'+2)}\sqrt{[j-m][j-m']}\ {}^q\mathfrak{D}^{j-\frac{1}{2}}_{m+\frac{1}{2},m'+\frac{1}{2}}(T),\end{aligned}$$

where we substituted for the q-WCG coefficients from (3.43). There are another four similar recurrence relations obtained by putting $j_1 = \frac{1}{2}$ in (5.23c) and we refer to Groza et al. [60] for details.

These recurrence relations can be used to express the ${}^q\mathfrak{D}$-matrices in terms of basic hypergeometric functions, as shown in [60]. It is not surprising that such expressions should exist, indeed we found in §4.4.1 that q-boson basis polynomials were expressible in terms of basic hypergeometric functions, and that these polynomials can be determined from the ${}^q\mathfrak{D}$-matrices acting on the vacuum state when the non-commuting elements of $T \in \mathfrak{M}_q(2)$ are realized in terms of q-boson operators. On the other hand, the explicit expression for ${}^q\mathfrak{D}(T)$ given in Definition 5.1 does not appear to be readily expressible as a basic hypergeometric function because of the noncommuting properties of the entries a, b, c, d. Nevertheless the recurrence relations for ${}^q\mathfrak{D}$ can be used to find the required expression; as an example consider the case $m \geqslant |m'|$, then:

$$\begin{aligned}{}^q\mathfrak{D}^j_{m,m'}(T) &= \left(\frac{[j+m]![j-m']!}{[j-m]![j+m']!}\right)^{\frac{1}{2}} q^{\frac{1}{4}(m+m')(m-m')}\frac{a^{m+m'}b^{m-m'}}{[m-m']!} \\ &\quad \times {}_2\phi_1\left(\begin{matrix} q^{-j+m} & q^{j+m+1} \\ & q^{m-m'+1}\end{matrix}\ ;\ q, -q^{\frac{1}{2}}bc\right). \end{aligned} \tag{5.24}$$

There are four such cases, each of which reduces to one of the special cases (5.10) at the extreme value of m or m'. The derivation of these expressions, in which the q-determinant is set equal to unity, uses relations between contiguous basic hypergeometric functions [60]. Similar expressions apply to $U_q(2)$ when the q-determinant is included as shown in (5.4).

The Wigner product law can be used to map the RTT relations, $RT_1T_2 = T_2T_1R$, into relations amongst the ${}^q\mathfrak{D}$-matrices. In order to demonstrate this we first require the following expression for the matrix elements of the R-matrix given in (2.20), p. 24:

$$\begin{aligned}\left(\mathcal{R}^{j_1j_2}\right)^{m_1,m_2}_{m_1',m_2'} &= \sum_{j,m} q^{-\frac{1}{2}\left(j_1(j_1+1)+j_2(j_2+1)-j(j+1)\right)}\ {}_qC^{j_1\ j_2\ j}_{m_1\ m_2\ m}\ {}_{q^{-1}}C^{j_1\ j_2\ j}_{m_1'\ m_2'\ m} \\ &= \sum_{j,m}(-)^{j_1+j_2-j} q^{-\frac{1}{2}\left(j_1(j_1+1)+j_2(j_2+1)-j(j+1)\right)}\ {}_qC^{j_1\ j_2\ j}_{m_1\ m_2\ m}\ {}_qC^{j_2\ j_1\ j}_{m_2'\ m_1'\ m},\end{aligned} \tag{5.25}$$

where we used (3.63b) in the last line. This expression, which is zero unless $m_1+m_2 = m_1' + m_2'$ and $m_1' - m_1 \geqslant 0$, was derived by Nomura [76] by first writing the q-$6j$ symbols as a sum over q-WCG coefficients, and then using the asymptotic limit of the q-$6j$ symbols (discussed in §3.6.1) to write a q-WCG coefficient as a sum over q-WCG coefficients, as we noted previously in (3.66), p. 101. Then the orthogonality properties of q-WCG coefficients lead to the formula (5.25). Now we may state [163]:

LEMMA **5.26** *The quantum rotation matrices* ${}^q\mathfrak{D}(T)$ *satisfy*

$$\sum_{m_1,m_2} \left(\mathcal{R}^{j_1j_2}\right)^{m_1,m_2}_{n_1,n_2}\ {}^q\mathfrak{D}^{j_1}_{m_1,m_1'}\ {}^q\mathfrak{D}^{j_2}_{m_2,m_2'} = \sum_{n_1',n_2'} {}^q\mathfrak{D}^{j_2}_{n_2,n_2'}\ {}^q\mathfrak{D}^{j_1}_{n_1,n_1'}\ \left(\mathcal{R}^{j_1j_2}\right)^{m_1',m_2'}_{n_1',n_2'}.$$

PROOF: Substitute on each side with the R-matrix in the form shown in (5.25) and then simplify by using (5.23a) on the left and (5.23b) on the right; each side is then seen to equal

$$\sum_{j,m,m_3} (-)^{j_1+j_2-j} q^{-\frac{1}{2}\left(j_1(j_1+1)+j_2(j_2+1)-j(j+1)\right)}\ {}_qC^{j_2\ j_1\ j}_{n_2\ n_1\ m}\ {}^q\mathfrak{D}^{j}_{m,m_3}(T)\ {}_qC^{j_1\ j_2\ j}_{m_1'\ m_2'\ m_3}.$$

□

As a final remark, although we have used the Wigner product law to define the ${}^q\mathfrak{D}$-matrices recursively by means of the recurrence relation (5.7) we could equally well derive the explicit form of the q-WCG coefficients from the ${}^q\mathfrak{D}$-matrices by substituting $j_3 = m_3$ into (5.23a) and using the following boundary value of the q-WCG coefficients:

$$\begin{aligned}{}_qC^{j_1\ j_2\ j}_{m_1\ m_2\ j} &= \delta_{j,m_1+m_2}(-)^{j_1-m_1} q^{\frac{1}{4}\left(-j_1(j_1+1)+j_2(j_2+1)-j(j+1)\right)+\frac{1}{2}m_1(j+1)} \\ &\quad\times\left(\frac{[j_1+m_1]![j_2+m_2]![j_1+j_2-j]![2j+1]!}{[j_1-m_1]![j_2-m_2]![j_1-j_2+j]![-j_1+j_2+j]![j_1+j_2+j+1]!}\right)^{\frac{1}{2}}.\end{aligned} \tag{5.27}$$

The calculation is simplified by taking the special case of the ${}^q\mathfrak{D}$-matrices in which either b or c is replaced by zero, as is consistent with the relations (5.2), and for which a and d now commute. Furthermore, the quantum rotation matrices reduce to monomials in the remaining elements of $\mathcal{A}$ which leads to an expression for the q-WCG coefficients in (5.23a) as a sum over a single index. This expression is in fact equivalent to the Racah form.

Chapter 6

Quantum Groups at Roots of Unity

The discussion of quantum groups in previous chapters has been based on the assumption that q is real and positive. If we permit q to take arbitrary complex values then the q-deformed universal enveloping algebra becomes complex, with non-unitary representations of little interest to physics. However, the special case where q is a complex root of unity is different. There can indeed be unitary representations and moreover this case appears in physical applications, such as in duality properties of conformal field theory (see for example Alvarez-Gaumé et al. [30, 75, 167]) in which q appears in the $SU(2)$ Witten-Wess-Zumino (WZW) model with the value $q^{\frac{1}{2}} = e^{\frac{i\pi}{k+2}}$, where the integer k is the level of the corresponding Kac-Moody algebra. Another example is the chiral Potts model (Date et al. [26]) and its generalizations (Bazhanov et al. [27]) in which the cyclic irreps of $\mathcal{U}_q(\mathfrak{sl}(2))$ play a significant role. Indeed, roots of unity appeared at an early stage in the development of quantum groups in the work of Sklyanin [168] in 1983 (in the trigonometric case), in the quantum group $\mathcal{U}_q(\mathfrak{su}(2))$ with $q = e^{2i\eta}$, in Sklyanin's notation.

As noted previously, for real positive q the representation theory of the quantum group is essentially the same as that of the corresponding Lie algebra, but at roots of unity new types of representations appear which have no classical analog. These are of two types, denoted $\mathcal{A}$ and $\mathcal{B}$, the first being deformations of the classical representations in which the raising/lowering generators are nilpotent, and the second being cyclic representations in which there are no states of highest or lowest weight, and in which the raising/lowering operators act injectively. In this chapter we discuss both types of irreps for $\mathcal{U}_q(\mathfrak{su}(2))$, and show how they may be constructed using the q-boson calculus.

A difficulty which arises at roots of unity is that since $\overline{q} = q^{-1}$, co-multiplication does not preserve the Hermiticity properties of the generators, and under the $\dagger$-operation co-multiplication Δ becomes the permuted co-multiplication $\overline{\Delta}$. Therefore, representations formed by co-multiplied unitary irreps are no longer unitary, and hence we investigate more generally finite-dimensional irreps of $\mathcal{U}_q(\mathfrak{sl}(2))$ which are not unitary. Nevertheless, we consider in detail the restrictions on the invariant labels required for unitary irreps of $\mathcal{U}_q(\mathfrak{su}(2))$. A further difficulty is that representations need not be completely reducible, that is, they need not be equivalent to the direct sum of irreps, but can contain indecomposable representations. (An **indecomposable representation** on a linear space $\mathfrak{V}$ is not irreducible and $\mathfrak{V}$ is not a direct sum of proper subspaces, each invariant under the quantum group.) Such representations appear in the tensor product of irreps, which we discuss briefly. One approach to this problem is to project from the tensor product only those representations which are fully reducible.

Following a discussion of properties of q-numbers and the construction of invariants of $\mathcal{U}_q(\mathfrak{sl}(2))$ at roots of unity in §6.1, we consider nilpotent and then cyclic representations. This is followed by an investigation of unitary irreps of $\mathcal{U}_q(\mathfrak{su}(2))$, and the consequent restrictions on the invariant labels. Then we determine representations of q-boson operators, not necessarily in a Fock space, and show in §6.3 how to construct all cyclic irreps of $\mathcal{U}_q(\mathfrak{sl}(2))$ using the q-boson calculus; in this way we determine all finite-dimensional irreps of $\mathcal{U}_q(\mathfrak{sl}(2))$, as classified by De Concini and Kac [169]. All representations of q-boson operators at roots of unity are finite dimensional and are either cyclic or nilpotent, which contrasts with the situation for generic q and the case $q = 1$, for which the boson operators are unbounded and representations are necessarily infinite dimensional. We investigate in §6.5 the extension of the Fock space so as to allow noninteger quanta and in which cyclic q-boson operators act. We conclude with some remarks in §6.7 on the application of the method of algebraic induction at roots of unity, a topic first introduced in Chapter 7 and which should therefore be perused before §6.7.

The new features we describe in this chapter are consequences of q being a root of unity and are not confined to $\mathcal{U}_q(\mathfrak{sl}(2))$, but are general to quantum groups.

6.1 The Special Linear Quantum Group for q a Root of Unity

We consider $\mathcal{U}_q(\mathfrak{sl}(2))$ and construct finite-dimensional irreps in a linear vector space without assuming the existence of an inner product. Let us now suppose[1]

$$q^{\frac{p}{2}} = 1, \tag{6.1}$$

where p is an integer with $p > 2$. We may choose $q^{\frac{1}{2}}$ to be a primitive root of unity:

$$q^{\frac{1}{2}} = \exp\left(\frac{2\pi i}{p}\right). \tag{6.2}$$

[1] With the convention used by several authors which chooses q in place of our $q^{\frac{1}{2}}$ one has $q^p = 1$.

We now encounter not just q-integers but q-numbers $[x]$ defined by[2]

$$[x] = \frac{q^{\frac{x}{2}} - q^{-\frac{x}{2}}}{q^{\frac{1}{2}} - q^{-\frac{1}{2}}},$$

which is real for complex q such that $|q| = 1$; suppose $q = e^{2i\phi}$, then $[x] = \frac{\sin \phi x}{\sin \phi}$. In particular, for q given by (6.2) we have

$$[x] = \frac{\sin\left(\frac{2\pi x}{p}\right)}{\sin\left(\frac{2\pi}{p}\right)}. \tag{6.3}$$

The q-number $[x]$ is periodic with period p:

$$[x+p] = q^{\frac{p}{2}}[x] + q^{-\frac{x}{2}}[p] = [x].$$

It is useful to note that

$$\left[x + \frac{p}{2}\right] = q^{\frac{p}{4}}[x] + q^{-\frac{x}{2}}\left[\frac{p}{2}\right] = q^{\frac{p}{4}}[x] = -[x],$$

where we used $q^{\frac{p}{4}} = -1$ in the last step, and so

$$\left[x + \frac{p}{2}\right]\left[y + \frac{p}{2}\right] = [x][y]. \tag{6.4}$$

It is helpful to keep in mind the graph of $[x]$, which is displayed in Figure 6.6; we note in particular that

$$\begin{aligned} &[x] > 0 \quad \text{for} \quad 0 < x < \tfrac{p}{2} \\ &[x] < 0 \quad \text{for} \quad \tfrac{p}{2} < x < p, \end{aligned} \tag{6.5}$$

with periodic extensions to all values of x. Evidently the q-numbers $[x]$ do not preserve the ordering of the real numbers, in contrast to the case $q \in \mathbb{R}^+$.

It follows from (6.3) that $[p] = 0$ and if p is even, that is $p = 2p'$ where $p' \in \mathbb{N}$, then $[p'] = 0$. In fact $[\frac{p}{2}] = 0$ for all $p \in \mathbb{Z}$, but the case where $[n]$ has an integer argument n is significant, since the dimension of cyclic irreps will be p for odd p and $p' = \frac{p}{2}$ for even p. It is convenient to define

$$p' \stackrel{\text{def}}{=} \begin{cases} p & \text{if } p \text{ is odd;} \\ \dfrac{p}{2} & \text{if } p \text{ is even,} \end{cases} \tag{6.7}$$

then we always have $q^{p'} = 1$, and $[p'] = 0$. The dimension of cyclic irreps will be p'.

---

[2]The affix q is suppressed.

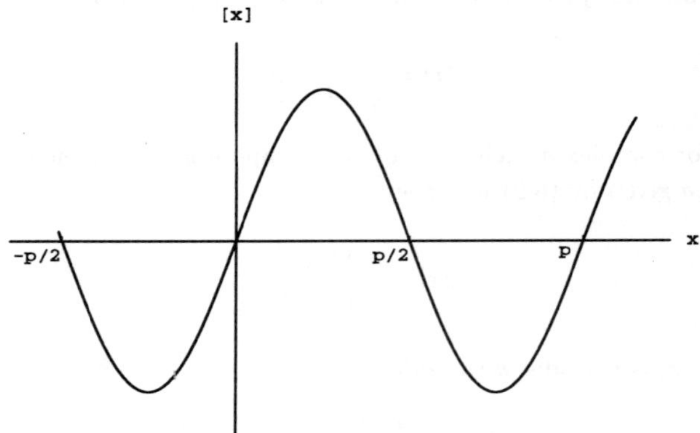

Figure 6.6: The graph of the q-number $[x]$ as a function of x when $q^{\frac{p}{2}} = 1$.

Let us also point out several properties of q-commuting elements at roots of unity. If $xy = q^{-\frac{1}{2}}yx$ then

$$xy^n = q^{-\frac{n}{2}}y^n x$$

and so y^p commutes with x, and similarly x^p commutes with y; also if p is even $y^{\frac{p}{2}}$ anticommutes with x, and similarly for $x^{\frac{p}{2}}$.

If we have elements a, b satisfying $ab = q^{-1}ba$, as occurs in applications of the q-binomial theorem 2.102 (p. 60) then using

$$(a+b)^n = a^n + [n]q^{\frac{1}{2}(n-1)}a^{n-1}b + \frac{[n][n-1]}{[2]}q^{n-2}a^{n-2}b^2 + \ldots + b^n,$$

we see that by putting $n = p'$:

$$(a+b)^{p'} = a^{p'} + b^{p'} \tag{6.8}$$

where p' is given by (6.7). It is sometimes useful to consider the indeterminate ratio of q-integers $[np']/[p']$ for integers n, and this is given by the limit:

$$\frac{[np']}{[p']} \stackrel{\text{def}}{=} \lim_{s\to p'} \frac{q^{\frac{1}{2}ns} - q^{-\frac{1}{2}ns}}{q^{\frac{s}{2}} - q^{-\frac{s}{2}}} = n\,. \tag{6.9}$$

More generally, we can assert that for all $x \in \mathbb{R}$ as $s \to p'$,

$$\frac{[xs]_q}{[s]_q} = [x]_{q^s} \longrightarrow x.$$

6.1.1 Invariants of $\mathcal{U}_q(\mathfrak{sl}(2))$ at Roots of Unity

For q a root of unity it is more appropriate to express the defining relations (2.1a, 2.1b) of $\mathcal{U}_q(\mathfrak{sl}(2))$ in terms of

$$K \stackrel{\text{def}}{=} q^{J_z}.$$

(This is because in contrast to the case $q \in \mathbb{R}^+$ where we can always recover J_z from $K = q^{J_z}$, J_z is not uniquely defined by K at roots of unity.) Hence we write (2.1a, 2.1b) in the form:

$$[J_+, J_-] = \frac{K - K^{-1}}{q^{\frac{1}{2}} - q^{-\frac{1}{2}}}, \quad KJ_\pm K^{-1} = q^{\pm 1} J_\pm. \tag{6.10}$$

The Casimir invariant is defined by:

$$C \stackrel{\text{def}}{=} J_- J_+ + \frac{Kq^{\frac{1}{2}} + K^{-1}q^{-\frac{1}{2}}}{(q^{\frac{1}{2}} - q^{-\frac{1}{2}})^2}$$

and commutes with K and $J_\pm$ (but differs by a constant from the definition (2.11)).

At roots of unity C is no longer the only invariant of $\mathcal{U}_q(\mathfrak{sl}(2))$. From the commutation relations (6.10) we find by induction on n:

$$[J_+, (J_-)^n] = [n]\frac{Kq^{\frac{1}{2}(n-1)} - K^{-1}q^{-\frac{1}{2}(n-1)}}{q^{\frac{1}{2}} - q^{-\frac{1}{2}}}(J_-)^{n-1}, \quad n \in \mathbb{N}, \tag{6.11}$$

and also

$$K(J_-)^n = q^{-n}(J_-)^n K, \quad n \in \mathbb{N}. \tag{6.12}$$

Now we put $n = p'$, where p' is defined by (6.7), into (6.11, 6.12) and using $[p] = 0 = [p']$ we find $[J_+, (J_-)^{p'}] = 0 = [K, (J_-)^{p'}]$. Hence $(J_-)^{p'}$ is an invariant, and similarly for $(J_+)^{p'}$. It also follows from (6.10) by induction on n that

$$K^n J_- = q^{-n} J_- K^n.$$

Again we put $n = p'$ and find that $K^{p'}$ is an invariant. We deduce:

Lemma 6.13 *The invariants of $\mathcal{U}_q(\mathfrak{sl}(2))$ at roots of unity are:*

$$C, \quad X = (J_+)^{p'}, \quad Y = (J_-)^{p'}, \quad Z = (K)^{p'}.$$

These four invariants are dependent since by induction we may prove:

$$(J_-)^n(J_+)^n = \prod_{i=1}^{n}\left(C - \frac{q^{\frac{1}{2}(2i-1)}K + q^{-\frac{1}{2}(2i-1)}K^{-1}}{\left(q^{\frac{1}{2}} - q^{-\frac{1}{2}}\right)^2}\right) \tag{6.14}$$

and by putting $n = p'$ we obtain a polynomial relation (discussed in [169, 170]) amongst X, Y, Z, C of the form

$$P(C) = (J_+)^{p'}(J_-)^{p'} + q^{\frac{p'}{2}} \frac{K^{p'} + K^{-p'}}{\left(q^{\frac{1}{2}} - q^{-\frac{1}{2}}\right)^{2p'}}, \tag{6.15}$$

where P is a polynomial of degree p'. This relation can be expressed explicitly in terms of Chebyshev polynomials, see Arnaudon [170], and [171] for the generalization to $\mathcal{U}_q(\mathfrak{sl}(n))$. Hence, irreps of $\mathcal{U}_q(\mathfrak{sl}(2))$ have three independent labels which we may take to be the eigenvalues x, y, z of X, Y, Z, with the Casimir invariant eigenvalue c taking one of the p' values allowed by (6.15).

The eigenvalues x, y, z may in general be complex, but we can rescale x, y by means of the following transformation, which leaves the defining relations (6.10) invariant:

$$J_- \to \omega J_-, \quad J_+ \to \omega^{-1} J_+, \quad \omega \in \mathbb{C}^\times. \tag{6.16}$$

In order to construct unitary representations of $\mathcal{U}_q(\mathfrak{su}(2))$ we assume that there exists an inner product, denoted (,), in the linear vector space $\mathfrak{V}$ carrying the irreducible representation with respect to which $(J_-)^\dagger = J_+$, $K^\dagger = K^{-1}$. In this case the eigenvalues x, y, z satisfy[3] $\overline{x} = y$ and $\overline{z} = z^{-1}$, as follows from $Z^\dagger = Z^{-1}$ and $X^\dagger = Y$, since

$$(Zv,\, v) = \overline{z}(v,\, v) = (v,\, Z^{-1}v) = z^{-1}(v,\, v)$$

for all $v \in \mathfrak{V}$, and similarly $\overline{x} = y$. In addition, we may assume that for unitary irreps both x and y are real, since we can transform any phase to zero by means of

$$J_- \to e^{i\psi} J_-, \quad J_+ \to e^{-i\psi} J_+, \tag{6.17}$$

which is a special case of (6.16), and which preserves the condition $(J_-)^\dagger = J_+$. Hence, in a unitary representation of $\mathcal{U}_q(\mathfrak{su}(2))$ we may choose phases such that the eigenvalues of $(J_-)^{p'}$ and $(J_+)^{p'}$ are equal and real. Unitary representations are therefore labelled by *two* independent invariant operators, which we can take to be one of $(J_-)^{p'}$ or $(J_+)^{p'}$, and C, and therefore by two real numbers.

In summary, the irreducible representations of $\mathcal{U}_q(\mathfrak{sl}(2))$ fall into the following two classes $\mathcal{A}$ and $\mathcal{B}$ respectively:

1. $(J_-)^{p'}$ and $(J_+)^{p'}$ have the zero eigenvalue, $(J_-)^{p'}v = 0 = (J_+)^{p'}v$ for all vectors v in the linear vector space $\mathfrak{V}$ carrying the irreducible representation, that is, J_- and J_+ are **nilpotent**. Consequently there exists a vector v_h of highest weight, and a vector v_l of lowest weight, with $J_-v_l = 0 = J_+v_h$ since otherwise, if J_-v is nonzero for all $v \in \mathfrak{V}$, then $(J_-)^{p'}v \neq 0$ in contradiction to the nilpotent property. Hence, there must exist a vector v_l such that $J_-v_l = 0$, and similarly

[3] Here $\overline{z}$ denotes the complex conjugate of z.

for the state of highest weight. These nilpotent representations have classical analogs, and can be constructed by specializing from general values of q to roots of unity.

2. The second class of representations are those for which in general both $(J_-)^{p'}$ and $(J_+)^{p'}$ have a nonzero eigenvalue, and in this case there can be no vectors of highest or lowest weight. These representations are called **cyclic** since we can generate all elements of $\mathfrak{V}$ by application of J_- to any one vector; hence J_- acts on $\mathfrak{V}$ injectively. For any nonzero vector $v_0 \in \mathfrak{V}$ we define the vectors $v_i = (J_-)^i v_0$ for $i = 1, 2, \ldots, p'-1$ and since $v_{p'} = (J_-)^{p'} v_0$ is proportional to v_0, further application of J_- generates the same set of vectors $v_i, i = 0, \ldots, p'-1$. These vectors span an irreducible representation space, and so the dimension of the irrep is p'. We can also include in this class representations for which only one of $(J_-)^{p'}$ or $(J_+)^{p'}$ has a nonzero eigenvalue and these can be obtained as special cases of cyclic representations, but these do not occur for unitary representations.

6.1.2 *Irreducible Nilpotent Representations of* $\mathcal{U}_q(\mathfrak{su}(2))$

At roots of unity the $(2j+1)$-dimensional representations of $\mathcal{U}_q(\mathfrak{su}(2))$ derived in §2.2 for generic q may be reducible, since for nilpotent irreps the invariants $(J_\pm)^{p'}$ have the zero eigenvalue and so irreps are of dimension p' or less.

For example if we let $p = 3$, for which $[3] = 0$ and $[2] = [2-3] = -1$, the four-dimensional matrices given in Example 2.17 (p. 22) become:

$$K = \begin{pmatrix} 1 & 0 & 0 & 0 \\ 0 & q^{\frac{1}{2}} & 0 & 0 \\ 0 & 0 & q^{-\frac{1}{2}} & 0 \\ 0 & 0 & 0 & 1 \end{pmatrix}, \quad J_+ = \begin{pmatrix} 0 & 0 & 0 & 0 \\ 0 & 0 & -1 & 0 \\ 0 & 0 & 0 & 0 \\ 0 & 0 & 0 & 0 \end{pmatrix}, \quad J_- = \begin{pmatrix} 0 & 0 & 0 & 0 \\ 0 & 0 & 0 & 0 \\ 0 & -1 & 0 & 0 \\ 0 & 0 & 0 & 0 \end{pmatrix},$$

which are evidently the direct sum of two trivial (1-dimensional) representations and the two-dimensional irrep. A difficulty related to the unitary requirement is apparent in the 3-dimensional representation in Example 2.17, for $p = 3$. Upon putting $[2] = -1$ in this matrix representation we see that the condition that J_-J_+ must have non-negative matrix elements, as follows from $(J_+)^\dagger = J_-$, is violated.

Let us construct nilpotent irreps using q-boson operators acting in a Fock space $\mathfrak{F}$. The inner product is given by

$$\langle n \,|\, n' \rangle = \langle 0 | \overline{a}^n a^{n'} | 0 \rangle,$$

and by induction we also determine $\overline{a}^n a^n = ([N+1])_n$, where we have used the shifted factorial notation defined in (2.122). Hence,

$$\langle 0 | \overline{a}^n a^n | 0 \rangle = ([1])_n = [n]!\,, \tag{6.18}$$

for non-negative integers n. By putting $n = p'$ and using $[p'] = 0$, we find that $\langle 0|\overline{a}^{p'}a^{p'}|0\rangle = \|a^{p'}|0\rangle\|^2 = 0$, showing that $a^{p'}|0\rangle = 0$, and hence $a^{p'}|n\rangle$ is the zero vector for any quantum number n. The operators a and $\overline{a}$ are therefore nilpotent in $\mathfrak{F}$ and so the dimension of $\mathfrak{F}$ is finite and equal to p'; no more than $p'-1$ quanta are allowed in any state.

REMARK **6.19** The nilpotent property $(a)^{p'} = 0 = (\overline{a})^{p'}$ is reminiscent of parafermion operators, however there does not appear to be a direct connection with q-boson operators at roots of unity. For one degree of freedom, parafermion creation and annihilation operators $\alpha, \overline{\alpha}$ satisfy

$$[\overline{\alpha}, [\alpha, \overline{\alpha}]] = 2\overline{\alpha},$$

representations of which can be classified by a positive integer P, the order of the parastatistics. The number operator is given by $N = \frac{1}{2}(P + [\alpha, \overline{\alpha}])$. In a Fock space, parafermion operators of order P satisfy $(\alpha)^{P+1} = 0 = (\overline{\alpha})^{P+1}$. Parafermions were introduced by Green [172] in 1953 and properties have been developed by Greenberg and Messiah [173], see also the monograph by Ohnuki and Kamefuchi [174]. We can realize parafermion operators of order P in terms of ordinary boson operators a, and hence in terms of q-boson operators a^q, acting in a Fock space with no more than P quanta by means of the following expressions:

$$\alpha = \sqrt{P - N + 1}\, a = \sqrt{\frac{N(P - N + 1)}{[N]}}\, a^q,$$

together with the Hermitean conjugate relations, where the last equation was obtained using (2.42). The nilpotent property appears as a consequence of repeated factors involving the number operator, as is clear if we calculate the following norm:

$$\|(\alpha)^n|0\rangle\|^2 = \prod_{i=1}^{n} i(P - i + 1).$$

Since the right hand side is zero for $n = P + 1$ we obtain $(\alpha)^{P+1}|0\rangle = 0$ and hence $(\alpha)^{P+1} = 0$ in the Fock space. In terms of q-boson operators this property is valid for generic q.

The definition of the q-boson adjoint operator requires special consideration since at roots of unity the norm of a q-boson state, given in (6.18), is not necessarily positive. A simple example is $p = 3$ for which $a^2|0\rangle$ has a norm of $[2] = -1$. For even p there is no difficulty since the norm of $a^n|0\rangle$ is $[n]!$ which is non-negative for $0 \leqslant n \leqslant \frac{p}{2}$, as is clear from Figure 6.6. For odd p we must redefine the adjoint of the q-boson operator in the way discussed in §6.4 for cyclic irreps, but which applies generally. With the required modification the norm in $\mathfrak{F}$ is again positive definite and the proof that the operator a^p is nilpotent in $\mathfrak{F}$ is valid when p is odd.

We construct the quantum group generators in the usual way as shown in (2.47) and consequently the group generators are nilpotent as follows from the same property of the q-boson operators. Basis states are as given in (2.48), p. 34, where the degree $2j$ is restricted to $2j \leqslant p'-1$. Matrix elements of the generators are given in Theorem 2.16 and satisfy the unitary property for even p. For odd p unitary irreps exist for certain values of j, with the conditions described more generally in §6.2.2 for the cyclic irreps. Considering now representations which may not be unitary, we can avoid square roots which could have negative arguments by normalizing the basis vectors $|jm\rangle$ of Theorem 2.16 (following Alvarez-Gaumé et al. [30] and Pasquier and Saleur [31]) according to

$$|jm\rangle \longrightarrow ([j+m]![j-m]!)^{-\frac{1}{2}}\,|jm\rangle.$$

Then the matrix elements of the generators are given by

$$\begin{aligned} K|jm\rangle &= \omega q^m |jm\rangle \\ J_+|jm\rangle &= \omega[j+m+1]\,|j,m+1\rangle \\ J_-|jm\rangle &= [j-m+1]\,|j,m-1\rangle, \end{aligned} \tag{6.20}$$

where we have included a discrete parameter $\omega = \pm 1$. We cannot necessarily set $\omega = 1$ by a choice of phase since ω is in fact determined by the invariant $z = (\omega)^{p'} = \pm 1$, which is the eigenvalue of $(K)^{p'}$; the sign ω is therefore also a representation label together with j (Lusztig [175]).

Let us consider tensor products of nilpotent representations of $\mathcal{U}_q(\mathfrak{sl}(2))$. Nilpotency is preserved under co-multiplication, essentially due to the q-binomial theorem (6.8) at roots of unity. Specifically, we have

$$\Delta(J_\pm)^n = \sum_{j=0}^{n} \begin{bmatrix} n \\ j \end{bmatrix} K^{-\frac{1}{2}(n-j)}(J_\pm)^j \otimes K^{\frac{1}{2}j}(J_\pm)^{n-j}, \quad n \in \mathbb{N}.$$

For $n = p'$ we see that the first and last terms vanish by nilpotency, and the other terms are zero because $[p'] = 0$. From (6.20) (up to a sign),

$$(J_\pm)^n|jm\rangle = ([j \pm m + 1])_n \,|j, m \pm n\rangle.$$

For $n = p'$ the shifted factorial vanishes since it contains the factor $[p'] = 0$. Denote by $T_\pm$ the operators

$$T_\pm \stackrel{\text{def}}{=} \frac{(J_\pm)^{p'}}{[p']!}, \tag{6.21}$$

then by the limiting procedure shown in (6.9) these operators have well-defined matrix elements.

Taking $p = 3$ as an example, the problem of indecomposable representations arises for the tensor product $\mathfrak{V}_{\frac{1}{2}} \otimes \mathfrak{V}_{\frac{1}{2}} \otimes \mathfrak{V}_{\frac{1}{2}}$ which for generic q decomposes into

$\mathfrak{V}_{\frac{3}{2}} \oplus 2\mathfrak{V}_{\frac{1}{2}}$. For $q^{\frac{3}{2}} = 1$, however, the weight states of the spin $\frac{3}{2}$ and one of the spin $\frac{1}{2}$ representations mix into an indecomposable representation. For example, the matrix elements (6.20) show that

$$J_-|\tfrac{3}{2},\tfrac{3}{2}\rangle = |\tfrac{3}{2},\tfrac{1}{2}\rangle, \quad J_+|\tfrac{3}{2},\tfrac{1}{2}\rangle = 0,$$

because $[3] = 0$. The state $|\frac{3}{2},-\frac{1}{2}\rangle$ is obtained from $|\frac{3}{2},\frac{1}{2}\rangle$ by application of J_-, but is then annihilated by a further application of J_-. However, the state $|\frac{3}{2},-\frac{3}{2}\rangle$ may be obtained from $|\frac{3}{2},\frac{3}{2}\rangle$ by application of T_-, defined in (6.21), and similarly with the operator T_+ one can obtain states not otherwise reached using J_+.

A general description has been given by Pasquier and Saleur [31], where the p-fold tensor product of irreps (for odd p) is shown to decompose into a sum of representations with labels j_k. States in each representation are obtained by application of J_- and T_-, some of which are annihilated by J_+. There exists a state not in this representation from which another representation j_{k-1} can be constructed by application of J_- and T_-. Mixing between these representations is accomplished with T_+. With this procedure representations are paired as far as possible and leads to the decomposition of the space of all states into a direct sum of representations which are indecomposable with respect to J_+ and T_+. A useful depiction of the possible states in the tensor product is the Bratteli diagram, and we refer to [31, 75] for further details.

In terms of the **q-dimension** given by $[2j+1]$ the representations mix in such a way that their total q-dimension is zero. The existence of indecomposable representations can be attributed to the fact that the Casimir invariant $[j][j+1]$ does not distinguish between representations labelled j mod p and $(-1-j)$ mod p. Under the transformation $j \to (-1-j) \bmod p$ we find that the q-dimension $[2j+1]$ changes sign, but that $j \to j \bmod p$ leaves $[2j+1]$ invariant.

Further discussion on the decomposition of tensor products (known in conformal field theory as fusion rules) has been given by Keller [176], Fröhlich and Kerler [32], and Arnaudon [170], wherein further references may be found.

6.2 Irreducible Cyclic Representations of $\mathcal{U}_q(\mathfrak{sl}(2))$

All finite dimensional representations of $\mathcal{U}_q(\mathfrak{sl}(2))$ have been classified by De Concini and Kac [169], and we rederive their results for type $\mathcal{B}$ irreps below in Theorem 6.22. Type $\mathcal{B}$ irreps all have dimension p' and do not have a classical analog, are cyclic, and are labelled by three continuous complex parameters which correspond to the eigenvalues of the three independent invariants. Let us denote these representations by $\mathcal{B}(x,y,x,c)$, where the labels x,y,z,c are the eigenvalues of the invariants X,Y,Z,C, and where c is one of the p values allowed by the identity (6.14) for $n=p$. Vectors within a basis for an irreducible representation are labelled by the eigenvalue of K.

THEOREM 6.22 *Every type $\mathcal{B}$ representation can be written, in some basis $\{v_i\}$ of $\mathfrak{V}$ where $0 \leqslant i \leqslant p'-1$, in the following form for some $\lambda \in \mathbb{C}^\times$, $a, b \in \mathbb{C}$:*

$$\begin{aligned} Kv_i &= \lambda q^{-i} v_i \quad \text{for} \quad i = 0, \ldots, p'-1 \\ J_- v_i &= v_{i+1} \quad \text{for} \quad i = 0, \ldots, p'-2 \\ J_- v_{p'-1} &= b v_0 \\ J_+ v_i &= M_i v_{i-1} \quad \text{for} \quad i = 1, \ldots, p'-1 \\ J_+ v_0 &= a v_{p'-1} \end{aligned} \tag{6.23}$$

where

$$M_i = ab + \frac{[i]\left(\lambda q^{-\frac{1}{2}(i-1)} - \lambda^{-1} q^{\frac{1}{2}(i-1)}\right)}{q^{\frac{1}{2}} - q^{-\frac{1}{2}}} \tag{6.24}$$

for $i = 1, \ldots p'-1$. The labels λ, a, b are related to the eigenvalues x, y, z, c of the invariants X, Y, Z, C by

$$x = a \prod_{i=1}^{p'-1} M_i, \quad y = b, \quad z = \lambda^{p'}, \quad c = ab + \frac{\lambda q^{\frac{1}{2}} + \lambda^{-1} q^{-\frac{1}{2}}}{\left(q^{\frac{1}{2}} - q^{-\frac{1}{2}}\right)^2}.$$

PROOF: Choose any eigenvector v_0 of K in $\mathfrak{V}$ with an eigenvalue $\lambda \in \mathbb{C}^\times$. Since $(K)^{p'}$ is the invariant Z with eigenvalue z, we have $(\lambda)^{p'} = z$, that is, λ is any p'th root of z. Now define the other vectors in the basis by $v_i = (J_-)^i v_0$ for $i = 1, \ldots, p'-1$. These vectors, which are all nonzero since J_- is injective, are also eigenvectors of K with eigenvalue λq^{-i}, as follows from (6.10). If we apply J_- to $v_{p'-1}$ we obtain the vector $(J_-)^{p'} v_0$, which equals $y v_0$ since $(J_-)^{p'}$ is an invariant with eigenvalue y. Hence the parameter b is in fact the invariant y. The vector $J_+ v_i$ for each $i = 1, \ldots, p'-1$ is also an eigenvector of K, with eigenvalue λq^{-i+1} (using again (6.10)), and so by irreducibility $J_+ v_i = M_i v_{i-1}$ for some coefficients M_i, $i = 1, \ldots p'-1$ to be determined. We consider separately $J_+ v_0$ which is an eigenvector of K with eigenvalue $\lambda q = \lambda q^{-p'+1}$, and so is equal to $a v_{p'-1}$ for some coefficient a. In order to determine the coefficients M_i, we first calculate the following matrix elements:

$$\begin{aligned} J_- J_+ v_i &= M_i v_i \quad \text{for} \quad i = 1, \ldots, p'-1 \\ J_- J_+ v_0 &= ab v_0 \\ J_+ J_- v_i &= M_{i+1} v_i \quad \text{for} \quad i = 0, \ldots, p'-2 \\ J_+ J_- v_{p'-1} &= ab v_{p'-1}. \end{aligned} \tag{6.25}$$

In order that the first commutation relation in (6.10) be satisfied, we must have

$$M_{i+1} - M_i = \frac{\lambda q^{-i} - \lambda^{-1} q^i}{q^{\frac{1}{2}} - q^{-\frac{1}{2}}} \tag{6.26}$$

for $i = 0, \ldots p'-2$, where we include the case $i = 0$ by defining $M_0 = ab$. In addition we must also satisfy, for the case $i = p'-1$,

$$ab - M_{p'-1} = \frac{\lambda q - \lambda^{-1} q^{-1}}{q^{\frac{1}{2}} - q^{-\frac{1}{2}}}$$

where we used $q^{p'} = 1$. The unique solution of these equations for M_i is precisely that given by (6.24).

The labels λ, a, b are determined by the invariants x, y, z, indeed, we have already seen that $\lambda^{p'} = z$ and $b = y$. By repeated application of J_+ we find that

$$(J_+)^{p'} v_0 = \left(a \prod_{i=1}^{p'-1} M_i \right) v_0 \tag{6.27}$$

which implies that $x = a \prod_{i=1}^{p'-1} M_i$. By repeated application of J_-, which commutes with $(J_+)^{p'}$, to both sides of (6.27) we find that (6.27) holds, as it must, with v_0 replaced by v_i. We may also determine the eigenvalue of the Casimir invariant by using the derived matrix elements to obtain c as stated. □

REMARK **6.28** 1. The notation of De Concini and Kac corresponds to ours according to $\varepsilon = q^{\frac{1}{2}}$, $E = J_+$, $F = J_-$, $\ell' = p'$.

2. The labels x, y, z, or λ, a, b do not distinguish inequivalent representations, since we may always rescale the invariants by means of the transformation (6.16).

3. The special case in which $b = 0$ and a is nonzero, but with $x = y = 0, z = \pm 1$ gives a representation which is indecomposable.

4. The representation (6.23) is irreducible if and only if one of the four following conditions is satisfied [170]: (a) $x \neq 0$ (b) $y \neq 0$ (c) $z \neq \pm 1$ and also (d) $c = 2\omega/(q^{\frac{1}{2}} - q^{-\frac{1}{2}})^2$ where $\omega = \pm 1$. Representations in the last category have $x = 0 = y$ and so are of type $\mathcal{A}$, that is nilpotent.

5. With a change of parametrization to complex invariants α, β according to the following equations, we can factorize the matrix element M_i:

$$\lambda = q^{\frac{1}{2}(\beta - \alpha)}, \quad ab = [\alpha][\beta + 1], \quad M_i = [\alpha + i][\beta + 1 - i].$$

These parameters appear naturally when we realize the generators in a polynomial space, see §6.6, and are considered in more detail in §6.2.2.

EXAMPLE **6.29** Let us write out the two simplest set of irrep matrices for $p = 4$ and $p = 3$, which are two and three dimensional respectively. Firstly for $p = 4$, when $q^{\frac{1}{2}} = i$, all cyclic irreps can be written as follows:

$$K = \begin{pmatrix} \lambda & 0 \\ 0 & -\lambda \end{pmatrix} \quad J_+ = \begin{pmatrix} 0 & M_1 \\ a & 0 \end{pmatrix} \quad J_- = \begin{pmatrix} 0 & b \\ M_1 & 0 \end{pmatrix}.$$

For $p = 3$, when $[2] = q^{\frac{1}{2}} + q^{-\frac{1}{2}} = -1$, all cyclic irreps are three-dimensional and take the form:

$$K = \begin{pmatrix} \lambda & 0 & 0 \\ 0 & \lambda q^{-1} & 0 \\ 0 & 0 & \lambda q^{\frac{1}{2}} \end{pmatrix} \quad J_+ = \begin{pmatrix} 0 & M_1 & 0 \\ 0 & 0 & M_2 \\ a & 0 & 0 \end{pmatrix} \quad J_- = \begin{pmatrix} 0 & 0 & b \\ 1 & 0 & 0 \\ 0 & 1 & 0 \end{pmatrix},$$

where M_1, M_2 are given by (6.24) or, in factorized form, by $M_i = [\alpha + i][\beta + 1 - i]$. We may verify in each case that the matrices $(J_+)^{p'}$, $(J_-)^{p'}$ and $(K)^{p'}$ are multiples of the identity matrix, and are therefore invariants of the irrep.

In the representations given by Theorem 6.22 the generators J_+ and J_- do not play a symmetric role, but this is partly a consequence of the normalization we chose for the vectors v_i. If instead we define vectors $u_i = (J_+)^{p'} u_0$ for $i = 1, \ldots, p' - 1$ (for some given u_0 which is an eigenfunction of K) then in place of (6.23) we have

$$\begin{aligned} Ku_i &= \lambda q^i u_i && \text{for } i = 0, \ldots, p' - 1 \\ J_+ u_i &= u_{i+1} && \text{for } i = 0, \ldots, p' - 2 \\ J_+ u_{p'-1} &= a u_0 && \\ J_- u_i &= M_i' u_{i-1} && \text{for } i = 1, \ldots, p' - 1 \\ J_- u_0 &= b u_{p'-1} && \end{aligned} \tag{6.30}$$

where M_i' is given by

$$M_i' = ab + \frac{[i] \left(\lambda^{-1} q^{-\frac{1}{2}(i-1)} - \lambda q^{\frac{1}{2}(i-1)} \right)}{q^{\frac{1}{2}} - q^{-\frac{1}{2}}}, \tag{6.31}$$

which is (6.24) with λ replaced by λ^{-1}. For these representations the invariants are given by $z = \lambda^{p'}$ (as before), $x = a$, $y = b \prod_{i=1}^{p'-1} M_i'$.

Tensor products of cyclic representations, and also cyclic and nilpotent representations, have been discussed by Arnaudon [170].

6.2.1 *Unitary Cyclic Representations of $\mathcal{U}_q(\mathfrak{su}(2))$*

We now specialise the results of the previous section to obtain unitary finite-dimensional irreducible cyclic representations of $\mathcal{U}_q(\mathfrak{su}(2))$, by imposing $\overline{z} = z^{-1}$ and $\overline{x} = y$ and determining all restrictions on the possible values of the complex invariants λ, a, b which occur in Theorem 6.22. In particular $K^\dagger = K^{-1}$ implies $\overline{\lambda} = \lambda^{-1}$ and so we may define

$$\lambda \stackrel{\text{def}}{=} q^\theta$$

for some real parameter θ which is unique for each λ provided we specify $0 \leqslant \theta < \frac{p}{2}$. If we now substitute $\lambda = q^\theta$ into the expression (6.24) we find that M_i can be written in the simpler form

$$M_i = ab + [i][2\theta + 1 - i]. \tag{6.32}$$

Here $i = 0, \ldots, p'-1$ but in fact we may allow i to take any integer value, since the right hand side of (6.32) is periodic in i with period p'.

There are also restrictions on the invariants a, b. For any $v \in \mathfrak{V}$ we have $(v,\ J_-J_+v) = (J_+v,\ J_+v) = \|J_+v\|^2 \geqslant 0$ and similarly $(v,\ J_+J_-v) = \|J_-v\|^2 \geqslant 0$. For cyclic irreps we require in fact strict positivity, and so we determine from (6.25) that each coefficient M_i must be real and positive: $M_i > 0$, for each $i = 0, \ldots, p'-1$. (We can include nilpotent irreps by allowing $M_i = 0$). The particular case $i = 0$ shows that ab must be real and positive. Since, as previously noted in §6.1.1, we can choose the eigenvalue $y = b$ of $Y = (J_-)^{p'}$ to be real, we may assert that a is also real. The eigenvalues of $(J_-)^{p'}$ and $(J_+)^{p'}$ are real and equal, and so we have from (6.27):

$$b = a \prod_{i=1}^{p'-1} M_i. \tag{6.33}$$

We can express each of a, b in terms of θ and a second invariant γ defined by

$$\gamma \stackrel{\text{def}}{=} ab.$$

Firstly,

$$M_i = \gamma + [i][2\theta + 1 - i],$$

for any integer i and hence, from (6.33), $b^2 = \prod_{i=0}^{p'-1} M_i$ (using $M_0 = ab$) and then $a = \gamma / b$, which expresses b, a in turn in terms of the invariants γ, θ.

The two real invariants γ, θ are further restricted by the positivity requirement $M_i > 0$, which means that we must have

$$\gamma > \max_{n \in \mathbb{Z}} [n][2\theta + 1 + n], \tag{6.34}$$

where we have allowed $n = -i$ to take any integer value. Hence we always require $\gamma > 0$, but if γ is sufficiently large (if $\gamma > \sin^{-2}(2\pi/p)$) then (6.34) imposes no restrictions on θ because $[n][2\theta+1+n]$, being a product of sin functions, is bounded above. For smaller values of γ, however, only certain values of θ satisfy (6.34) and this is investigated in more detail below when we choose a certain form for γ.

Let us write the cyclic irreps in a form which is manifestly unitary by defining the following vectors, in the ket notation:

$$|\gamma, \theta, m\rangle = \left(\prod_{i=p'-m+1}^{p'} M_i \right)^{\frac{1}{2}} v_{p'-m}, \quad m = 1, \ldots, p'.$$

The matrix elements of the generators of $\mathcal{U}_q(\mathfrak{su}(2))$ in this basis are given by:

$$\begin{aligned} K\,|\gamma, \theta, m\rangle &= q^{\theta+m}\,|\gamma, \theta, m\rangle \\ J_+\,|\gamma, \theta, m\rangle &= \sqrt{\gamma - [2\theta + 1 + m][m]}\,|\gamma, \theta, m+1\rangle \\ J_-\,|\gamma, \theta, m\rangle &= \sqrt{\gamma - [2\theta + m][m-1]}\,|\gamma, \theta, m-1\rangle. \end{aligned}$$

Here $m = 1, \ldots, p'$, and we have written the matrix elements $\sqrt{M_{p'-m}}$ and $\sqrt{M_{p'-m+1}}$ in the form shown by using periodicity in p'. We have also used (6.33) to obtain the second equation for $m = p'$ and the last equation for $m = 1$, where we defined $|\gamma, \theta, p'+1\rangle = |\gamma, \theta, 1\rangle$, and $|\gamma, \theta, 0\rangle = |\gamma, \theta, p'\rangle$. It is straightforward to show directly from these equations that the commutation relations (6.10) are satisfied. For this we use the familiar identity

$$[a][b-c] + [b][c-a] + [c][a-b] = 0, \tag{6.35}$$

with $a = 2\theta + m, b = -m, c = -1$. The basis vectors in any irrep are labelled by $m = 0, \ldots, p'-1$, but we may allow m to take any integer value by periodic extension, that is, we define $|\gamma, \theta, m\rangle$ for all $m \in \mathbb{Z}$ by means of $|\gamma, \theta, m\rangle = |\gamma, \theta, m+p'\rangle$. This is possible because the matrix elements of the generators in this basis are also periodic in m, with period p'. The matrix elements are also invariant under the replacement $\theta \to \theta + \frac{p}{2}$, so we may allow θ to take any real value.

We may summarize the results in the following lemma:

LEMMA **6.36** *All unitary cyclic irreps of $\mathcal{U}_q(\mathfrak{su}(2))$, where $q^{\frac{p}{2}} = 1$, are labelled by two real numbers γ, θ satisfying*

$$\gamma > \max_{n \in \mathbb{Z}} [n][2\theta + 1 + n].$$

The dimension of the irrep is p' and the basis vectors are denoted $|\gamma, \theta, m\rangle$, where $m \in \mathbb{Z}$, and the matrix elements of the generators are given by

$$\begin{aligned} K\,|\gamma,\theta,m\rangle &= q^{\theta+m}\,|\gamma,\theta,m\rangle \\ J_+\,|\gamma,\theta,m\rangle &= \sqrt{\gamma - [2\theta+1+m][m]}\,|\gamma,\theta,m+1\rangle \\ J_-\,|\gamma,\theta,m\rangle &= \sqrt{\gamma - [2\theta+m][m-1]}\,|\gamma,\theta,m-1\rangle. \end{aligned}$$

The invariants take the values:

$$x = y = \left(\prod_{i=0}^{p'-1} (\gamma - [2\theta+1+i][i])\right)^{\frac{1}{2}}, \quad z = q^{\theta p'}, \quad c = \gamma + \frac{q^{\theta+\frac{1}{2}} + q^{-\theta-\frac{1}{2}}}{\left(q^{\frac{1}{2}} - q^{-\frac{1}{2}}\right)^2}.$$

EXAMPLE **6.37** The two and three dimensional irreps take the following form, firstly for $p = 4$:

$$K = \begin{pmatrix} q^\theta & 0 \\ 0 & -q^\theta \end{pmatrix} \quad J_+ = \begin{pmatrix} 0 & \sqrt{\gamma+[2\theta]} \\ \sqrt{\gamma} & 0 \end{pmatrix} \quad J_- = \begin{pmatrix} 0 & \sqrt{\gamma} \\ \sqrt{\gamma+[2\theta]} & 0 \end{pmatrix},$$

and secondly for $p = 3$:

$$K = \begin{pmatrix} q^\theta & 0 & 0 \\ 0 & q^{\theta+1} & 0 \\ 0 & 0 & q^{\theta+\frac{1}{2}} \end{pmatrix} \quad J_+ = \begin{pmatrix} 0 & 0 & \sqrt{\gamma+[2\theta]} \\ \sqrt{\gamma} & 0 & 0 \\ 0 & \sqrt{\gamma+[2\theta-1]} & 0 \end{pmatrix},$$

with J_- the transpose of J_+. By using the particular properties of q-numbers for each case, we may verify that the defining relations (6.10) are satisfied.

6.2.2 Factorized Matrix Elements

We can write the matrix elements shown in Lemma 6.36 in a simpler form which is very suggestive for the purpose of generalizing these results to higher unitary quantum groups. Let us replace the invariants γ, θ by two new real invariants α, β satisfying:

$$2\theta = \beta - \alpha, \quad \gamma = [\alpha][\beta + 1]. \tag{6.38}$$

(These equations do not determine α, β uniquely from γ, θ, neither can all possible invariants γ, θ be expressed in this form if α and β are real, since $[\alpha][\beta+1]$ is bounded.) The basis vectors are now denoted $|\alpha, \beta, m\rangle$ and we find that the matrix elements in Lemma 6.36 can be factorized to read

$$\begin{aligned} K\,|\alpha,\beta,m\rangle &= q^{\frac{1}{2}(\beta-\alpha)+m}\,|\alpha,\beta,m\rangle \\ J_+\,|\alpha,\beta,m\rangle &= \sqrt{[\alpha-m][\beta+m+1]}\,|\alpha,\beta,m+1\rangle \\ J_-\,|\alpha,\beta,m\rangle &= \sqrt{[\alpha-m+1][\beta+m]}\,|\alpha,\beta,m-1\rangle. \end{aligned} \tag{6.39}$$

These matrix elements follow from those in Lemma 6.36 by using (6.35) with $a = \beta + 1, b = \alpha - m, c = -m$. This substitution also displays a symmetry between the invariants α and β, which take real values, but do not label the irreps uniquely. The eigenvalues of C and $K^{p'}$ are given by

$$c = \left[\tfrac{1}{2}(\alpha+\beta)\right]\left[\tfrac{1}{2}(\alpha+\beta)+1\right], \quad z = q^{\frac{1}{2}(\beta-\alpha)}.$$

We can also express the eigenvalues of $(J_-)^{p'}$ and $(J_+)^{p'}$ in a simple form. Let us recall the notation of the shifted factorial $([x])_n$ defined in (2.122), p. 67, for any $x \in \mathbb{R}$ and for both positive and negative integers. A property of $([x])_n$ we will use frequently is

$$([x])_n = [x+n-1]([x])_{n-1}$$

which holds for both positive and negative n. Another useful property which is valid at roots of unity is

$$([x])_{p'} = ([x+n])_{p'} \tag{6.40}$$

for any integer n and $x \in \mathbb{R}$; in other words the "rising" product is circular in the sense that the last factor can be viewed as preceding the first. We also note that as a result $([n])_{p'} = 0$ whenever n is an integer, since then $([n])_{p'} = ([0])_{p'} = 0$.

The invariants $(J_-)^{p'}$ and $(J_+)^{p'}$ have a common eigenvalue which, from (6.39), is given by

$$x = y = \sqrt{([\alpha])_{p'}([\beta])_{p'}}.$$

We have yet to consider the restrictions placed on the invariants α, β by the positivity requirement (6.34). The matrix elements (6.39) show that this requirement takes the form

$$[\alpha - m][\beta + m + 1] > 0, \quad m \in \mathbb{Z}. \tag{6.41}$$

(We need to impose this only for $m = 1, \ldots, p'$, however, since this product is periodic with period p' we may allow m to take all integer values). Eqn. (6.41) imposes restrictions on the possible values of α and β which we summarize in Lemma 6.44 below.

DEFINITION **6.42** *Define the* **greatest integer function** *of $x \in \mathbb{R}$, denoted $[\![x]\!]$, by*

$$[\![x]\!] \stackrel{\text{def}}{=} n \text{ if and only if } n \leqslant x < n+1, \quad \text{where } n \in \mathbb{Z}.$$

Some properties of this function are:

$$\begin{aligned} [\![n]\!] &= n \quad \text{if and only if } n \in \mathbb{Z}, \\ [\![x+n]\!] &= [\![x]\!] + n \quad \text{for all } n \in \mathbb{Z}, \\ [\![x]\!] + [\![-x]\!] &= -1 \quad \text{for all } x \notin \mathbb{Z}. \end{aligned}$$

It follows from the last equation that for x noninteger, $[\![x]\!] + [\![n-x]\!] = n - 1$ for all $n \in \mathbb{Z}$. For any $x \in \mathbb{R}$ now define for notational convenience:

$$x' \stackrel{\text{def}}{=} x - [\![x]\!], \tag{6.43}$$

then $0 \leqslant x' < 1$, and if x is noninteger, $0 < x' < 1$.

LEMMA **6.44** *(i) Let $p > 2$ be an even integer. Then $[\alpha - m][\beta + m + 1] > 0$ for any nonintegers α, β, and for all $m \in \mathbb{Z}$, if and only if*

$$[\![\alpha \bmod p]\!] + [\![\beta \bmod p]\!] = \tfrac{1}{2}(p-4).$$

(ii) Let $p > 2$ be an odd integer. Then $[\alpha - m][\beta + m + 1] > 0$ for all $m \in \mathbb{Z}$ if and only if either

$$0 < \alpha', \ \beta' < \tfrac{1}{2} \quad \text{and} \quad [\![\alpha \bmod p]\!] + [\![\beta \bmod p]\!] = \tfrac{1}{2}(p-3), \tag{6.45}$$

or

$$\tfrac{1}{2} < \alpha', \ \beta' < 1 \quad \text{and} \quad [\![\alpha \bmod p]\!] + [\![\beta \bmod p]\!] = \tfrac{1}{2}(p-5).$$

REMARK **6.46** We have considered strict positivity of $[\alpha - m][\beta + m + 1]$ since any zero values would imply that the representation is nilpotent, and we consider these separately. Since we have defined $\alpha' = \alpha - [\![\alpha]\!]$, and $\beta' = \beta - [\![\beta]\!]$, the condition that α, β be noninteger could also be stated $0 < \alpha', \ \beta' < 1$.

PROOF: It is convenient to assume that $0 \leqslant \alpha, \beta \leqslant p$ for we may extend the results to all α, β using periodicity. It is also sufficient to prove positivity for only p' consecutive integers m, rather than all $m \in \mathbb{Z}$, because the product $[\alpha - m][\beta + m + 1]$ is periodic in m, with period p' and hence we consider the product for $m = M, M + 1, \dots, M + p' - 1$, for some M to be chosen. We proceed by listing the values of m for which $[\alpha - m] > 0$, and then requiring that $[\beta + m + 1] > 0$ for these same values, and similarly for the negative values of $[\alpha - m]$, for which $[\beta + m + 1]$ must also be negative.

Consider first even p. We have

$$[\alpha - m] = [\alpha' + [\![\alpha]\!] - m] > 0 \tag{6.47}$$

for $m = [\![\alpha]\!] + 1 - \frac{p}{2}, \dots, [\![\alpha]\!]$. Since these consecutive values of m range over exactly one half period $\frac{p}{2}$ of $[\alpha - m]$ we need only verify (6.47) at the endpoint values $m = [\![\alpha]\!] + 1 - \frac{p}{2}$ and $m = [\![\alpha]\!]$. This follows easily if we recall the definition (6.3) of $[x]$ and the inequalities (6.5), and so we see that there can be at most one sign change for this range of values of m. In order to verify (6.47) we need to show that $0 < \alpha - m < \frac{p}{2}$ for each m. Taking now $m = [\![\alpha]\!] + 1 - \frac{p}{2}$ we find $0 < \alpha' - 1 + \frac{p}{2} < \frac{p}{2}$, since $0 < \alpha' < 1$ and $p > 2$, and hence $[\alpha' - 1 + \frac{p}{2}] > 0$, and similarly for $m = [\![\alpha]\!]$ we have $0 < \alpha' < \frac{p}{2}$. We now impose $[\beta' + [\![\beta]\!] + m + 1] > 0$ for $m = [\![\alpha]\!] + 1 - \frac{p}{2}, \dots, [\![\alpha]\!]$, which requires $0 < \beta' + [\![\beta]\!] + m + 1 < \frac{p}{2}$ for the same values of m. Again, it is sufficient to impose these conditions for the two endpoint values of m. These inequalities are satisfied if and only if

$$\frac{p}{2} - 1 - \beta' < [\![\alpha]\!] + [\![\beta]\!] + 1 < -\beta' + \frac{p}{2}.$$

This inequality allows one and only one solution (for any $0 < \beta' < 1$):

$$[\![\alpha]\!] + [\![\beta]\!] + 1 = \tfrac{1}{2}(p - 2). \tag{6.48}$$

We conclude that the product $[\alpha - m][\beta + m + 1]$ is strictly positive for $m = [\![\alpha]\!] + 1 - \frac{p}{2}, \dots, [\![\alpha]\!]$ if and only if the condition (6.48) is satisfied. Since the range of m considered comprises a complete period of $[\alpha - m][\beta + m + 1]$, we have in fact verified that $[\alpha - m][\beta + m + 1] > 0$ for *all* integers m if and only if (6.48) is satisfied. (For values of m not considered explicitly, whenever $[\alpha - m] < 0$ we also have $[\beta + m + 1] < 0$ provided (6.48) is satisfied). Hence we have proved part (i) of the lemma.

Consider now the case where p is odd, and let us write $p = 2k + 1$ for some integer $k \geqslant 1$. We have

$$[\alpha - m] = [\alpha' + [\![\alpha]\!] - m] > 0$$

for $m = [\![\alpha]\!] - k, [\![\alpha]\!] - k + 1, \dots, [\![\alpha]\!]$, provided $0 < \alpha' < \frac{1}{2}$. To prove this, put $m = [\![\alpha]\!] - i$ for $i = 0, \dots, k$, then we have $0 < \alpha' + i (= \alpha - m) < \frac{1}{2} + k = \frac{p}{2}$ and so $[\alpha - m] > 0$. Now we impose $[\beta + m + 1] > 0$ for the same set of m values, that is, $0 < [\![\beta]\!] + \beta' + m + 1 < \frac{p}{2}$ for $m = [\![\alpha]\!] - i$, $i = 0, \dots, k$. The endpoint values of m show that we must have

$$k - \beta' < [\![\alpha]\!] + [\![\beta]\!] + 1 < -\beta' + k + \tfrac{1}{2}.$$

This inequality allows only one solution,

$$[\![\alpha]\!] + [\![\beta]\!] + 1 = k \tag{6.49}$$

but only if $0 < \beta' < \frac{1}{2}$. (If $\frac{1}{2} < \beta' < 1$ and $0 < \alpha' < \frac{1}{2}$ there is no solution and $[\alpha - m][\beta + m + 1]$ is not positive for all integers m). It may now be checked that $[\beta + m + 1] > 0$ for all $m = [\![\alpha]\!] - i, i = 0, \ldots, k$.

Since we must consider a full period of p consecutive values for m, we now let $m = [\![\alpha]\!] + i$, $i = 1, \ldots, k$. We find $-\frac{p}{2} < \alpha - m < 0$ for each such m, since we have $-\frac{p}{2} < -k < \alpha' - k \leqslant \alpha' - i (= \alpha - m) \leqslant \alpha' - 1 < 0$. Hence $[\alpha - m] < 0$ for these values of m. Also $\frac{p}{2} < \beta + m + 1 < p$ for these same values of m, since $\frac{p}{2} = k + \frac{1}{2} < \beta' + k + 1 (= \beta + m + 1) < 2k + 1 = p$ (putting $m = [\![\alpha]\!] + 1$ and using (6.49)). For $m = [\![\alpha]\!] + k$ we find $\frac{p}{2} = k + \frac{1}{2} < \beta' + 2k (= \beta + m + 1) < 2k + 1 = p$ and in each case we obtain $[\beta + m + 1] < 0$. Hence for p consecutive values of m we have shown that $[\alpha - m][\beta + m + 1] > 0$ provided that (6.49) holds with $0 < \alpha', \beta' < \frac{1}{2}$, and so (6.45) is verified.

The remaining case we need to consider is $\frac{1}{2} < \alpha' < 1$. We have

$$[\alpha - m] = [\alpha' + [\![\alpha]\!] - m] > 0$$

for $m = [\![\alpha]\!] - k + 1, [\![\alpha]\!] - k + 2, \ldots, [\![\alpha]\!]$, provided $\frac{1}{2} < \alpha' < 1$. For $m = [\![\alpha]\!] - k + 1$ we have $0 \leqslant k - 1 < \alpha' + k - 1 (= \alpha - m) < k < \frac{p}{2}$ and similarly for $m = [\![\alpha]\!]$. Hence $[\alpha - m] > 0$ for these values of m. We require $[\beta + m + 1] > 0$ for the same set of m values, that is, $0 < \beta + \beta' + m + 1 < \frac{p}{2}$. Hence we must have

$$k - 1 - \beta' < [\![\alpha]\!] + [\![\beta]\!] + 1 < -\beta' + k + \tfrac{1}{2},$$

which allows *two* solutions, $[\![\alpha]\!] + [\![\beta]\!] + 1 = k$ or $k - 1$. Only the second possibility is allowed, as we shall see by consideration of the remaining values, $m = [\![\alpha]\!] + 1, \ldots, [\![\alpha]\!] + k + 1$. We find $[\alpha - m] < 0$ for these values and, using the same method as before, $\frac{p}{2} < \beta + m + 1 < p$ provided

$$k - \tfrac{1}{2} - \beta' < [\![\alpha]\!] + [\![\beta]\!] + 1 < k - \beta.'$$

The only solution is $[\![\alpha]\!] + [\![\beta]\!] + 1 = k - 1$, provided $\frac{1}{2} < \beta' < 1$. This now proves the remaining equation of the lemma. □

This result essentially determines the values on the real axis for which a product of two sin functions is positive.

EXAMPLE **6.50** A simple example is the case $p = 4$ for which the unitary irreps of $\mathcal{U}_q(\mathfrak{su}(2))$ are:

$$K = \begin{pmatrix} q^{\frac{1}{2}(\beta-\alpha)} & 0 \\ 0 & -q^{\frac{1}{2}(\beta-\alpha)} \end{pmatrix} \quad J_+ = \begin{pmatrix} 0 & \sqrt{[\alpha+1][\beta]} \\ \sqrt{[\alpha][\beta+1]} & 0 \end{pmatrix},$$

with J_- the transpose of J_+. Positivity of the arguments of the square roots, namely $[\alpha][\beta+1] > 0$, $[\alpha+1][\beta] > 0$, is equivalent to the condition $[\![\alpha \bmod 4]\!] + [\![\beta \bmod 4]\!] = 0$, which means that the only allowed invariants α, β are those for which $[\![\alpha]\!] + [\![\beta]\!]$ is some integer multiple of 4.

6.2.3 Analytic Extension of $U(2)$ Representations

In the form (6.39) we can see that the matrix elements can be deduced rather simply from the unitary irreps of $\mathcal{U}_q(\mathfrak{u}(2))$ found previously for $q \in \mathbb{R}^+$. This comes about in the following way. As Example 2.76 shows, the irreps of $\mathcal{U}_q(\mathfrak{u}(2))$ are given in the Gel'fand-Weyl basis by:

$$\begin{aligned} K\left|\begin{smallmatrix} m_{11} & & m_{22} \\ & m_{11} & \end{smallmatrix}\right\rangle &= q^{m_{11}-\frac{1}{2}(m_{22}-m_{12})}\left|\begin{smallmatrix} m_{12} & & m_{22} \\ & m_{11} & \end{smallmatrix}\right\rangle \\ J_+\left|\begin{smallmatrix} m_{12} & & m_{22} \\ & m_{11} & \end{smallmatrix}\right\rangle &= \sqrt{[m_{12}-m_{11}][m_{11}+1-m_{22}]}\left|\begin{smallmatrix} m_{12} & & m_{22} \\ & m_{11}+1 & \end{smallmatrix}\right\rangle \\ J_-\left|\begin{smallmatrix} m_{12} & & m_{22} \\ & m_{11} & \end{smallmatrix}\right\rangle &= \sqrt{[m_{12}-m_{11}+1][m_{11}-m_{22}]}\left|\begin{smallmatrix} m_{12} & & m_{22} \\ & m_{11}-1 & \end{smallmatrix}\right\rangle \end{aligned} \tag{6.51}$$

where the entries m_{ij} are non-negative integers. For $\mathcal{U}_q(\mathfrak{su}(2))$ with generic q these matrix elements for integers m_{12}, m_{22} also form irreducible representations, but ones which are inequivalent only for different values of $m_{12} - m_{22}$. If we allow the labels m_{ij} to take noninteger values we still obtain a representation of $\mathcal{U}_q(\mathfrak{u}(2))$, but one which is infinite-dimensional, since the action of the raising or lowering generators no longer terminates and so there are no states of highest or lowest weight.

Now let us relabel the integers m_{ij} in the following way: put

$$m_{12} = \alpha, \quad m_{22} = \beta, \quad m_{11} = m,$$

then we find that at roots of unity the matrix elements (6.51) are identical with (6.39)! In other words, we can allow the entries m_{ij} in (6.51) to take noninteger values but despite this the representations of $\mathcal{U}_q(\mathfrak{su}(2))$ at roots of unity remain finite-dimensional because of the periodicity properties of the matrix elements. Furthermore, representations with different values of $\alpha \bmod p'$ and $\beta \bmod p'$ are now inequivalent even for $\mathcal{U}_q(\mathfrak{su}(2))$, because at roots of unity there is a second invariant (as stated in Lemma 6.13) which distinguishes otherwise equivalent representations.

The fact that representations of $\mathcal{U}_q(\mathfrak{su}(2))$ can be obtained at roots of unity from the Gel'fand-Weyl basis for $\mathcal{U}_q(\mathfrak{u}(2))$ has been observed by Arnaudon and Chakrabarti [177] and extended to all unitary quantum groups. The defining commutation relations for $\mathcal{U}_q(\mathfrak{su}(n))$ (and hence $\mathcal{U}_q(\mathfrak{sl}(n))$) are given in Definition 2.65 (p. 44) and may be expressed in terms of $k_i \stackrel{\text{def}}{=} q^{\frac{1}{2}h_i}$. At roots of unity we have, besides the usual $n-1$ Casimir invariants, also the following invariants (assuming now that p is odd):

$$(k_i)^p, \quad (e_i)^p, \quad (f_i)^p \tag{6.52}$$

as may be verified in the same way as for $\mathcal{U}_q(\mathfrak{sl}(2))$. For example, we find:

$$(k_i)^n e_i = e_i (k_i)^n q^n, \quad (k_i)^n e_{i\pm1} = e_{i\pm1} (k_i)^n q^{-\frac{n}{2}}, \quad n \in \mathbb{N}$$

and by putting $n = p$ we infer that $(k_i)^p$ commutes with all e_j, and similarly with all f_j. In considering $(e_i)^p$ we firstly note that

$$k_i (e_i)^n = (e_i)^n k_i q^n, \quad k_i (e_{i\pm1})^n = (e_{i\pm1})^n k_i q^{-\frac{n}{2}}$$
$$[(e_i)^n, f_i] = [n](e_i)^{n-1} \frac{k_i q^{\frac{1}{2}(n-1)} - k_i^{-1} q^{-\frac{1}{2}(n-1)}}{q^{\frac{1}{2}} - q^{-\frac{1}{2}}}, \quad n \in \mathbb{N}$$

and so $(e_i)^p$ commutes with all k_j and f_j. Secondly, we must also verify that $(e_i)^p$ commutes with all e_j and this is done using the Serre relations (2.67). It is helpful to use an identity which is stated later in (7.86), p. 243, and by putting $A = e_i$, $B = e_{i\pm1}$, $n = p$ in this identity we conclude that $(e_i)^p$ is an invariant of $\mathcal{U}_q(\mathfrak{sl}(n))$ at roots of unity.

Representations of $\mathcal{U}_q(\mathfrak{sl}(n))$ are classified therefore by the $n-1$ Casimir invariants and also the $3(n-1)$ dependent invariants (6.52). It is possible to have "partially periodic" representations in which the generators e_i, f_j can be either injective or nilpotent. Matrix elements may be determined by extending the Gel'fand-Weyl labels to complex values in a way similar to the $n = 2$ case, and we refer to [177] for further details.

REMARK **6.53** It appears at first that one does not obtain all cyclic representations of $\mathcal{U}_q(\mathfrak{sl}(2))$ by the device of extending the irrep matrix elements of $\mathcal{U}_q(\mathfrak{u}(2))$ to non-integer values of the labels, since the label γ defined in §6.2.1 by $\gamma = ab$ is in general unbounded above, but is now restricted to take the value $\gamma = [\alpha][\beta + 1]$ as shown in (6.38), which *is* bounded above for real α, β. However, by allowing the Gel'fand-Weyl labels to take complex values one can analytically continue the $\mathcal{U}_q(\mathfrak{u}(2))$ matrix elements to obtain all finite-dimensional $\mathcal{U}_q(\mathfrak{sl}(2))$ cyclic representations at roots of unity.

6.3 q-Boson Operator Construction of Representations

We can construct all irreducible representations of $\mathcal{U}_q(\mathfrak{sl}(2))$ at roots of unity with the help of q-boson operators and, by imposing Hermiticity properties, also construct the unitary irreps of $\mathcal{U}_q(\mathfrak{su}(2))$. This requires us to reconsider the definition of inner product in the space $\mathfrak{V}$ carrying the irreps, in order to avoid a metric which is not positive definite. For irreps which are not necessarily unitary we may construct all finite-dimensional irreps of $\mathcal{U}_q(\mathfrak{sl}(2))$ with q-boson operators in which a and $\overline{a}$ are not necessarily related as adjoint operators. It is interesting to note that the q-boson algebra admits cyclic representations which are finite dimensional, in contrast with the situation for generic q and also the case $q = 1$, for which the boson operators are unbounded, when representations are necessarily infinite dimensional.

When q is a root of unity it is more appropriate[4] to express the q-boson algebra in terms of

$$L \stackrel{\text{def}}{=} q^{\frac{N}{2}},$$

rather than in terms of N, the number operator, which is not uniquely determined by L. However, operators such as $[N]$ are expressible in terms of L and L^{-1} and will also appear in our development. We may write the defining relations (2.26, 2.27) as

$$\overline{a}a = q^{\frac{1}{2}}a\overline{a} + L^{-1}, \tag{6.54a}$$

$$La L^{-1} = q^{\frac{1}{2}}a, \qquad L\overline{a}L^{-1} = q^{-\frac{1}{2}}\overline{a}. \tag{6.54b}$$

The restricted q-boson algebra given by (2.33), p. 28, takes the form:

$$\begin{aligned} \overline{a}a &= \frac{q^{\frac{1}{2}}L - q^{-\frac{1}{2}}L^{-1}}{q^{\frac{1}{2}} - q^{-\frac{1}{2}}} \\ a\overline{a} &= \frac{L - L^{-1}}{q^{\frac{1}{2}} - q^{-\frac{1}{2}}}, \end{aligned} \tag{6.55}$$

together with (6.54b).

Now we determine irreducible representations of the q-boson algebra and use them to construct the cyclic representations of $\mathcal{U}_q(\mathfrak{sl}(2))$, assuming that p is odd so that we obtain irreducible representations; for even p the representations are reducible and this case is considered separately in §6.3.1. For generic q the matrix representations of the q-boson operators are infinite-dimensional, as shown in (2.37), p. 29, but at roots of unity irreducible matrix representations are of dimension p'.

Firstly, we find the invariants of the q-boson algebra. From (6.54a) it follows by induction on n that

$$\overline{a}a^n = q^{\frac{n}{2}}a^n\overline{a} + L^{-1}q^{\frac{1}{2}(n-1)}[n]a^{n-1}. \tag{6.56}$$

For $n = p$ it follows from $[p] = 0$ and $q^{\frac{p}{2}} = 1$ that

$$[\overline{a}, a^p] = 0 = [a, a^p]$$

and, also from (6.54b),

$$La^nL^{-1} = q^{\frac{n}{2}}a^n, \quad L\overline{a}^nL^{-1} = q^{-\frac{n}{2}}\overline{a}^n, \tag{6.57}$$

showing that $[L, a^p] = 0 = [L, \overline{a}^p]$ (this also implies that $[N]$ commutes with a^p and $\overline{a}^p$). Hence both a^p and $\overline{a}^p$ commute with all elements of the q-boson algebra. Furthermore, we have

$$L^n a L^{-n} = q^{\frac{n}{2}}a, \quad L^n\overline{a}L^{-n} = q^{-\frac{n}{2}}\overline{a}, \tag{6.58}$$

which shows that $[L^p, a] = 0 = [L^p, \overline{a}]$. We deduce:

[4]Just as discussed for J_z previously.

Lemma 6.59 *The invariants of the q-boson algebra for $q^{\frac{p}{2}} = 1$ are a^p, $\overline{a}^p$, L^p.*

When (6.55) is satisfied these three invariant operators are dependent, as is implied by the identity:

$$(\overline{a})^n (a)^n = [N+1][N+2]\dots[N+n], \quad n \in \mathbb{N},$$

with $n = p$. This relation between the invariants is similar to that implied by the relation (6.14) satisfied by the generators and the Casimir invariant of $\mathcal{U}_q(\mathfrak{su}(2))$.

We now construct irreducible cyclic representations of the q-boson algebra by allowing the operators $a, \overline{a}, L$ to act in a linear vector space $\mathfrak{V}$ (which need not be a Fock space), and simultaneously diagonalizing the invariants $a^p, \overline{a}^p, L^p$, and also L, since it commutes with the three invariants. The representations which we seek are those for which a^p and $\overline{a}^p$ have nonzero eigenvalues and so are cyclic, when all vectors in the representation (of dimension p) can be generated from any nonzero vector.

Lemma 6.60 *Irreducible representations of the q-boson algebra (6.54b) are labelled by three invariants $\ell \in \mathbb{C}^\times$ and $\mu, \nu \in \mathbb{C}$, and for odd p are given in some basis $v_0, v_1, \dots, v_{p-1}$ by:*

$$\begin{aligned} Lv_i &= \ell q^{\frac{i}{2}} v_i && \text{for } i = 0, \dots, p-1 \\ av_i &= v_{i+1} && \text{for } i = 0, \dots, p-2 \\ av_{p-1} &= \mu v_0 && \\ \overline{a}v_i &= \left(\ell^{-1}[i] + q^{\frac{i}{2}}\mu\nu\right) v_{i-1} && \text{for } i = 1, \dots, p-1 \\ \overline{a}v_0 &= \nu v_{p-1}. && \end{aligned}$$

The eigenvalues of the invariants $(L)^p$, $(a)^p$, $(\overline{a})^p$ are given respectively by

$$\ell^p, \quad \mu, \quad \left(\nu \prod_{i=1}^{p-1} \ell^{-1}[i] + q^{\frac{i}{2}}\mu\nu\right).$$

Proof: Let $v_0 \in \mathfrak{V}$ and suppose that $Lv_0 = \ell v_0$, where $\ell \in \mathbb{C}^\times$. We have $L^p v_0 = \ell^p v_0$ and hence ℓ is any p-th root of the invariant eigenvalue of L^p. Now define the p vectors $v_1, \dots, v_{p-1}$ by

$$v_i = a^i v_0, \quad i = 1, \dots, p-1,$$

then we find

$$\begin{aligned} av_i &= v_{i+1}, \quad i = 0, \dots, p-2 \\ av_{p-1} &= \mu v_0, \end{aligned}$$

where $\mu \in \mathbb{C}$ is the eigenvalue of a^p. Each v_i is an eigenvector of L, as follows from (6.54b):

$$Lv_i = \ell q^{\frac{i}{2}} v_i, \quad i = 0, \dots, p-1.$$

Furthermore $\overline{a}v_i$ is also an eigenvector of L, with $L(\overline{a}v_i) = \ell q^{\frac{1}{2}(i-1)}(\overline{a}v_i)$ and hence, since the representation is irreducible in which case the basis vectors are uniquely labelled by the eigenvalues of L, we must have

$$\begin{aligned}\overline{a}v_i &= A_i v_{i-1}, \quad i = 1, \ldots, p-1 \\ \overline{a}v_0 &= \nu v_{p-1},\end{aligned}$$

for some set of coefficients $A_i, \nu \in \mathbb{C}$. From the matrix elements

$$\begin{aligned}\overline{a}av_i &= A_{i+1}v_i, \quad i = 0, \ldots, p-2 \\ a\overline{a}v_i &= A_i v_i, \quad i = 1, \ldots, p-1 \\ a\overline{a}v_{p-1} &= \mu\nu v_{p-1}, \\ \overline{a}av_0 &= \mu\nu v_0\end{aligned}$$

we can determine an expression for each $A_i, i = 1, \ldots, p-1$, by using the defining relations (6.54a), and we find that the following recurrence relations must be satisfied:

$$\begin{aligned}A_{i+1} - q^{\frac{1}{2}}A_i &= \ell^{-1}q^{-\frac{i}{2}}, \quad i = 1, \ldots, p-2 \\ A_1 &= \ell^{-1} + q^{\frac{1}{2}}\mu\nu, \\ A_{p-1} &= -\ell^{-1} + q^{\frac{1}{2}}\mu\nu,\end{aligned}$$

where in the last equation we used $q^{\frac{p}{2}} = 1$. The unique solution is

$$A_i = \ell^{-1}[i] + q^{\frac{i}{2}}\mu\nu, \quad i = 0, \ldots, p-1, \tag{6.61}$$

where we have defined $A_0 = \mu\nu$ (indeed we may define A_i for all integers i by (6.61) using the periodicity of $[i]$). □

We may specialize these results to the case where the q-boson algebra is restricted to satisfy (6.55), for which the three invariants in Lemma 6.59 become dependent. In this case the matrix elements for $\overline{a}$ are replaced by

$$\overline{a}v_i = \left(\frac{\ell q^{\frac{i}{2}} - \ell^{-1}q^{-\frac{i}{2}}}{q^{\frac{1}{2}} - q^{-\frac{1}{2}}}\right) v_{i-1}, \quad i = 1, \ldots, p-1,$$

and the invariants ℓ, μ, ν are related by

$$\mu\nu = \frac{\ell - \ell^{-1}}{q^{\frac{1}{2}} - q^{-\frac{1}{2}}}. \tag{6.62}$$

We observe that ℓ, μ, ν do not uniquely specify the representations, because the defining relations for the q-boson algebra are invariant to the transformations

$$a \to \omega a, \quad \overline{a} \to \omega^{-1}\overline{a}, \quad \omega \in \mathbb{C}^{\times}.$$

EXAMPLE 6.63 For $p = 3$, when $q^{\frac{3}{2}} = 1$ and $[2] = -1$, the q-boson representation matrices are all three-dimensional and are given by:

$$L = \begin{pmatrix} \ell & 0 & 0 \\ 0 & \ell q^{\frac{1}{2}} & 0 \\ 0 & 0 & \ell q \end{pmatrix} \quad a = \begin{pmatrix} 0 & 0 & \mu \\ 1 & 0 & 0 \\ 0 & 1 & 0 \end{pmatrix} \quad \overline{a} = \begin{pmatrix} 0 & \ell^{-1} + q^{\frac{1}{2}}\mu\nu & 0 \\ 0 & 0 & -\ell^{-1} + q\mu\nu \\ \nu & 0 & 0 \end{pmatrix},$$

and satisfy (6.54a, 6.54b). When in addition (6.55) must be satisfied then (6.62) holds and $\overline{a}$ is now represented by

$$\overline{a} = \begin{pmatrix} 0 & \dfrac{q^{\frac{1}{2}}\ell - q^{-\frac{1}{2}}\ell^{-1}}{q^{\frac{1}{2}} - q^{-\frac{1}{2}}} & 0 \\ 0 & 0 & \dfrac{q\ell - q^{-1}\ell^{-1}}{q^{\frac{1}{2}} - q^{-\frac{1}{2}}} \\ \nu & 0 & 0 \end{pmatrix}.$$

We can construct representations of $\mathcal{U}_q(\mathfrak{sl}(2))$, and hence of $\mathcal{U}_q(\mathfrak{su}(2))$, by expressing the generators in terms of q-boson operators as before, shown in (2.47). The two sets of q-boson operators act in the space $\mathfrak{V}_1 \otimes \mathfrak{V}_2$ of dimension p^2 spanned by the basis vectors $v_i \otimes v_j$, $i, j = 0, \ldots, p-1$, which are eigenstates of the six q-boson invariants $a_1^p, \overline{a}_1^p, L_1^p$ and $a_2^p, \overline{a}_2^p, L_2^p$, as well as L_1, L_2. Since we now employ the restricted q-boson algebra (6.55) there are in fact only four independent invariants. Let $\mathfrak{V}$ be the p-dimensional subspace of $\mathfrak{V}_1 \otimes \mathfrak{V}_2$ spanned by $w_i = v_i \otimes v_i$, $i = 1, \ldots, p-1$, then a direct calculation using the matrix elements of Lemma 6.60 enables us to derive the matrix elements of the generators $J_\pm, K$, which we find are the same as those determined directly in Theorem 6.22, after a suitable normalization of the basis vectors. Hence, we obtain all the irreps listed in Theorem 6.22, and have therefore constructed all finite dimensional cyclic irreps of $\mathcal{U}_q(\mathfrak{sl}(2))$. We carry out the details of this calculation in the remainder of this section.

We may express the matrix elements in Lemma 6.60 in many ways, for example by generating a basis using $\overline{a}$ instead of a. Hence, for any given $u_0 \in \mathfrak{V}$, we define $u_i = \overline{a}^i u_0$ for $i = 1, \ldots, p-1$, and then

$$\begin{aligned} Lu_i &= \ell q^{-\frac{i}{2}} u_i && \text{for} \quad i = 0, \ldots, p-1 \\ \overline{a}u_i &= u_{i+1} && \text{for} \quad i = 0, \ldots, p-2 \\ \overline{a}u_{p-1} &= \nu u_0 \\ au_i &= A'_i u_{i-1} && \text{for} \quad i = 1, \ldots, p-1 \\ au_0 &= \mu u_{p-1} \end{aligned} \tag{6.64}$$

where

$$A'_i = -q^{-\frac{1}{2}}\ell^{-1}[i] + q^{-\frac{i}{2}}\mu\nu, \quad i = 0, \ldots, p-1.$$

For the restricted q-boson algebra (6.55) we have in addition

$$\mu\nu = \frac{\ell q^{\frac{1}{2}} - \ell^{-1}q^{-\frac{1}{2}}}{q^{\frac{1}{2}} - q^{-\frac{1}{2}}}, \tag{6.65}$$

and then

$$A'_i = \frac{\ell q^{-\frac{1}{2}(i-1)} - \ell^{-1} q^{\frac{1}{2}(i-1)}}{q^{\frac{1}{2}} - q^{-\frac{1}{2}}}, \quad i = 0, \ldots, p-1.$$

We express the generators of $\mathcal{U}_q(\mathfrak{sl}(2))$ as bilinear operators in two commuting sets $\{a_1, \overline{a}_1, L_1\}$ and $\{a_2, \overline{a}_2, L_2\}$ of q-boson operators, each of which act in spaces $\mathfrak{V}_1, \mathfrak{V}_2$, and each satisfying the restricted q-boson algebra (6.55). The quantum group generators are given by

$$J_+ = a_1 \overline{a}_2, \quad J_- = a_2 \overline{a}_1, \quad K = L_1 L_2^{-1}, \tag{6.66}$$

and the invariants are

$$(J_+)^p = (a_1)^p (\overline{a}_2)^p, \quad (J_-)^p = (a_2)^p (\overline{a}_1)^p, \quad K^p = (L_1)^p (L_2)^{-p}, \tag{6.67}$$

and so the eigenfunctions of the q-boson invariants are also eigenfunctions of the $\mathcal{U}_q(\mathfrak{sl}(2))$ invariants.

Let us choose the representation for $\{a_1, \overline{a}_1, L_1\}$ to be that given in (6.64), and so we put $\ell = \ell_1, \mu = \mu_1, \nu = \nu_1$; for $\{a_2, \overline{a}_2, L_2\}$ we choose the irrep as given by Lemma 6.60 and let $\ell = \ell_2, \mu = \mu_2, \nu = \nu_2$. Let $\mathfrak{V}$ be the p-dimensional subspace of $\mathfrak{V}_1 \otimes \mathfrak{V}_2$ spanned by $w_i = u_i \otimes v_i$, $i = 1, \ldots, p-1$ where u_i, v_i are the eigenvectors in (6.64) and Lemma 6.60 respectively. A direct calculation shows that the matrix elements of the generators $J_\pm, K$ are given precisely by Theorem 6.22 in which M_i is given by (6.24) with

$$\lambda = \ell_1 \ell_2^{-1}, \quad a = \mu_1 \nu_2, \quad b = \nu_1 \mu_2.$$

(We also replace p' by p and v_i by w_i). We find that M_i is factorized, as expected from the factorized form (6.66) for the generators:

$$M_i = A'_i A_i = \frac{\left(\ell_1 q^{-\frac{1}{2}(i-1)} - \ell_1^{-1} q^{\frac{1}{2}(i-1)}\right)\left(\ell_2 q^{\frac{i}{2}} - \ell_2^{-1} q^{-\frac{i}{2}}\right)}{\left(q^{\frac{1}{2}} - q^{-\frac{1}{2}}\right)^2}, \tag{6.68}$$

where we used

$$ab = (\mu_1 \nu_1)(\mu_2 \nu_2) = \frac{(\ell_1 q^{\frac{1}{2}} - \ell_1^{-1} q^{-\frac{1}{2}})(\ell_2 - \ell_2^{-1})}{\left(q^{\frac{1}{2}} - q^{-\frac{1}{2}}\right)^2},$$

as follows from (6.65) and (6.62).

We have therefore constructed all cyclic representations of $\mathcal{U}_q(\mathfrak{sl}(2))$ using the algebra (6.55). Although there are four independent invariants of this q-boson algebra (ℓ_1, ℓ_2 and one of each of μ_1, ν_1 and μ_2, ν_2) the representations of $\mathcal{U}_q(\mathfrak{sl}(2))$ are labelled by three invariants; this is because only the combination $\ell_1 \ell_2^{-1}$ appears in this construction, that is, representations with different ℓ_1, ℓ_2 are inequivalent only for different values of $\ell_1 \ell_2^{-1}$. In fact we have constructed representations of $\mathcal{U}_q(\mathfrak{gl}(2))$, in which a fourth generator H can be defined by $H = L_1 L_2$, and which commutes with $J_\pm, K$ as given in (6.66). We can distinguish representations with different values of ℓ_1, ℓ_2 by use of both K and H.

6.3.1 Cyclic Representations for Even p

Since the representations of $\mathcal{U}_q(\mathfrak{sl}(2))$ which we have constructed are of dimension p, they are irreducible only for odd p. Suppose now that p is even, then irreducible representations have dimension $p' = \frac{p}{2}$ and we can investigate this case by returning to the q-boson algebra and putting $n = p'$ into each of (6.56),(6.57) and (6.58). Using $q^{\frac{p}{4}} = -1$ we determine the following anticommutators:

$$\begin{aligned} \{\overline{a}, a^{p'}\} = & \; 0 \; = \{a, \overline{a}^{p'}\} \\ \{L, a^{p'}\} = & \; 0 \; = \{L, \overline{a}^{p'}\} \\ \{L^{p'}, a\} = & \; 0 \; = \{L^{p'}, \overline{a}\}. \end{aligned} \tag{6.69}$$

Hence, for even p the operators $(a)^{p'}$, $(\overline{a})^{p'}$, $(L)^{p'}$ are no longer invariants of the q-boson algebra, and so we need to determine a new set of invariants. In order to do this we first take as elements of the q-boson algebra, not $a, \overline{a}, L$ as before, but instead

$$a, \; \overline{a}L, \; L^2, \tag{6.70}$$

(where we could have chosen $\overline{a}L^{-1}$ instead of $\overline{a}L$). We can choose these elements because it is possible to write the defining relations (6.54b), together with the additional restrictions (6.55) if necessary, entirely in terms of the elements (6.70). For example (6.54a, 6.54b) can be written:

$$\begin{aligned} (\overline{a}L)a &= qa(\overline{a}L) + q^{\frac{1}{2}}, \\ L^2 a L^{-2} &= qa, \\ L^2(\overline{a}L)L^{-2} &= q^{-1}(\overline{a}L). \end{aligned}$$

The operators $(a)^{p'}$, $(\overline{a}L)^{p'}$, $(L^2)^{p'}$ mutually commute with a, $\overline{a}L$, L^2, as follows from (6.69), and also

$$(\overline{a}L)^n a = q^n a(\overline{a}L)^n + q^{\frac{n}{2}}[n](\overline{a}L)^{n-1},$$

which is valid for all $n \in \mathbb{N}$. Hence the invariants are:

$$a^{p'}, \; (\overline{a}L)^{p'}, \; (L^2)^{p'}. \tag{6.71}$$

We construct representations of the q-boson algebra as before, by choosing some vector $v_0 \in \mathfrak{V}$ which is an eigenvector of the invariants and also the commuting operator L^2:

$$L^2 v_0 = \ell^2 v_0 \tag{6.72}$$

where $\ell^2 \in \mathbb{C}^\times$. Define the basis vectors by $v_i = a^i v_0$ for $i = 1, \ldots, p'-1$, each of which is also an eigenvector of L^2. We find $av_{p'-1} = a^{p'}v_0 = \mu v_0$, where μ is the invariant eigenvalue of $a^{p'}$. The matrix elements of $\overline{a}L$ are determined by noting that $(\overline{a}L)v_i$ must be proportional to v_{i-1} (since each of these vectors is an eigenvector of L^2 with eigenvalue $q^{i-1}\ell^2$), and by using the q-boson algebra to solve for the coefficient.

In summary, we find that for even p:

LEMMA 6.73 *Irreducible representations of the q-boson algebra are labelled by three invariants $\ell^2 \in \mathbb{C}^\times, \mu, \nu \in \mathbb{C}$, and for even p are given in some basis $v_0, v_1, \ldots, v_{p'-1}$ by*

$$\begin{aligned} L^2 v_i &= \ell^2 q^i v_i && \text{for } i = 0, \ldots, p'-1 \\ av_i &= v_{i+1} && \text{for } i = 0, \ldots, p'-2 \\ av_{p'-1} &= \mu v_0 && \\ (\overline{a}L)v_i &= \left(q^i\mu\nu + q^{\frac{i}{2}}[i]\right) v_{i-1} && \text{for } i = 1, \ldots, p'-1 \\ (\overline{a}L)v_0 &= \nu v_{p'-1}\,. && \end{aligned}$$

The eigenvalues of the invariants $(L^2)^{p'}$, $(a)^{p'}$, $(\overline{a}L)^{p'}$ are given respectively by

$$(\ell^2)^{p'}, \quad \mu, \quad \nu \prod_{i=1}^{p'-1} \left(q^i\mu\nu + q^{\frac{i}{2}}[i]\right).$$

For the restricted q-boson algebra (6.55) the three invariants ℓ^2, μ, ν satisfy

$$\ell^2 - 1 = (q^{\frac{1}{2}} - q^{-\frac{1}{2}})\mu\nu.$$

EXAMPLE 6.74 Let us select $p = 4$, for which $q^{\frac{1}{2}} = i$, then the two-dimensional irrep matrices are:

$$L^2 = \begin{pmatrix} \ell^2 & 0 \\ 0 & -\ell^2 \end{pmatrix} \quad a = \begin{pmatrix} 0 & \mu \\ 1 & 0 \end{pmatrix} \quad \overline{a}L = \begin{pmatrix} 0 & i - \mu\nu \\ \nu & 0 \end{pmatrix},$$

with $\ell^2 = 1 + 2i\mu\nu$ for the restricted q-boson algebra.

6.4 Hermitean Adjoints of q-Boson Operators

Since we wish to construct unitary representations of $\mathcal{U}_q(\mathfrak{su}(2))$ by means of the simpler constructs of q-boson operators, we must impose Hermiticity conditions on the q-boson algebra which determine a suitable inner product in the space $\mathfrak{V}$ in which these operators act, and as a result of which the generators satisfy the correct Hermiticity properties. We have already observed in §6.2.1 that a positive definite metric imposes restrictions, such as (6.34), on the possible invariants which can be admitted.

For positive real q there is a natural definition of the inner product in a Fock space which generalizes the $q = 1$ case, and which leads to the required Hermiticity properties of the quantum algebra generators as we described in §2.4. Specifically we imposed the conditions $a^\dagger = \overline{a}$ and $N^\dagger = N$ for each set of q-bosons and, as a consequence, the $\mathcal{U}_q(\mathfrak{su}(2))$ generators satisfied the Hermiticity properties $(J_-)^\dagger = J_+$, $(J_z)^\dagger = J_z$. The inner product defines a norm, as (2.36) shows, where the norm is positive definite because the q-integers $[n]$ are positive for all $n > 0$ *provided q is real.*

At roots of unity it is necessary to redefine the adjoint q-boson operator, essentially because q-integers $[n]$ are no longer positive for $n > 0$. For example, if $p = 3$ we have $[2] = -1$ and so the vector $a^2|0\rangle$ has negative norm in a Fock space. We cannot define $a^\dagger = \overline{a}$ as before but may still impose $L^\dagger = L^{-1}$ for each set of q-bosons, which is consistent with $N^\dagger = N$, remembering that $\overline{q} = q^{-1}$. Hence, $\overline{\ell} = \ell^{-1}$ and we may define a real angle ϕ according to:

$$\ell \stackrel{\text{def}}{=} q^{\frac{\phi}{2}},$$

which is unique provided $0 \leqslant \phi < p$.

In order to investigate the possibly indefinite metric, consider irreps of the restricted q-boson algebra when p is odd, discussed in §6.3. According to (6.62) we have $\mu\nu = [\phi]$, and from the matrix elements in Lemma 6.60 we find:

$$a\overline{a}\, v_i = [\phi + i]v_i, \quad \overline{a}a\, v_i = [\phi + i + 1]v_i, \tag{6.75}$$

for all $i = 0, \ldots, p-1$. We again observe the potential difficulty mentioned above, of a metric that is not positive definite if $a^\dagger = \overline{a}$, because then

$$\|\overline{a}v_i\|^2 = (\overline{a}v_i, \overline{a}v_i) \stackrel{?}{=} (v_i, a\overline{a}v_i) = [\phi + i](v_i, v_i) \geqslant 0$$

for all $i = 0, \ldots, p-1$. However, there is no real ϕ for which the inequalities $[\phi + i] \geqslant 0$ are all satisfied! This is readily seen by noting that $[x]$ changes sign at least once over any interval of length p for any $x \in \mathbb{R}$.

Hence we must redefine the q-boson adjoint operators. First define the **signum function** $\varepsilon(x)$ by:

$$\varepsilon(x) \stackrel{\text{def}}{=} \begin{cases} 1 & \text{for } x \geqslant 0, \\ -1 & \text{for } x < 0. \end{cases}$$

We have therefore $x\varepsilon(x) = |x|$ and will also use $(\varepsilon(x))^2 = 1$, for all $x \in \mathbb{R}$.

Now define the adjoint of the q-boson operator by:

$$(a)^\dagger = \overline{a}\, \varepsilon([N]), \quad (\overline{a})^\dagger = (\varepsilon([N]))a, \tag{6.76}$$

where $[N]$ may be expressed in terms of L, namely,

$$[N] = \frac{L - L^{-1}}{q^{\frac{1}{2}} - q^{-\frac{1}{2}}}.$$

The property $N^\dagger = N$ implies that the $\dagger$-operation is involutive.

The q-boson algebra admits this definition of the adjoint, that is, "$\dagger$" is an involutive anti-automorphism of the relations (6.54a, 6.54b) and also of the relations (6.55). We find that $(a\overline{a})^\dagger = a\overline{a}$, where we use $(\varepsilon(x))^2 = 1$ and the property that any function, even if discontinuous, of N commutes with N. The effect of the definitions (6.76) is essentially to insert the appropriate sign in any norm so as to ensure the positive definite property; for example we now have:

$$\|av_i\|^2 = (v_i, (a)^\dagger a v_i) = |[\phi + i + 1]|(v_i, v_i) \geqslant 0$$

for any $\phi \in \mathbb{R}$ and all $i = 0, \ldots, p-1$.

Now define the generators of the quantum group in a slightly different way than before:

$$J_+ = a_1 a_2^\dagger = a_1 \overline{a}_2 \varepsilon([N_2]), \quad J_- = a_2 a_1^\dagger = a_2 \overline{a}_1 \varepsilon([N_1]), \tag{6.77}$$

with $K = L_1 L_2^{-1}$ as before, and for which $(J_-)^\dagger = J_+$. We must investigate the conditions under which the defining relations of the quantum group are satisfied. Clearly, $KJ_\pm K^{-1} = q^{\pm 1} J_\pm$, and also

$$[J_+, J_-] = [N_1][N_2 + 1]\,\varepsilon([N_1])\,\varepsilon([N_2 + 1]) - [N_2][N_1 + 1]\,\varepsilon([N_2])\,\varepsilon([N_1 + 1])$$

which should equal $[2J_z] = [N_1][N_2 + 1] - [N_2][N_1 + 1]$. This implies that we must have:

$$\varepsilon([N_1])\,\varepsilon([N_2 + 1]) = I = \varepsilon([N_2])\,\varepsilon([N_1 + 1]), \tag{6.78}$$

where I is the identity operator. This equation imposes restrictions on the linear vector space $\mathfrak{V}$ which carries the irrep, and indeed we shall see with an explicit realization that these restrictions are precisely those given in (6.41) and analysed in Lemma 6.44 for the representations labelled by α, β.

With the definition (6.76) we can construct representations of the q-boson algebra which lead to unitary representations of $\mathcal{U}_q(\mathfrak{su}(2))$. In the same way as for the generators, where we found that $(J_\pm)^{p'}$ have eigenvalues which are real and equal, the eigenvalues of a^p and $(a^\dagger)^p$ may be chosen by a choice of phase to be real and therefore equal. From Lemma 6.60 (substituting (6.62) and $\ell = q^{\frac{\phi}{2}}$), we determine the common eigenvalue to be:

$$\mu = \nu\varepsilon([\phi])\left|([\phi + 1])_{p-1}\right|,$$

which together with $\mu\nu = [\phi]$ determines each of μ, ν as a function of ϕ. Hence, when the operators satisfy the Hermiticity properties (6.76) there is a single invariant ϕ of the q-boson algebra. We can write the matrix elements in Lemma 6.60 in a manifestly Hermitean form by renormalizing the basis vectors according to the definition

$$|\phi, m\rangle = \left(\left|([\phi + 1])_{p-1}\right|\right)^{-\frac{1}{2}} v_m \tag{6.79}$$

for $m = 0, \ldots, p-1$, and then we find

$$\begin{aligned} L\,|\phi, m\rangle &= q^{\frac{1}{2}(\phi+m)}\,|\phi, m\rangle \\ a\,|\phi, m\rangle &= \sqrt{|[\phi+m+1]|}\,|\phi, m+1\rangle \\ a^\dagger\,|\phi, m\rangle &= \sqrt{|[\phi+m]|}\,|\phi, m-1\rangle. \end{aligned} \tag{6.80}$$

Since the matrix elements are periodic in m with period p, we may extend this periodicity to the basis vectors, and therefore define $|\phi, m+p\rangle = |\phi, m\rangle$ for all $m \in \mathbb{Z}$ (as was done for the $\mathcal{U}_q(\mathfrak{su}(2))$ basis vectors). It is straightforward to verify directly that the defining relations (6.54b) and (6.55) are satisfied, and that the invariants L^p and a^p have the respective eigenvalues: $q^{\frac{1}{2}p\phi}$, $\sqrt{\left|(([\phi])_p\right|}$.

EXAMPLE **6.81** For $p = 3$ the representation matrices are three-dimensional and are labelled by ϕ which we may take to be any real number:

$$L = \begin{pmatrix} q^{\frac{\phi}{2}} & 0 & 0 \\ 0 & q^{\frac{1}{2}(\phi+1)} & 0 \\ 0 & 0 & q^{\frac{1}{2}(\phi+2)} \end{pmatrix} \quad a = \begin{pmatrix} 0 & 0 & \sqrt{|[\phi]|} \\ \sqrt{|[\phi+1]|} & 0 & 0 \\ 0 & \sqrt{|[\phi+2]|} & 0 \end{pmatrix},$$

with $a^\dagger$ the transpose of a. These matrices satisfy $aa^\dagger = |[N]|$ and $a^\dagger a = |[N+1]|$ correctly.

Now let us construct unitary representations of $\mathcal{U}_q(\mathfrak{su}(2))$ using the generators (6.77) which act in the space $\mathfrak{V}_1 \otimes \mathfrak{V}_2$ spanned by the vectors $|\beta, m\rangle \otimes |\alpha, m'\rangle$, where β, α denote the invariants ϕ_1, ϕ_2 for $\mathfrak{V}_1, \mathfrak{V}_2$ respectively. Choose the subspace $\mathfrak{V}$ spanned by the vectors $|\alpha, \beta, m\rangle = |\beta, m\rangle \otimes |\alpha,\, p-m\rangle$, $m = 0, \ldots, p-1$, which are eigenstates of L_1, L_2 and consequently also of $[N_1], [N_2]$:

$$\begin{aligned} \left[N_1\right] |\alpha,\, \beta,\, m\rangle &= [\beta+m]|\alpha,\, \beta,\, m\rangle \\ \left[N_2\right] |\alpha,\, \beta,\, m\rangle &= [\alpha-m]|\alpha,\, \beta,\, m\rangle. \end{aligned} \tag{6.82}$$

In order that the generators (6.77) satisfy the $\mathcal{U}_q(\mathfrak{su}(2))$ algebra (6.10) the invariants α, β must be such as to satisfy (6.78). This requires

$$\varepsilon([\beta+m])\,\varepsilon([\alpha-m+1]) = 1 = \varepsilon([\beta+m+1])\,\varepsilon([\alpha-m]), \tag{6.83}$$

for $m = 0, \ldots, p-1$ (and hence for all $m \in \mathbb{Z}$), which is equivalent to the conditions $[\alpha-m][\beta+m+1] \geqslant 0$ for all $m \in \mathbb{Z}$. These are the inequalities (6.41) required for positivity and which are analyzed explicitly in Lemma 6.44. Assuming now that α, β satisfy these inequalities, we find that the generators (6.77) satisfy the quantum group algebra with the appropriate Hermiticity properties, and we may calculate the matrix elements from those of the q-boson operators (6.80). We obtain exactly the matrix elements (6.39), for example:

$$\begin{aligned} J_+|\alpha,\beta,m\rangle &= a_1 a_2^\dagger|\beta,m\rangle\otimes|\alpha,\,p-m\rangle \\ &= \sqrt{|[\beta+m+1][\alpha+p-m]|}\;|\beta,m+1\rangle\otimes|\alpha,\,p-m-1\rangle \\ &= \sqrt{[\alpha-m][\beta+m+1]}\;|\alpha,\beta,m+1\rangle. \end{aligned} \tag{6.84}$$

Again, we remark that these representations are irreducible for odd p. For even p we use the q-boson representations given in Lemma 6.73 in order to construct irreducible representations.

6.5 Cyclic q-Boson Operators in a Fock Space

In §6.4 we determined representations of q-boson operators acting in an abstract vector space and we now realize this space as a Fock space $\mathfrak{F}$, which consists of polynomials in a acting on a vacuum state $|0\rangle$, and may be relevant to physical applications of the cyclic irreps of $\mathcal{U}_q(\mathfrak{su}(2))$. The vacuum state is assumed to be unique and satisfies $\overline{a}|0\rangle = 0$. The vectors in $\mathfrak{F}$ will be denoted $|n\rangle$ and are eigenstates of the number operator N, with eigenvalue n (the "number of quanta") where $n \in \mathbb{N}$. In order to realize the vectors $|\phi,m\rangle$ defined in (6.79) as states in a Fock space, however, we must allow noninteger quanta ϕ and the unitary representations of $\mathcal{U}_q(\mathfrak{su}(2))$ may then be constructed as in §6.4.

We have seen already in §6.1.2 that the q-boson operators a and $\overline{a}$ are nilpotent in a Fock space carrying an integer number of quanta, and hence one cannot construct cyclic representations of $\mathcal{U}_q(\mathfrak{su}(2))$ in such a space. Let us extend the meaning of the Fock space, however, by constructing states with noninteger quanta. We discuss two ways of achieving this goal, firstly by introducing noninteger powers of q-boson operators and secondly by realizing q-boson operators in terms of boson operators as in (2.43), p. 32, for which the q-boson and boson vacua may differ. The need for the construction of noninteger quanta can be anticipated in view of the fact that the cyclic representations given in (6.39) are labelled by nonintegers α, β.

We wish to form the states $|\alpha\rangle$, where α is any real number, by allowing the operator a^α to act on the vacuum. We need not be precise about the meaning of noninteger powers of operators, which are best defined in specific contexts; for example, fractional derivatives (pseudo-differential operators) can be defined using Fourier transforms. However, we will assume various properties of states with noninteger quanta and operators with noninteger powers, which form part of the definition of these operators, and which we shall use to manipulate them. Consider the state $|\alpha\rangle = \mathcal{N}^{-\frac{1}{2}}a^\alpha|0\rangle$ for $\alpha\in\mathbb{R}$ (where $\mathcal{N}$ is the normalization), and which is an eigenstate of L:

$$L|\alpha\rangle = q^{\frac{\alpha}{2}}|\alpha\rangle.$$

(Hence, $[N]\ |\alpha\rangle = [\alpha]\ |\alpha\rangle$.) We assume that there exists an inner product, with respect to which the states $|\alpha\rangle$ are orthonormal:

$$\langle\beta|\alpha\rangle = \delta_{\alpha,\beta}.$$

We can change the number of quanta by one unit, by applying the operator a or $\overline{a}$ to the state $|\alpha\rangle$. Specifically,

$$a|\alpha\rangle = \sqrt{|[\alpha+1]|}\ |\alpha+1\rangle, \quad a^\dagger\ |\alpha\rangle = \sqrt{|[\alpha]|}\ |\alpha-1\rangle, \tag{6.85}$$

where the matrix elements are obtained using $a^\dagger a = [N+1]$. Hence we obtain the matrix elements of $a^{p'}$ and $(a^\dagger)^{p'}$:

$$a^{p'}|\alpha\rangle = \sqrt{\left|([\alpha])_{p'}\right|}\ |\alpha+p'\rangle, \quad (a^\dagger)^{p'}|\alpha\rangle = \sqrt{\left|([\alpha])_{p'}\right|}\ |\alpha-p'\rangle. \tag{6.86}$$

where we used the property (6.40). For integer $\alpha = n$, these equations reduce to $a^{p'}|n\rangle = 0 = (a^\dagger)^{p'}|n\rangle$.

We restrict our consideration now to odd p only, for which the Fock space states are eigenvectors of both L and a^p. Whereas the properties (6.85) and (6.86) generalize readily from the case in which α takes integer values, we cannot assume that a^α commutes with a for noninteger values of α, for if we do, it follows from $a^{p}\ |n\rangle = 0$ that $a^p|\alpha\rangle = \mathcal{N}^{-\frac{1}{2}} a^\alpha a^p|0\rangle = 0$ for any non-negative α in contradiction to (6.86).

A similar observation can be made about the operator $([N])_p$, which has the property that it commutes with a^n for any integer n, but not with a^α for noninteger α. This follows from

$$a^\alpha([N])_p = ([N-\alpha])_p a^\alpha$$

where for integer α the right hand side is equal to $([N])_p a^\alpha$ by (6.40), but not for noninteger values.

In order to investigate further the properties of the invariant a^p, we define the operators

$$A \overset{\text{def}}{=} a^p\left(\left|([N])_p\right|\right)^{-\frac{1}{2}}, \quad A^\dagger \overset{\text{def}}{=} \left(\left|([N])_p\right|\right)^{-\frac{1}{2}} (a^\dagger)^p, \tag{6.87}$$

which satisfy

$$A^\dagger A = I = AA^\dagger,$$
$$[A, a^n] = 0 = [A, (a^\dagger)^n] = [A^\dagger, a^n] = [A^\dagger, (a^\dagger)^n], \quad n \in \mathbb{N},$$

as follows from

$$(a^\dagger)^n a^n = |([N+1])_n|\,. \tag{6.88}$$

Therefore A and $A^\dagger$ are bounded operators with unit norm. Since a^α commutes with $A^\dagger A = 1$ for all real α, and commutes with A for all integer values of α, we *assume* that

$$[A, a^\alpha] = 0 = [A^\dagger, a^\alpha] \tag{6.89}$$

for all real α. We can regard this assumption as part of the definition of the noninteger power a^α and will see that it leads to conclusions consistent with (6.88).

From (6.89) we determine that A and $A^\dagger$ are each represented by the identity matrix in the Fock space of states with fractional quanta. We have

$$(a^\dagger)(A|0\rangle) = A((a^\dagger)|0\rangle) = 0,$$

and so, by uniqueness of the vacuum, we have $A|0\rangle = c|0\rangle$ for some constant $c \in \mathbb{C}$. Now $A^\dagger A = 1$ implies $c = 1$ up to a phase which we can choose to be zero, and so we have $A|0\rangle = |0\rangle$. An arbitrary state $|\alpha\rangle$ is proportional to $a^\alpha|0\rangle$ and hence, using now the assumption of commutativity, we deduce that $A|\alpha\rangle = |\alpha\rangle$ for all real α. Similarly $A^\dagger|\alpha\rangle = |\alpha\rangle$ for all real α. From the definition of A in (6.87), this means that

$$a^p|\alpha\rangle = \left(\left|([N])_p\right|\right)^{\frac{1}{2}}|\alpha\rangle = \left(\left|([\alpha])_p\right|\right)^{\frac{1}{2}}|\alpha\rangle = (a^\dagger)^p|\alpha\rangle, \tag{6.90}$$

which accords with (6.86). The number of quanta on each side of (6.90) is measured by $[N]$ (or L) and is equal to $[\alpha + p] = [\alpha]$. By comparing (6.90) with (6.86) we see that the states we have constructed are cyclic:

$$|\alpha\rangle = |\alpha + p\rangle, \tag{6.91}$$

and it is this property which enables us to construct the cyclic representations of $\mathcal{U}_q(\mathfrak{su}(2))$.

If we write $\alpha = [\![\alpha]\!] + \alpha'$ as defined in (6.43) then since $a^{[\![\alpha]\!]}$ does not commute with $a^{\alpha'}$, we must specify an order for these operators. We adopt the convention that for any $\alpha \in \mathbb{R}$, $a^\alpha|0\rangle \stackrel{\text{def}}{=} a^{[\![\alpha]\!]}a^{\alpha'}|0\rangle$, that is, we place the integer powers to the left of any noninteger powers. The normalization of the state a^α is determined from $\mathcal{N} = \langle 0|(a^\dagger)^\alpha a^\alpha|0\rangle$. For integer α this gives $\mathcal{N} = [\alpha]!$ and for noninteger α we can generalize this to $\mathcal{N} = |\Gamma_q(\alpha + 1)|$, where the q-analog of the Euler gamma function has been discussed in §2.8.2.

The second means by which noninteger q-boson quanta may be interpreted is the realization of q-boson operators in terms of boson operators given in (2.43), namely[5]:

$$a^q = \sqrt{\frac{|[N+\alpha]_q|}{N}}\,a, \quad (a^q)^\dagger = \overline{a}\sqrt{\frac{|[N+\alpha]_q|}{N}}, \quad N^q = N + \alpha, \quad \alpha \in \mathbb{R},$$

where a and $\overline{a}$ are boson creation and annihilation operators and $N = a\overline{a}$. These q-boson operators satisfy $a^q(a^q)^\dagger = |[N^q]|$, $(a^q)^\dagger a^q = |[N^q + 1]|$ as required. We now create states in the boson Fock space as polynomials in the q-boson operators built on the *boson* vacuum $|0\rangle$ annihilated by $\overline{a}$, and which carries a q-quantum number of α, that is, $[N^q]|0\rangle = [\alpha]|0\rangle$. We identify this boson vacuum state with the state

[5] For the remainder of this section we distinguish q-boson operators with the affix q.

$|\alpha, 0\rangle$ appearing in the matrix elements (6.80) in which we have put $\phi = \alpha$. The states $|\alpha, m\rangle$ are proportional to $a^m|0\rangle$ and carry a q-quantum number of $\alpha + m$, and we may re-derive the matrix elements (6.80). Effectively we are realizing the cyclic representations of the q-boson algebra in a space of polynomials of one variable, and we return to this construction in §6.6 where we discuss cyclic irreps of $\mathcal{U}_q(\mathfrak{sl}(2))$ in a polynomial space.

6.5.1 Unitary Cyclic Representations in a Fock Space

We can construct unitary cyclic representations of $\mathcal{U}_q(\mathfrak{su}(2))$ in an extended Fock space containing noninteger quanta α, β, where the possible values of these quanta is restricted by the condition (6.83). The explicit basis states, denoted $|\alpha, \beta, m\rangle$, are:

$$|\alpha, \beta, m\rangle = \mathcal{N}_m^{-\frac{1}{2}}(a_1)^{\beta+m}(a_2)^{\alpha-m}|0\rangle, \tag{6.92}$$

where $\mathcal{N}_m$ is the normalization, and are eigenstates of L_1, L_2 and hence also $[N_1], [N_2]$ as shown in (6.82). The invariants $(a_1)^p, (a_1^\dagger)^p$ and $(a_2)^p, (a_2^\dagger)^p$ take the respective eigenvalues:

$$(a_1)^p,\ (a_1^\dagger)^p \longrightarrow \sqrt{\left|([\alpha])_p\right|}, \quad (a_2)^p,\ (a_2^\dagger)^p \longrightarrow \sqrt{\left|([\beta])_p\right|},$$

as follows from (6.90). Although the states (6.92) have only the exponents $\alpha - m$ and $\beta + m$, we can actually distinguish four labels, these being essentially the integer and noninteger components of each exponent, and these arise because we have two additional invariants $(L_1)^p$ and $(L_2)^p$ which have the respective eigenvalues $q^{\frac{p\beta}{2}}$, $q^{\frac{p\alpha}{2}}$. The states

$$|\alpha, \beta, m_1, m_2\rangle = \mathcal{N}^{-\frac{1}{2}}(a_1)^{\beta+m_1}(a_2)^{\alpha-m_2}|0\rangle, \tag{6.93}$$

with the four labels α, β, m_1, m_2 span a space which carries irreducible cyclic representations of $\mathcal{U}_q(\mathfrak{u}(2))$. We recall that the meaning of the powers in the states (6.92) or (6.93) is that noninteger powers are written to the right of the integer powers.

We can now see that the states (6.92) are also eigenfunctions of the invariants $(J_+)^p$ and K^p, with nonzero eigenvalues, which are given by (respectively):

$$\sqrt{\left|([\alpha])_p([\beta])_p\right|}, \quad q^{\frac{1}{2}p(\beta-\alpha)}.$$

The label m is determined from the eigenvalue $q^{m+\frac{1}{2}(\beta-\alpha)}$ of K.

The normalization $\mathcal{N}_m$ is calculated from

$$\mathcal{N}_m = \langle 0|(a_2^\dagger)^\alpha(a_1^\dagger)^\beta(a_2^\dagger)^{-m}(a_2)^{-m}(a_1^\dagger)^m(a_1)^m(a_1)^\beta(a_2)^\alpha|0\rangle$$

and by using

$$(a_1^\dagger)^m(a_1)^m = \left|([N_1+1])_m\right|, \quad (a_2^\dagger)^{-m}(a_2)^{-m} = \frac{1}{\left|([N_2+1-m])_m\right|},$$

which follow from (6.88) and properties of the number operators, we obtain the dependence of the normalization on m:

$$\begin{aligned}\mathcal{N}_m &= \langle 0|(a_2^\dagger)^\alpha (a_1^\dagger)^\beta \frac{|(N_1+1])_m|}{|([N_2+1-m])_m|}(a_1)^\beta (a_2)^\alpha|0\rangle \\ &= \frac{|([\beta+1])_m|}{|([\alpha+1-m])_m|}\mathcal{N}_0, \qquad (6.94)\end{aligned}$$

where $\mathcal{N}_0$ is independent of m.

From the action (6.85) of the q-boson operators on the orthonormal basis (6.92) we can calculate the matrix elements of the generators $J_\pm, K$ and we derive precisely the matrix elements (6.39), with α and β restricted as in Lemma 6.44. The basis vectors are cyclic as indicated in (6.91), that is, the labels of these vectors depend only on $\alpha \bmod p$, $\beta \bmod p$ and $m \bmod p$. These representations are irreducible provided p is odd.

6.6 Cyclic Representations in a Space of Polynomials

In §2.4.3 we derived a realization of the generators of $\mathcal{U}_q(\mathfrak{sl}(2))$, in terms of a single complex variable z, which is related to the q-analog of fractional linear transformations on the complex plane (see particularly Lemma 2.55, p. 39). This construction appears naturally also in the method of algebraic induction considered in Chapter 7, and we now discuss the application of this approach to cyclic irreps of $\mathcal{U}_q(\mathfrak{sl}(2))$.

The realization of the generators of $\mathcal{U}_q(\mathfrak{sl}(2))$, for generic q, takes the form:

$$J_+ = D_z, \quad J_- = z[2j-N], \quad K = q^{j-N} = q^j L^{-2} \qquad (6.95)$$

where j may in general be taken to be a complex number, or an invariant operator, and N is the number operator which we can identify with $z\partial/\partial z$, and the finite difference operator D_z is defined by (2.39), p. 30. Two properties which we will frequently use are the q-boson relations, expressed in operator form,

$$zD_z = [N], \quad D_z z = [N+1]. \qquad (6.96)$$

It is straightforward to verify directly that the generators given by (6.95) do in fact satisfy the correct relations by using (6.96) and (6.35) with $a = N, b = 2j - N, c = -1$. The (unnormalized) basis states in a finite-dimensional irrep are $1, z, z^2, \dots z^{2j}$, showing that the dimension of the representation is $2j+1$ and that j is a non-negative half-integer.

Let us extend this construction to the cyclic irreps of $\mathcal{U}_q(\mathfrak{sl}(2))$ at roots of unity. The operator $J_+ = D_z$ acting in a space of polynomials in z is always nilpotent, for we have

$$(D_z)^p z^n = [n][n-1]\dots[n-p+1]z^{n-p} = ([n-p+1])_p z^{n-p} = 0$$

for all integers n, where we used $([n-p+1])_p = ([1])_p = [p]! = 0$ as follows from the cyclic property $[x+p] = [x]$ and $[p] = 0$. The operator D_z is therefore also nilpotent when acting in a space of functions expressible as a Laurent series expansion about $z = 0$. In order to construct the cyclic representations we must either allow nonpolynomial states and noninteger powers of z, or generalize the realization (6.95) for J_-. Since we have discussed the former strategy in the previous section and since for finite-dimensional representations it is natural to have polynomial basis vectors, let us implement the latter approach by finding a realization in which the generators $J_\pm$ are injective in the p-dimensional space spanned by the monomials $1, z, \ldots, z^{p-1}$.

We first write $J_+ = D_z = z^{-1}[N]$ by using (6.96), and then generalize the expressions for $J_\pm$ by putting

$$J_+ = z^{-1}[N+\alpha], \quad J_- = z[\beta - N], \tag{6.97}$$

where $\alpha, \beta \in \mathbb{C}$. The commutation relations for $\mathcal{U}_q(\mathfrak{sl}(2))$ are satisfied (using (6.35) with $a = N+\alpha$, $b = \beta - N$, $c = -1$) provided we define $K = q^{\frac{1}{2}(\beta-\alpha)-N}$. The generator $J_+ = z^{-1}[N+\alpha]$ is no longer nilpotent when acting on polynomials, provided α is noninteger, in fact

$$\left(J_+\right)^p z^n = ([n+\alpha-p+1])_p z^{n-p} = ([\alpha])_p z^{n-p}$$

and similarly for J_+.

Let us also consider the cyclic property of the basis vectors, which requires that we identify the polynomials z^n and z^{n+p}. We observe that the generators $J_\pm, K$ commute with z^p: $[J_\pm, z^p] = 0 = [K, z^p]$. Now we separate the monomials z^j into equivalence classes by defining the equivalence relation

$$z^j \sim z^k \quad \text{if} \quad j = k \bmod p. \tag{6.98}$$

Denote the set of all such equivalence classes by $\mathfrak{P}$, which is a finite set of elements which can be denoted $\{1, z, \ldots, z^{p-1}\}$. We now construct basis states from the elements of $\mathfrak{P}$, in effect letting $\mathfrak{P}$ consist of all monomials and setting $z^p = 1$. In order for operators on $\mathfrak{P}$ to be well defined, they must have the same action on any representative of an equivalence class, and this is guaranteed for operators which commute with z^p, in particular the generators.

The basis vectors are given by $v_m = \mathcal{N}_m^{-\frac{1}{2}} z^{p-m}$, where $\mathcal{N}_m$ is defined in (6.94), and where m takes the values $0, 1, \ldots, p-1$ or, equivalently, we let m take all integer values and the cyclic property means that the basis vectors are labelled by $m \bmod p$. Choosing now α, β to be real parameters satisfying the inequalities (6.41), we find that the matrix elements are as given in (6.39) for unitary irreps of $\mathcal{U}_q(\mathfrak{su}(2))$.

The cyclic representations of $\mathcal{U}_q(\mathfrak{sl}(2))$ determined from the realization (6.97) are reducible if p is even, but we can include this case by rewriting the realization such that the generators each commute with $z^{p'}$. This can be done in several ways by rearranging the factors involving the number operator, for example:

$$J_+ = z^{-1}[\beta - N + 1][\alpha + N], \quad J_- = z, \quad K = q^{\frac{1}{2}(\beta-\alpha)-N}, \tag{6.99}$$

where $\alpha, \beta \in \mathbb{C}$, and then each generator commutes with $z^{p'}$ (using (6.4)). The equivalence in (6.98) is modified by imposing instead $z^{p'} \sim b$ where $b \in \mathbb{C}$, and the basis vectors may be chosen to be simply $v_i = z^i$, $i = 0, \ldots, p' - 1$. The matrix elements of the generators of $\mathcal{U}_q(\mathfrak{sl}(2))$ are exactly as given in Theorem 6.22 upon identifying

$$\lambda = q^{\frac{1}{2}(\beta-\alpha)}, \quad ab = [\alpha][\beta + 1], \quad M_i = [\alpha + i][\beta + 1 - i].$$

As observed also in (6.68) in a similar situation, the substitution in terms of the complex parameters α, β factorizes the matrix elements M_i.

6.7 Algebraic Induction at Roots of Unity

The method of algebraic induction discussed in Chapter 7 provides a means of constructing the representations of the quantum unitary groups recursively[6]. Let us investigate how this can be done for the cyclic representations of $\mathcal{U}_q(\mathfrak{su}(2))$.

For generic q the induced basis vectors are constructed from those of the subgroup $U(1)$ by means of the formula (7.10), given in terms of the Gel'fand-Weyl labels. A difficulty which arises at roots of unity is that the q-exponential function in (7.10) can no longer be defined by its power series expansion, because of the appearance of the zero $[p] = 0$ in the factorial denominator (this can be avoided if the argument z is replaced by a nilpotent operator, but this does not occur for cyclic representations). Hence, we must replace the q-exponential by another holomorphic function $E(z)$.

The particular property of the q-exponential function which we wish to extend to $E(z)$ is that it is an eigenfunction of the generator J_+, as the case of generic q in (7.9) shows. Hence, we seek a function $E_\alpha(z)$ such that

$$z^{-1}[N + \alpha]E_\alpha(zJ_+) = J_+ E_\alpha(zJ_+), \tag{6.100}$$

where J_+ is an abstract $\mathcal{U}_q(\mathfrak{su}(2))$ generator, and where we have chosen the realization (6.97) of the generators which leads to cyclic irreps for odd p. For nonunitary irreps we may take α to be a complex number, and have indicated the dependence of $E(z)$ on α. Again, we impose the equivalence relation $z^p \sim 1$. If we expand $E_\alpha(z)$ in a Laurent series then by means of this equivalence we can regard $E_\alpha(z)$ as a polynomial in z of degree $p - 1$, and therefore write

$$E_\alpha(zJ_+) = \sum_{i=0}^{p-1} a_i z^i (J_+)^i,$$

[6] **It is convenient to discuss the application of algebraic induction to quantum groups at roots of unity in this chapter, however we assume some knowledge of Chapter 7, in particular §7.4.**

for some complex coefficients a_i. We determine these coefficients by substituting this polynomial into (6.100) and find $a_i = 1/([\alpha+1])_i$ (where we use the notation of the shifted factorial), provided that (formally) J_+ satisfies $(J_+)^p = ([\alpha])_p$. Hence, we define:

$$E_\alpha(z) \stackrel{\text{def}}{=} \sum_{i=0}^{p-1} \frac{z^i}{([\alpha+1])_i}.$$

Now let us outline the method of algebraic induction at roots of unity applied to $\mathcal{U}_q(\mathfrak{u}(2))$. We define induced vectors analogously to those in (7.10):

$$|\alpha,\beta,m\rangle_{\text{ind}} = \langle\alpha,\beta,0|\ E_\alpha(zJ_+)\ |\alpha,\beta,m\rangle\ \otimes |\alpha,\beta,0\rangle,$$

where $|\psi\rangle = |\alpha,\beta,0\rangle$ is a fixed vector carrying an irrep of $U(1)\times U(1) \subset \mathcal{U}_q(\mathfrak{u}(2))$ which spans the intrinsic space, and has the properties (in the notation of §7.4)

$$\mathcal{E}_{11}|\psi\rangle = \beta|\psi\rangle, \quad \mathcal{E}_{22}|\psi\rangle = \alpha|\psi\rangle,$$

and $|\alpha,\beta,m\rangle$ is a generic vector in a cyclic irrep of $\mathcal{U}_q(\mathfrak{su}(2))$. The polynomial

$$\mathcal{P}(z) = \langle\alpha,\beta,0|\ E_\alpha(zJ_+)\ |\alpha,\beta,m\rangle,$$

which we may regard as a coherent state operator, carries the invariant labels α,β (which in general are complex numbers) as well as the integer label m, which enumerates the p vectors spanning the irrep carrier space. The induced action of the generators is defined in the same way as for generic q by:

$$\Gamma(J)|\alpha,\beta,m\rangle_{\text{ind}} \stackrel{\text{def}}{=} \langle\alpha,\beta,0|\ E_\alpha(zJ_+)J\ |\alpha,\beta,m\rangle\ \otimes |\alpha,\beta,0\rangle,$$

where J denotes a $\mathcal{U}_q(\mathfrak{su}(2))$ generator.

We now compute the explicit form of the operators $\Gamma(J)$ and find:

Lemma **6.101** *The $\mathcal{U}_q(\mathfrak{su}(2))$ generators J have a right action on the induced vectors $|\alpha,\beta,m\rangle_{\text{ind}}$ with $\Gamma(J)$ given by*

$$\Gamma(J_+) = z^{-1}[N+\alpha], \quad \Gamma(J_-) = z[\beta - N], \quad \Gamma(K) = q^{\frac{1}{2}(\beta-\alpha)-N},$$

where N is the number operator $z\partial/\partial z$.

Proof: The expression for $\Gamma(K)$ follows from

$$Kq^{-N}z^n(J_+)^n = z^n(J_+)^nK$$

and the matrix element $\langle\alpha,\beta,0|K = q^{\frac{1}{2}(\beta-\alpha)}\langle\alpha,\beta,0|$. The expression for $\Gamma(J_+)$ follows from the constructed property (6.100) of $E_\alpha(zJ_+)$, and the eigenvalue relation $(J_+)^p \to ([\alpha])_p$. Finally, the expression for $\Gamma(J_-)$ is derived using firstly the identity

$$(zJ_+)^nJ_- = J_-(zJ_+)^n + z[n][2J_z - n + 1](zJ_+)^{n-1}, \quad n\in\mathbb{N},$$

from which follows

$$E_\alpha(zJ_+)J_- = J_-E_\alpha(zJ_+) + z\frac{[N+1][2J_z - N]}{[\alpha + N + 1]}E_\alpha(zJ_+).$$

Next, we use the identity (6.35) with $a = \beta - N$, $b = -N - 1$, $c = \alpha$, and the matrix element

$$\langle\alpha,\beta,0|J_- = z[\alpha][\beta+1][\alpha+N+1]^{-1}\langle\alpha,\beta,0|$$

which follows from

$$\langle\alpha,\beta,0|J_-J_+ = [\alpha][\beta+1]\langle\alpha,\beta,0|$$

to complete the proof. (In the cyclic case for which there is no vector of highest weight we can choose any fixed vector $|\psi\rangle$ in order to form the coherent state $\mathcal{P}(z)$, and matrix elements on this vector are determined from the eigenvalues of the Casimir invariant and K.) □

The method of algebraic induction has also been applied to quantum groups at roots of unity by Links and Zhang [178], and Schnizer [179] has described the construction of quantum group irreps for rank n from those of a quantum group of rank $n-1$.

Chapter 7

Algebraic Induction of Quantum Group Representations

In previous chapters we have discussed irreducible representations of $\mathcal{U}_q(\mathfrak{su}(2))$ in detail, classifying all such representations and deriving matrix elements, and have also extended these considerations to tensor operators. We have explicitly constructed the irreps in a Fock space, and presented some results for general unitary quantum groups, specifically, the construction of symmetric irreps (§2.5.1). In principle, this construction can be extended to include all irreps of $\mathcal{U}_q(\mathfrak{u}(n))$ by following the example of the boson calculus applied to $U(n)$, as summarized in the monograph [103] (see also the review by Louck [108]).

In this chapter we describe a different approach to the construction of irreps of $\mathcal{U}_q(\mathfrak{u}(2))$ which admits a relatively simple generalization to arbitrary n. This method of algebraic induction is recursive in that one assumes that all irreps of of the subgroup $\mathcal{U}_q(\mathfrak{u}(n-1)) \times U(1)$ have already been constructed and is explicit to the extent that the basis vectors spanning an irrep space for $\mathcal{U}_q(\mathfrak{u}(n))$ are expressed in terms of those for $\mathcal{U}_q(\mathfrak{u}(n-1)) \times U(1)$. Furthermore, the method extends to tensor operators and provides a powerful means of expressing properties of $\mathcal{U}_q(\mathfrak{u}(n))$ tensor operators in terms of those for $\mathcal{U}_q(\mathfrak{u}(n-1))$; this leads to significant properties such as limits and certain matrix elements which may expressed in terms of those for the subgroup.

7.1 Introduction and Summary

We begin with a general description of algebraic induction in §7.2, indicating its relation to the inducing methods of Borel and Weil, and proceed in §7.3 to $U(2)$ for which the method is sufficiently simple that we may demonstrate its application by direct calculation; we assume the existence of $U(1) \times U(1)$ irreps and induce those of $U(2)$ using "vector coherent states" which are holomorphic functions of complex variables, and are called "coherent" because they may equivalently be constructed as exponentials in boson creation operators. Following this we are able to write down the extension to $\mathcal{U}_q(\mathfrak{u}(2))$ in §7.4 without difficulty.

Before generalizing the $\mathcal{U}_q(\mathfrak{u}(2))$ construction to all $\mathcal{U}_q(\mathfrak{u}(n))$, which turns out to be rather complex, we first review in §7.5 the application of this approach to $U(n)$ in order to highlight the more significant aspects of the construction and to indicate where difficulties might arise in the extension to the quantum case. Then in §7.6 we turn to the general case of $\mathcal{U}_q(\mathfrak{u}(n))$ and construct all its irreps, assuming full knowledge of those of $\mathcal{U}_q(\mathfrak{u}(n-1)) \times U(1)$. Included is a description of the $\mathcal{U}_q(\mathfrak{u}(n))$ realization thus obtained, and a derivation which expresses the basis vectors in a different form using the coupling coefficients of the subgroup, which generalizes a similar form for $U(n)$.

In §7.7 we turn again to a particular case, $n = 3$, and discuss algebraic induction in detail for both $U(3)$ and its quantum extension $\mathcal{U}_q(\mathfrak{u}(3))$. This has a twofold purpose: firstly, it demonstrates explicitly how the construction may be implemented in a case which is considerably more complex than $n = 2$ but is still sufficiently simple that details can be worked out by direct calculation, and as such may be perused with profit by the reader before the case of general n is considered in §7.6. Secondly, the details of the $\mathcal{U}_q(\mathfrak{u}(3))$ construction reveal an aspect of quantum groups which we frequently highlight, the relationship to and implications for the theory of special functions, principally basic hypergeometric functions. The equality of two specific forms of the basis vectors in the Borel-Weil approach turns out to imply special function identities of considerable generality and complexity. For $U(3)$ we shall derive, using only group theoretic properties, a transformation of ${}_6F_5$ hypergeometric functions into ${}_3F_2$ functions, themselves related to the $SU(2)$ Wigner-Clebsch-Gordan coefficients, which is a special case of an important identity found by Whipple in 1926. For $\mathcal{U}_q(\mathfrak{u}(3))$ this identity generalizes to one which relates a very well-poised basic hypergeometric function ${}_7\phi_6$ to a terminating balanced ${}_3\phi_2$ series, and is a special case of a general identity due to Watson, which can also be derived using Bailey's Lemma (Andrews [180]). (We actually use this identity to verify a form of the basis vectors deduced from general considerations, but conversely we could regard this known form as implying the existence of the identity). Hence, quantum groups can again be seen to supply an algebraic understanding of q-analog identities, generalizing those of the classical functions.

7.2 The Algebraic Borel-Weil Construction

In physical applications of symmetry groups, both classical and quantum, there are two basic technical problems that require solution: the construction of all unitary irreps of the group, and the construction of all tensor operators acting on the model space for this group. A general technique for the construction of representations was developed by Frobenius; this is the method of induced representations, which constructs a representation of G from a representation of a subgroup H. Gel'fand and Naimark [181], and also Mackey [182] have extended the method of induced representations to all Lie groups, but the problem of irreducibility did not have an optimal resolution in this work.

For compact Lie groups a complete resolution was given in the famous work of Borel and Weil [183], who demonstrated that all unitary irreps of a compact simple Lie group G could be constructed by induction from a character (a one dimensional unitary irrep) of the Cartan subgroup T, the maximal toroid. To every character of T one associates a holomorphic line bundle G/T which carries a natural G-action. The holomorphic sections of this homogeneous holomorphic line bundle are irreps of G with highest weight given by the character of T.

More precisely, the Borel-Weil method, applied to representations of a compact simple Lie group G, constructs a line bundle over the homogeneous space G/T, where T is the maximal torus of G. This coset space G/T can be made into a complex manifold, as can be seen from the fact that $G/T \cong G_c/B^+$, where G_c is the complexified group G and B^+ is the Borel subgroup. (We will be considering $U(n)$ below for which B^+ is the subgroup of upper triangular matrices.) To every character χ of the torus T one associates a homogeneous holomorphic line bundle L_χ over G/T; this uses the fact that every homomorphism $\chi : T \to \mathbb{C}^\times$ extends uniquely to a holomorphic homomorphism $\chi : B^+ \to \mathbb{C}^\times$ so that one may define the associated line bundle $L_\chi \overset{\text{def}}{=} G_c \times_{B^+} \mathbb{C}$ (which denotes the quotient of $G_c \times \mathbb{C}$ by the equivalence relation $(gb, \xi) = (g, \chi(b)\xi)$ for all $b \in B^+$). The group G_c acts on the line bundle L_χ and hence on its cross-sections; this action descends to an action by the group G. The Borel-Weil theorem asserts that if χ is a dominant weight then the space of holomorphic sections of L_χ gives an irrep of G with dominant weight χ.

The work of Borel and Weil is an elegant resolution of the problem of constructing all unitary irreps, but this method is heavily topological and coordinate-free and hence relatively inaccessible to physicists. In point of fact physicists (Holstein and Primakoff [96]) had actually discovered the essence of the Borel-Weil (BW) construction at least a decade before the work of Borel and Weil, but this was not generally realized until an equivalent, purely algebraic procedure, known as the "vector coherent state" method, was developed much later by Rowe, Hecht, LeBlanc and others [184, 133, 185]. This algebraic version of the BW technique, also called the algebraic induction method, was extended to quantum groups by Biedenharn and Lohe [186, 187, 188].

The algebraic induction method is particularly well adapted to standard techniques in physics, since the complex manifolds in the BW construction can be easily identified with boson operator spaces (Fock spaces) thereby combining the utility of the Jordan map with an operator version of the BW induction procedure. The algebraic induction procedure is therefore of great utility in resolving the second technical problem, the construction of all tensor operators. This has been shown for classical [131] and also quantum groups [186] and will be discussed below briefly in §7.8.

The unitary group $U(n)$ is particularly well suited to the algebraic induction technique, because of the existence of the subgroup chain

$$U(n) \supset U(n-1) \times U(1) \supset \ldots \supset \overbrace{U(1) \times \ldots \times U(1)}^{n \text{ times}} \overset{\text{def}}{=} T.$$

In our extension of this construction to the quantum group $\mathcal{U}_q(\mathfrak{u}(n))$ we will employ a recursive approach which differs from the BW procedure in that the induction will be from an irrep of the subgroup $\mathcal{U}_q(\mathfrak{u}(n)) \times U(1)$ rather than from a character of the maximal torus T. The base manifold is then $\mathcal{U}_q(\mathfrak{u}(n))/(\mathcal{U}_q(\mathfrak{u}(n-1)) \times U(1))$ which, however, still has a complex structure.

The recursive procedure assumes a prior construction of irreps of $\mathcal{U}_q(\mathfrak{u}(n-1))$ itself by the same method, and likewise for each step in the induction procedure. Appealing to Mackey's theorem [182] on "induction-in-stages" one sees that this recursive approach is, despite the seeming difference, equivalent in fact to the BW construction itself.

7.3 Algebraic Induction for the Classical Group $U(2)$

We will illustrate the method of algebraic induction by considering first the elementary (classical) $U(2)$ example. The generalization to $\mathcal{U}_q(\mathfrak{u}(2))$ will then be carried out.

For $U(2)$ the maximal Abelian (Cartan) subgroup is $U(1) \times U(1)$, generated by the Weyl operators $\mathcal{E}_{11}$ and $\mathcal{E}_{22}$. The induction process assumes the existence of an irrep of $U(1) \times U(1)$ carried by a single vector $|\psi\rangle$, which we label by the eigenvalues of $\mathcal{E}_{11}$ and $\mathcal{E}_{22}$. Explicitly, we take the vector $|\psi\rangle$ to be

$$|\psi\rangle = \left|\begin{matrix} m_{12} & & m_{22} \\ & m_{12} & \end{matrix}\right\rangle, \tag{7.1}$$

where we use a Gel'fand-Weyl pattern (discussed in §2.6) to denote $|\psi\rangle$ as the highest weight vector in the irrep $[m_{12}, m_{22}]$, which is the irrep to be induced; $|\psi\rangle$ has, by definition, the properties:

$$\mathcal{E}_{11}|\psi\rangle = m_{12}|\psi\rangle, \quad \mathcal{E}_{22}|\psi\rangle = m_{22}|\psi\rangle. \tag{7.2}$$

REMARK **7.3** Since we have restricted our attention to unitary irreps of compact groups, it will be seen at this stage of our construction that we have actually made a choice in (7.1) by requiring $|\psi\rangle$ to be a highest weight vector. Clearly (for a compact group) we could equally well have chosen a lowest weight vector. The irrep so constructed will be the same in either case, but the specific techniques will differ (corresponding to "subtraction" vs. "addition" of angular momentum in the present case) with many concomitant sign differences. We will in fact choose a different sign convention in §7.6 when developing the method for $\mathcal{U}_q(\mathfrak{u}(n))$. In the extension of the method of algebraic induction to the construction of tensor operators, the two choices can lead to distinct results.

The algebraic induction method augments the vector $|\psi\rangle$ by tensoring it with a polynomial $\mathcal{P}(a)$ of the boson operator a, acting on the boson vacuum ket $|0\rangle$, to yield a general induced state $|(m)\rangle_{\text{ind}}$ in the irrep $[m_{12}, m_{22}]$. In other words, we form the direct product space $\mathfrak{F} \otimes \{|\psi\rangle\}$, where $\mathfrak{F}$ denotes the Fock space in which the boson operator a acts. A general induced state in $\mathfrak{F} \otimes \{|\psi\rangle\}$ is then given by

$$|(m)\rangle_{\text{ind}} = \left|\begin{matrix} m_{12} & & m_{22} \\ & m_{11} & \end{matrix}\right\rangle_{\text{ind}} \stackrel{\text{def}}{=} \mathcal{P}(a)\,|0\rangle \otimes \left|\begin{matrix} m_{12} & & m_{22} \\ & m_{12} & \end{matrix}\right\rangle.$$

The operator-valued polynomial $\mathcal{P}(a)$, which implicitly carries the m_{11} label, has a characteristic form; it is a matrix element of a coherent state operator:

$$\mathcal{P}(a) \stackrel{\text{def}}{=} \left\langle\begin{matrix} m_{12} & & m_{22} \\ & m_{12} & \end{matrix}\right| e^{aE_{12}} \left|\begin{matrix} m_{12} & & m_{22} \\ & m_{11} & \end{matrix}\right\rangle, \tag{7.4}$$

where E_{12} is the Weyl generator (the raising operator) in $\mathfrak{u}(2)$ with a well-defined action on the Gel'fand-Weyl vectors $|(m)\rangle$ in the bra-ket. Note that $\mathcal{P}$ is a polynomial in aE_{12} because only one term (which depends on m_{11}) in the series representation of $e^{aE_{12}}$ contributes to the matrix element. (E_{12} acts to the right as a raising generator, but to the left as a lowering generator; the notation (,) for inner product is perhaps more appropriate than the bra-ket notation in this context).

The Weyl generators E_{ij}, $(i,j = 1,2)$ have an induced right action on the vectors $|(m)\rangle_{\text{ind}}$. Let us denote by $\Gamma(E_{ij})$ the generators which act in the space spanned by these induced vectors $|(m)\rangle_{\text{ind}}$. (Hence Γ maps from $\mathfrak{g} = \mathfrak{u}(2)$ to the space of operators spanned by $\mathcal{E}_{11}, \mathcal{E}_{22}$ and polynomials in the boson creation and destruction operators.) The induced action of the generators is then defined by:

$$\Gamma(E_{ij}) \left|\begin{matrix} m_{12} & & m_{22} \\ & m_{11} & \end{matrix}\right\rangle_{\text{ind}} \stackrel{\text{def}}{=} \left\langle\begin{matrix} m_{12} & & m_{22} \\ & m_{12} & \end{matrix}\right| e^{aE_{12}} E_{ij} \left|\begin{matrix} m_{12} & & m_{22} \\ & m_{11} & \end{matrix}\right\rangle |0\rangle \otimes \left|\begin{matrix} m_{12} & & m_{22} \\ & m_{12} & \end{matrix}\right\rangle.$$

We may now compute the explicit form of the operators $\Gamma(E_{ij})$ and we find [131]:

LEMMA **7.5**

$$\begin{aligned} \Gamma(E_{11}) &= \mathcal{E}_{11} - a\overline{a} \\ \Gamma(E_{22}) &= \mathcal{E}_{22} + a\overline{a} \\ \Gamma(E_{12}) &= \overline{a} \\ \Gamma(E_{21}) &= a(\mathcal{E}_{11} - \mathcal{E}_{22} - a\overline{a}). \end{aligned}$$

PROOF: The first three relations are easily verified, for example,

$$\overline{a}e^{aE_{12}}|0\rangle = e^{aE_{12}}E_{12}|0\rangle$$

so that $\Gamma(E_{12})|(m)\rangle_{\text{ind}} = \overline{a}|(m)\rangle_{\text{ind}}$. To determine $\Gamma(E_{21})$ we use

$$e^{aE_{12}}E_{21}e^{-aE_{12}} = E_{21} + a(E_{11} - E_{22}) - a^2(E_{12}),$$

which, upon using (7.4), yields the required expression for $\Gamma(E_{21})$. □

It can be determined directly [186] that the generators $\Gamma(E_{ij})$, $i,j=1,2$ obey the commutation relations of $U(2)$, that is, the mapping Γ is a Lie algebra homomorphism, and that the induced vectors $|(m)\rangle_{\text{ind}}$ carry the irrep $[m_{12}, m_{22}]$ of $U(2)$. But note that these vectors, although orthogonal, do not have unit norm with respect to the Fock space inner product; instead, they are normalized such that the action of the induced generators $\Gamma(E_{ij})$ yields the standard matrix elements for the Cartan-Weyl generators E_{ij} of $U(2)$ acting on an orthonormal Gel'fand-Weyl basis. In other words, we redefine the Fock space inner product by means of a simple scale factor, with respect to which the generators are Hermitean.

The polynomial $\mathcal{P}(a)$ can be easily evaluated, yielding the following explicit form for the induced vectors:

$$\left|\begin{smallmatrix} m_{12} & & m_{22} \\ & m_{11} & \end{smallmatrix}\right\rangle_{\text{ind}} = K\frac{a^{m_{12}-m_{11}}}{\sqrt{(m_{12}-m_{11})!}}|0\rangle \otimes \left|\begin{smallmatrix} m_{12} & & m_{22} \\ & m_{12} & \end{smallmatrix}\right\rangle, \tag{7.6}$$

where

$$K = \left(\frac{(m_{12}-m_{22})!}{(m_{11}-m_{22})!}\right)^{\frac{1}{2}}.$$

This factor K ensures that the generators have the standard matrix elements, and K^2 has the further significance of defining a Kähler metric in a holomorphic manifold (see for example Wallach [189]). The boson operator a may equivalently be considered as a complex variable when realized according to $a \to z, \bar{a} \to d/dz$, so that "holomorphic" has the standard meaning (this realization of boson operators is discussed in §2.4.1, p. 29).

REMARK **7.7** The BW procedure emphasizes another aspect of the function $\mathcal{P}(a)$, that of an equivalence relation on the associated bundle, which merits mention here. As defined in (7.4), this function admits a *left* group action, for $g \in U(1)\times U(1)$, such that

$$\begin{aligned} \mathcal{P}(ga) &= \left\langle\begin{smallmatrix} m_{12} & & m_{22} \\ & m_{12} & \end{smallmatrix}\right| g e^{aE_{12}} \left|\begin{smallmatrix} m_{12} & & m_{22} \\ & m_{11} & \end{smallmatrix}\right\rangle \\ &= \chi(g^{-1})\mathcal{P}(a), \end{aligned} \tag{7.8}$$

where χ is the character of $U(1)\times U(1)$ carried by $|\psi\rangle$ as defined in (7.1). It is clear that this left action is equivalently given by the action of the Weyl generators $\mathcal{E}_{11}, \mathcal{E}_{22}$ acting on $|\psi\rangle$. The relation (7.8) is essential in the more abstract geometric BW approach.

This treatment of algebraic induction for $U(2)$ is equivalent to the construction of vector coherent states for angular momentum as discussed, for example, by Hecht [133, §1.1, §1.2]. Coherent states $|z\rangle$ may be defined (ignoring a z-dependent normalization) by the relation

$$|z\rangle = e^{za}|0\rangle, \quad z \in \mathbb{C},$$

where a is a boson creation operator, and obey the property, characteristic of coherent states, $\overline{a}|z\rangle = z|z\rangle$. (We discuss coherent states and their q-analogs in §8.1.1, p. 255). Vector coherent states are constructed in a similar way, namely

$$|z\rangle = e^{zJ_-}|j,j\rangle,$$

where $J\_$ is the lowering generator of $SU(2)$ and $|j,j\rangle$ is the state of highest weight in some irrep of $SU(2)$. Hence, state vectors $|\phi\rangle$ are mapped into the holomorphic functions

$$\langle z|\phi\rangle = \langle j,j|\, e^{zJ_+}\, |\phi\rangle,$$

which are essentially the matrix elements (7.4). The factor K above, which ensures that irreps of $SU(2)$ are unitary, may be promoted to an operator $\mathbf{K}$ with which we perform a similarity transformation on the non-unitary generators to obtain unitary irreps. Essentially, we use $\mathbf{K}$ to redefine the inner product. We return to properties of this operator when we generalize these methods to $\mathcal{U}_q(\mathfrak{u}(2))$ in the next section.

7.4 Algebraic Induction for the Quantum Group $\mathcal{U}_q(\mathfrak{u}(2))$

To proceed to the quantum group $\mathcal{U}_q(\mathfrak{u}(2))$ is now relatively straightforward. The ordinary exponential in (7.4) is generalized to the q-exponential $\exp_q$ defined by its series expansion in Definition 2.114 (p. 64), and is a holomorphic function on the complex plane. (We extend the domain of $\exp_q$ to commuting operators in order to construct the q-exponential of a q-boson operator acting on $|0\rangle$, but with the realization discussed in §2.4.1 this is equivalent to $\exp_q$ defined on $\mathbb{C}$). One property amongst several which we require is the derivative property of Lemma 2.115 (p. 65) which, in terms of q-boson operators, reads:

$$\overline{a}^q \exp_q(\alpha a^q)|0\rangle = \alpha \exp_q(\alpha a^q)|0\rangle, \tag{7.9}$$

for real- or operator-valued α.

The carrier space for all unitary irreps of $\mathcal{U}_q(\mathfrak{u}(2))$ is then spanned by the induced vectors:

$$\begin{aligned} \left|\begin{smallmatrix} m_{12} & & m_{22} \\ & m_{11} & \end{smallmatrix}\right\rangle_{\text{ind}} &= \left\langle\begin{smallmatrix} m_{12} & & m_{22} \\ & m_{12} & \end{smallmatrix}\right| \exp_q\left(a^q E_{12}\right) \left|\begin{smallmatrix} m_{12} & & m_{22} \\ & m_{11} & \end{smallmatrix}\right\rangle |0\rangle \otimes \left|\begin{smallmatrix} m_{12} & & m_{22} \\ & m_{12} & \end{smallmatrix}\right\rangle \\ &= K_q \frac{(a^q)^{m_{12}-m_{11}}}{\sqrt{[m_{12}-m_{11}]!}} |0\rangle \otimes \left|\begin{smallmatrix} m_{12} & & m_{22} \\ & m_{12} & \end{smallmatrix}\right\rangle, \end{aligned} \tag{7.10}$$

where

$$K_q = \left(\frac{[m_{12}-m_{22}]!}{[m_{11}-m_{22}]!}\right)^{\frac{1}{2}}. \tag{7.11}$$

The expression (7.10) is a straightforward q-analog of the $q = 1$ expression, and is obtained by assuming the known matrix elements of the $\mathcal{U}_q(\mathfrak{u}(2))$ generators, given in Example 2.76 (p. 48). Just as for $q = 1$, the q-boson monomial is normalized using the Fock space inner product, and the K_q-factor ensures that the generators $\Gamma(E_{ij})$ have the correct matrix elements for unitary irreps of $\mathcal{U}_q(\mathfrak{u}(2))$.

The induced action of the generators is defined analogously to $q = 1$ by:

$$\Gamma(E_{ij})\left|\begin{smallmatrix} m_{12} & & m_{22} \\ & m_{11} & \end{smallmatrix}\right\rangle_{\text{ind}} \stackrel{\text{def}}{=} \left\langle\begin{smallmatrix} m_{12} & & m_{22} \\ & m_{12} & \end{smallmatrix}\right| \exp_q(aE_{12})E_{ij}\left|\begin{smallmatrix} m_{12} & & m_{22} \\ & m_{11} & \end{smallmatrix}\right\rangle |0\rangle \otimes \left|\begin{smallmatrix} m_{12} & & m_{22} \\ & m_{12} & \end{smallmatrix}\right\rangle.$$

LEMMA 7.12 *The $\mathcal{U}_q(\mathfrak{u}(2))$ generators $\{E_{ij}, (i,j = 1,2)\}$ have a right action on the induced vectors (7.10) with the realizations $\Gamma(E_{ij})$ given by*

$$\begin{aligned} \Gamma(E_{11}) &= \mathcal{E}_{11} - N \\ \Gamma(E_{22}) &= \mathcal{E}_{22} + N \\ \Gamma(E_{12}) &= \overline{a}^q \\ \Gamma(E_{21}) &= a^q[\mathcal{E}_{11} - \mathcal{E}_{22} - N], \end{aligned}$$

where N is the number operator for the q-boson operators defined in (2.27) and, as before, $\mathcal{E}_{11}$, $\mathcal{E}_{22}$ are the generators of the inducing subgroup $U(1) \times U(1)$.

PROOF: The first three relations follow as in Lemma 7.5, for example, the commutator $[E_{11} - N, a^q E_{12}] = 0$ implies $[E_{11} - N, \exp_q(a^q E_{12})] = 0$, which in turn implies

$$(E_{11} - N)\exp_q(a^q E_{12})|0\rangle = \exp_q(a^q E_{12})E_{11}|0\rangle,$$

and hence,

$$\Gamma(E_{11})|(m)\rangle_{\text{ind}} = (\mathcal{E}_{11} - N)|(m)\rangle_{\text{ind}}.$$

Similarly $[E_{22} + N, a^q E_{12}] = 0$ leads to the second relation. The proof of the last relation requires the identity

$$[E_{12}^n, E_{21}] = [n][E_{11} - E_{22} - n + 1]E_{12}^{n-1}, \quad n \in \mathbb{N}$$

which may be proved by induction on n using

$$E_{12}^n[E_{11} - E_{22}] = [E_{11} - E_{22} - 2n]E_{12}^n,$$

and the ubiquitous identity (2.12) with $a = 1, b = E_{11} - E_{22} - n + 1, c = n + 1$. Hence, after postmultiplying by $(a^q)^n/[n]!$, summing over n, and letting the result act on $|0\rangle$, we obtain

$$E_{21}\exp_q(a^q E_{12})|0\rangle = \exp_q(a^q E_{12})E_{21}|0\rangle - [E_{11} - E_{22} - N + 1]a^q \exp_q(a^q E_{12})|0\rangle.$$

The last relation follows now by using that fact that E_{21} annihilates the highest weight bra-vector and that $E_{ii} \to \mathcal{E}_{ii}$ on the same vector. □

LEMMA **7.13** *The map* Γ *is an isomorphism of the quantum group* $\mathfrak{g} = \mathcal{U}_q(\mathfrak{u}(2))$.

This result has already been proved in Lemma 2.56 (p. 39) in an apparently different context. We obtained there essentially the above realization of $\mathcal{U}_q(\mathfrak{su}(2))$ in terms of $a^q, \overline{a}^q$ by means of a mapping from a homogeneous to a projective space.

For the same reason as for $q = 1$, the K_q-factor in (7.11) is necessary because the generators $\Gamma(E_{ij})$ are not Hermitean with respect to the q-boson inner product, as is evident from the expressions for $\Gamma(E_{ij})$ in Lemma 7.12. However, following Rowe et al. [184] and Hecht [133] (for $q = 1$), we may redefine the inner product so that the generators become Hermitean. This redefinition can be accomplished most simply by means of a similarity transformation with an operator $\mathbf{K}$, that is, we define new generators

$$\gamma(E_{ij}) = \mathbf{K}^{-1}\Gamma(E_{ij})\mathbf{K},$$

where $\mathbf{K}$ can be chosen to be Hermitean with respect to the Fock space, $\mathbf{K} = \mathbf{K}^\dagger$, and commutes with $\Gamma(E_{11})$ and $\Gamma(E_{22})$. The requirement $\gamma(E_{12}) = \gamma(E_{21})^\dagger$ reduces to $\mathbf{K}^2\,\Gamma(E_{12})^\dagger = \Gamma(E_{21})\,\mathbf{K}^2$, that is,

$$\mathbf{K}^2\, a^q = a^q[\mathcal{E}_{11} - \mathcal{E}_{22} - N]\,\mathbf{K}^2.$$

It is helpful to introduce an operator Λ, which depends on N, according to the equation

$$[\Lambda,\, a^q] = a^q[\mathcal{E}_{11} - \mathcal{E}_{22} - N]$$

and which therefore satisfies $\Lambda(N+1) - \Lambda(N) = [\mathcal{E}_{11} - \mathcal{E}_{22} - N]$. (We could, if we wish, solve for Λ as a function of N in closed form by using the summation identity (2.101) on p. 59). If we now take matrix elements of the relation

$$\mathbf{K}^2 a^q = [\Lambda,\, a^q]\mathbf{K}^2$$

between normalized q-boson (monomial) states of quanta $n+1$ and n, where $n = m_{12} - m_{11}$, tensored with the vector $|\psi\rangle$ in (7.1), we find

$$\left(\frac{K_q(n+1)}{K_q(n)}\right)^2 = \Lambda_{n+1} - \Lambda_n = [m_{12} - m_{22} - n]$$

where $K_q(n+1), \Lambda_n$ are the eigenvalues of $\mathbf{K}$ and Λ respectively, and where we have substituted the eigenvalues of $\mathcal{E}_{11}$ and $\mathcal{E}_{22}$ as shown in (7.2). Solving for K_q and putting $n = m_{12} - m_{22}$ gives the same value of K_q in (7.11) above, which we found there by *assuming* the form of the matrix elements of the generators. In contrast, by normalizing the q-boson monomials using the operator $\mathbf{K}$ we have constructed a unitary irrep of $\mathcal{U}_q(\mathfrak{u}(2))$, and the matrix elements of the generators E_{ij}, as given in Example 2.76 (p. 48), may then be *derived.*

The K_q-factor defined in (7.11) has the significance, for $q = 1$, of defining a metric in the Kähler manifold, and hence it is likely that this construction for the q-analog case allows one to define the metric for an analog to a "q-Kähler manifold" for the quantum group $\mathcal{U}_q(\mathfrak{u}(2))$.

REMARK **7.14** It is significant that in this construction all irrep vectors of $\mathcal{U}_q(\mathfrak{u}(2))$ appear as monomials, involving only a single q-boson a^q, in contrast to the two q-bosons involved in the construction of $\mathcal{U}_q(\mathfrak{su}(2))$ (see(2.47)), or the four q-bosons in (2.59). This suggests that the explicit construction of all unitary irreps of $\mathcal{U}_q(\mathfrak{u}(n))$ can be achieved with only $n-1$ q-bosons, and this is indeed correct; the corresponding recursive approach is an algebraic generalization of the Borel-Weil result. This generalization, for the classical Lie group $U(n)$, corresponds to constructing holomorphic sections of a bundle whose fibers carry irreps of the subgroup $U(n-1) \times U(1)$. The base manifold is now $U(n)\big/(U(n-1) \times U(1))$ having $(n-1)$ complex dimensions.

7.5 The Algebraic Induction Construction for the Classical Unitary Groups

Having described in detail the method of algebraic induction (the BW method) applied to $\mathfrak{u}(2)$ and its quantum group analog $\mathcal{U}_q(\mathfrak{u}(2))$, we may turn to the generalization to all unitary quantum groups but, because of the complexity of this construction, we find it helpful to first review the methods as applied to the classical unitary groups. In this generalization to $U(n)$, developed by Hecht, Le Blanc and Biedenharn [133, 131], we show how to construct the matrix element of a coherent state operator corresponding to (7.4) for $\mathfrak{u}(2)$, by means of suitable raising generators, and hence are able to write down the general induced state $|(m)\rangle_{\text{ind}}$. We also determine the induced action of the generators and their explicit form, and express the states in this induced irrep in terms of $U(n)$ coupling coefficients. These results all generalize to the unitary quantum groups.

In standard notation, the generators of the Lie algebra $\mathfrak{u}(n)$ are $\{E_{ij}; 1 \leqslant i, j \leqslant n\}$, and obey the commutation relations

$$[E_{ij}, E_{kl}] = \delta_{jk}E_{il} - \delta_{il}E_{kj}.$$

We choose the subgroup $U(n-1) \times U(1)$ from which to induce irreps of $U(n)$, and accordingly separate $\mathfrak{u}(n)$ into four subsets:

(a) the $\mathfrak{u}(n-1)$ subalgebra[1]: $\{C_{\alpha\beta}\} = \{E_{\alpha\beta};\ 1 \leqslant \alpha, \beta \leqslant n-1\}$,

(b) the $\mathfrak{u}(1)$ subalgebra: $\{W\} = \{E_{nn}\}$,

(c) the raising operators: $\{A_\alpha\} = \{E_{\alpha n};\ 1 \leqslant \alpha \leqslant n-1\}$, and

(d) the lowering operators: $\{B_\alpha\} = \{E_{n\alpha};\ 1 \leqslant \alpha \leqslant n-1\}$.

[1] **We adopt the convention in this section that Latin indices run from 1 to n while Greek indices run from 1 to $n-1$.**

Algebraic induction focuses on a special subset of the vectors in the space carrying the irrep $[m]$, namely, the vectors (μ) annihilated by the raising operators $\{A_\alpha\}$. Since $A_\alpha|(m)\rangle = 0$ implies that $W|(m)\rangle = m_{nn}|(m)\rangle$, it follows that the vectors in this subset have maximal labels in row $n-1$, that is, $m_{i,n-1} = m_{i,n}$. This special subset of vectors — which span a linear vector space which we call the **intrinsic space** — corresponds exactly to vectors carrying the irrep $[m_{1n}\cdots m_{n-1,n}]\times[m_{nn}]$ of the subgroup $U(n-1)\times U(1)$. With algebraic induction we construct the same irrep $[m]$ but over a new basis consisting of vectors from the intrinsic space, multiplied by holomorphic vector functions which we construct as boson polynomials acting on the vacuum ket. (For $U(2)$ the intrinsic space was spanned by the vector denoted $|\psi\rangle$ in §7.3.)

To construct this new carrier space, let us define an induced vector of the new basis by

$$|(m)\rangle_{\mathrm{ind}} \stackrel{\mathrm{def}}{=} \sum_{(\mu)} \langle(\mu)|e^{a\cdot A}|(m)\rangle\ |0\rangle \otimes |(\mu)\rangle, \tag{7.15}$$

where the summation is over the vectors (μ) of the intrinsic space, with $W|(\mu)\rangle = m_{nn}|(\mu)\rangle$, and where

$$a\cdot A \stackrel{\mathrm{def}}{=} \sum_{\alpha=1}^{n-1} a_\alpha A_\alpha.$$

Here, $\{a_\alpha\}$ are boson creation operators commuting with all $\{E_{ij}\}$, and acting on the vacuum ket $|0\rangle$. This notation is designed to make clear that vectors in the new basis are labelled by the same labels (m) as in the old basis, but that the new basis is built on the direct product of boson ket vectors with the subset of vectors of the original basis that comprise the intrinsic space. The boson ket vectors span the factor space $SU(n)/U(n-1)$.

Let us now demonstrate that the basis $\{|(m)\rangle_{\mathrm{ind}}\}$ carries exactly the same irrep as before. For each $g \in U(n)$ we define the action:

$$g|(m)\rangle = \sum_{(m')} \mathfrak{D}^{[m]}_{(m'),(m)}(g)\ |(m')\rangle, \tag{7.16}$$

with $g \to \mathfrak{D}(g)$ defining the irrep $[m]$ on the original basis $\{|(m)\rangle\}$. To show that the new basis (7.15) carries this same irrep, $\mathfrak{D}(g)$, we define an action by

$$g\circ|(m)\rangle_{\mathrm{ind}} \stackrel{\mathrm{def}}{=} \sum_{(\mu)} \langle(\mu)|e^{a\cdot A}g|(m)\rangle\ |0\rangle \otimes |(\mu)\rangle. \tag{7.17}$$

It follows from (7.16) that

$$g\circ|(m)\rangle_{\mathrm{ind}} = \sum_{(m')} \mathfrak{D}^{[m]}_{(m'),(m)}(g)|(m')\rangle_{\mathrm{ind}}, \tag{7.18}$$

which establishes the desired result.

Consider now the action of the Lie algebra $\mathfrak{u}(n)$ on the basis $\{|(m)\rangle_{\text{ind}}\}$. From (7.17) we see that

$$\begin{aligned} E \circ |(m)\rangle_{\text{ind}} &= \sum_{(\mu)} \langle(\mu)|\ e^{a\cdot A} E\ |(m)\rangle\ |0\rangle \otimes |(\mu)\rangle \\ &= \sum_{(\mu)} \langle(\mu)|\ (e^{a\cdot A} E e^{-a\cdot A})\ e^{a\cdot A}\ |(m)\rangle\ |0\rangle \otimes |(\mu)\rangle, \end{aligned}$$

where $E \in \mathfrak{u}(n)$. Using the Baker-Campbell-Hausdorff identity to evaluate $e^{a\cdot A} E e^{-a\cdot A}$ as E runs over the generators E_{ij}, and denoting

$$E \circ |(m)\rangle_{\text{ind}} \stackrel{\text{def}}{=} \Gamma(E)|(m)\rangle_{\text{ind}},$$

we find [131]:

LEMMA 7.19 *The following operators form a realization of $\mathfrak{u}(n)$, that is, Γ is an isomorphism of the Lie algebra of $U(n)$:*

$$\begin{aligned} \Gamma(A_\alpha) &= \bar{a}_\alpha \\ \Gamma(C_{\alpha\beta}) &= \mathcal{E}_{\alpha\beta} - a_\beta \bar{a}_\alpha \\ \Gamma(B_\alpha) &= a_\gamma \mathcal{E}_{\gamma\alpha} - a_\alpha \mathcal{E}_{nn} - a_\alpha a_\gamma \bar{a}_\gamma \\ \Gamma(W) &= \mathcal{E}_{nn} + a_\gamma \bar{a}_\gamma, \end{aligned}$$

where the indices γ are summed in each of the last two lines and where, for clarity, we have denoted by $\mathcal{E}_{\alpha\beta}$ the operators $E_{\alpha\beta}$ acting on the intrinsic subspace $\{|(\mu)\rangle\}$.

As previously explained, we could equally well have expressed these boson operator relations by their equivalents in terms of complex variables in order to emphasize that these generators act on a base manifold of holomorphic functions. The relations shown in Lemma 7.19 reduce to those already found for $U(2)$ in Lemma 7.5.

REMARK 7.20 1. The form of the isomorphism $\mathfrak{g} \to \Gamma(\mathfrak{g})$ in Lemma 7.19 is quite important for further developments. Note that the generators $\Gamma(C_{\alpha\beta})$ and $\Gamma(W)$ in this lemma consist of the direct sum of intrinsic generators $\mathcal{E}_{\alpha\beta}$ and $\mathcal{E}_{nn}$ of $U(n-1) \otimes U(1)$ and the Jordan map for $U(n-1)$ on the n-boson basis[2] $-\bar{a}_\alpha \in [\dot{0}, -1]$ with conjugate $a_\beta \in [1, \dot{0}]$ in $U(n-1)$ (here, the notation $a_\beta \in [1, \dot{0}]$ means that the set of operators $\{a_\beta\}$ transforms as an irreducible tensor operator with respect to the irrep $[1, \dot{0}]$). These two realizations of $U(n-1)$ commute, so that vectors in $|(m)\rangle_{\text{ind}}$ are determined by $U(n-1)$ vector couplings. As we will show below in Lemma 7.23 (see in particular (7.25)), this form for the

[2] We use $\dot{0} = 0, \ldots, 0$ to denote repeated zeros.

isomorphism given by Γ also implies that the generic eigenvector has an m_{nn} dependence residing solely in the factor K.

The implication of these observations is that the isomorphism is of such a form as to allow the construction $U(n) \to U(n-1) \times H_{n-1}$ to be easily implemented (where H_{n-1} denotes the Heisenberg-Weyl algebra of dimension $n-1$). Physically this result is that the curved Kähler manifold limits to a flat manifold. There are many applications of this observation, principally to limits of the matrix elements of tensor operators (the limit being $m_{nn} \to -\infty$), as indicated in §7.8. The purpose of this remark, however, is to demonstrate intuitively the origin of the limit structure in the form of the isomorphism given in Lemma 7.19. All of these results have a quantum group extension.

2. In the defining relation (7.15), the matrix elements that occur, $\langle(\mu)|e^{a\cdot A}|(m)\rangle$, are polynomials in the variables $\{a_\alpha\}$, since $(a\cdot A)^k|(m)\rangle$ vanishes for sufficiently large k. The form of this matrix element is familiar from coherent state investigations, but the construction in (7.15) is more general, since the expansion is over the vectors of the intrinsic space as well as over vectors in the "coherent state" basis. Note also that $\{\overline{a}_\alpha\}$, like $\{A_\alpha\}$, carries the $U(n-1)$ irrep $[1\,\dot{0}]$. Similarly $\{a_\alpha\}$ belongs to $[\dot{0},-1]$ in $U(n-1)$.

Let us now give a more explicit form for the vectors $|(m)\rangle_{\text{ind}}$. Using the fact that $\{a_\alpha\} \in [\dot{0},-1]$ in $U(n-1)$, we see that the direct product $[\dot{0},-1] \times \cdots \times [\dot{0},-1]$ of w such irreps defines uniquely the irrep $[\dot{0},-w]$ as polynomials in $\{a_\alpha\}$ of degree w. Let us denote the vectors of this irrep, normalized in the standard (boson) way, by

$$\left|\begin{matrix}[\dot{0},-w]\\(\lambda)\end{matrix}\right\rangle \overset{\text{def}}{=} B^{[\dot{0},-w]}_{(\lambda)}(a)|0\rangle, \tag{7.21}$$

where $[\dot{0},-w]$ is an irrep in $U(n-1)$, (λ) denotes the associated Gel'fand-Weyl $U(n-2)$ pattern, and $B(a)$ is a normalized boson polynomial in $\{a_\alpha\}$. The induced vectors having prescribed $U(n-1)\otimes U(1)$ labels are then:

$$\left[\left|\begin{matrix}[\dot{0},-w]\\(\cdot)\end{matrix}\right\rangle \times \left|\begin{matrix}m_{1n}\ldots m_{n-1,n}\\(\cdot)\end{matrix}\right\rangle\right]_{(\nu)} \otimes |m_{nn}\rangle, \tag{7.22}$$

where we have denoted the $U(n-1)$ vector coupling — yielding the $U(n-1)$ state $|(\nu)\rangle$ — by the bracket $[\;]_{(\nu)}$, with $(\cdot)$ denoting quantum numbers summed over in effecting the coupling.

Acting on this set of vectors with the raising operator $\{\overline{a}_\alpha\}$ successively lowers the degree of the polynomial $B(a)$ to zero, so that the highest weight vectors of the $U(n)$ irrep are given by the vectors of the intrinsic space. It follows that the $U(n)$ irrep is $[m]$, with these labels being supplied by the intrinsic state vectors alone.

Although the states (7.22) are normalized in $U(n-1)\otimes U(1)$, they are not necessarily normalized correctly in $U(n)$. To obtain the proper relative normalization, we recall that both bases $\{|(m)\rangle\}$ and $\{|(m)\rangle_{\text{ind}}\}$ carry the same representation. Eqn. (7.18) shows that:

$$\Gamma(E)|(m)\rangle = \sum_{(m')} \langle (m')|\, E\, |(m)\rangle\ |(m')\rangle_{\text{ind}}$$

where $\langle (m')|\, E\, |(m)\rangle$ are the standard matrices of the generators of $U(n)$. Hence, by direct computation, one obtains the following result (Le Blanc and Biedenharn [131]):

LEMMA **7.23** *An explicit form for the induced vectors of a general $U(n)$ irrep is given by:*

$$|(m)\rangle_{\text{ind}} = (-1)^{\varphi} K(m) \left[\left| \begin{matrix} [\dot{0}, -w] \\ (\cdot) \end{matrix} \right\rangle \times \left| \begin{matrix} [\mu] \\ (\cdot) \end{matrix} \right\rangle \right]_{(m)_{n-1}} \otimes |m_{nn}\rangle, \tag{7.24}$$

where

(a) $|{}^{[\dot{0},-w]}_{\ \cdot}\rangle$ *is defined by the normalized boson polynomial, (7.21), belonging to the $U(n-1)$ irrep $[\dot{0}, -w]$.*

(b) *The label* $w \overset{\text{def}}{=} \sum_{i=1}^{n-1} (m_{in} - m_{i,n-1})$.

(c) $\left|{}^{[\mu]}_{\ \cdot}\right\rangle \otimes |m_{nn}\rangle$ *denotes an intrinsic state vector carrying the $U(n-1)\times U(1)$ irrep $[m_{1n} m_{2n} \cdots m_{n-1,n}] \times [m_{nn}]$.*

(d) *the bracket $[\cdots \times \cdots]$, as in (7.22), denotes $U(n-1)$ vector coupling, and $(m)_{n-1}$ denotes the lowest $n-1$ rows of the $U(n)$ Gel'fand-Weyl pattern (m).*

(e) *The normalization factor $K(m)$ is given by:*

$$K(m) = \left(\prod_{\alpha=1}^{n-1} \frac{(p_{\alpha n} - p_{nn} - 1)!}{(p_{\alpha,n-1} - p_{nn})!} \right)^{\frac{1}{2}}, \tag{7.25}$$

where $p_{ij} \overset{\text{def}}{=} m_{ij} + j - i$.

Since $K(m)$ involves only the invariant labels of $U(n)$ and $U(n-1)$ we can promote $K(m)$ to a Hermitean operator, $\mathbf{K}$, with matrix elements $K(m)$. The generators of $U(n-1)\times U(1)$ (namely, $\Gamma(E_{\alpha\beta})$ and $\Gamma(E_{nn})$) commute with $\mathbf{K}$.

(f) *The phase factor $(-1)^{\varphi}$ is defined from (7.22) by the phase conventions of the standard realization of $U(n)$ (see [131, Eqn. (2.16)] and [190])).*

(g) *The orthonormal dual vectors,* ${}_{\rm ind}\langle(m)|$, *are Hermitean conjugates to (7.24) but with* $\mathbf{K}$ *replaced by* $\mathbf{K}^{-1}$. *(The inverse* $\mathbf{K}^{-1}$ *is well-defined as a consequence of the betweenness constraints on the* m_{ij}.*)*

For $U(2)$ the complicated result (7.24) reduces to the expression (7.6), and the boson polynomial defined in (7.21) reduces to a monomial, coupled with a trivial $U(1)$ coupling coefficient, with $w = m_{12} - m_{11}$, which explains the great simplicity of the algebraic induction method for $U(2)$.

In §7.7 we return to the expression (7.24) for $n = 3$ and verify it explicitly using properties of generalized hypergeometric functions. In this case the coupling coefficient is a standard $SU(2)$ Wigner-Clebsch-Gordan coefficient, and we obtain precisely the expression (7.24), even with identical phase factors!

REMARK **7.26** Although the BW construction achieves nothing new in terms of $U(n)$ representations alone, its novelty and importance lie in a different direction, since one sees that the construction yields a new realization, $\Gamma(E)$, of $\mathfrak{u}(n)$ and more importantly, a new construction of $U(n)$ carrier spaces defined entirely within the context of $U(n-1)$ vector addition coefficients. In other words, this construction allows us to obtain $U(n)$ results with $U(n-1)$ methods, and is accordingly a most useful technique in a recursive approach to the set of all $U(n)$ groups and their associated operators.

Now let us point out several aspects of algebraic induction which require particular attention when we generalize to $\mathcal{U}_q(\mathfrak{u}(n))$. In order to form the induced vectors (7.15) we can replace the exponential by its q-analog, $\exp_q$, however it is not immediately clear which operators will play the role of the raising generators $\{A_\alpha\}$. The proper choice is the set of commuting Cartan-Weyl generators defined by (2.70a,2.70b), p. 45, or their equivalents. Using an extension of the addition law for the q-exponential function given in Lemma 2.116 (p. 65), we can write the q-analog of the coherent states (7.15) either as a single q-exponential of an operator $\mathcal{O}$, or as a product of q-exponential factors. The boson operators are generalized of course to q-boson operators.

In calculating the realization $\Gamma(E), E \in \mathcal{U}_q(\mathfrak{u}(n))$, the q-analog of Lemma 7.19, we do not have at our disposal the q-analog of the Baker-Campbell-Hausdorff identity, however, we may proceed as in the example of $\mathcal{U}_q(\mathfrak{u}(2))$, by determining operator identities as in Lemma 7.12. Finally, the alternative form of the induced vectors given in Lemma 7.23 also generalizes using properties of quantum group coupling coefficients.

7.6 Extension of Algebraic Induction to the Unitary Quantum Groups

The q-analog generalization of the BW method, to be developed in this section, corresponds — as we have discussed for the classical group $U(n)$ — to constructing holomorphic sections of an associated bundle whose fibers carry irreps of the subgroup $U(n-1)\times U(1)$. This is in contrast (as mentioned in §7.2) to the standard BW construction where the fibers carry one-dimensional irreps of the maximal torus. The base manifold is accordingly $U(n)/(U(n-1)\times U(1))$ having $n-1$ complex dimensions. Such a procedure is well adapted to a recursive approach to the unitary group $U(n)$ since at each stage in the construction one needs only explicit results for a simpler group known by construction from the previous step, that is to say $U(n-1)$ for $U(n)$, $U(n-2)$ for $U(n-1),\ldots$ etc. For $U(2)$ itself one has, of course, the original construction.

This approach is well known in the physics literature for classical groups, quite in contrast to the quantum group analog where the construction had not been obtained prior to the works of Biedenharn and Lohe [186]. It is clear from this recursive approach that the original BW result itself, extended to quantum groups, can be easily obtained by substituting explicit prior results for all recursive steps although, to be sure, the final result involving $\frac{1}{2}n(n-1)$ boson operators will be very complicated.

Our construction begins by assuming, as the recursion hypothesis, the existence and knowledge of a realization of the generators of $\mathcal{U}_q(\mathfrak{u}(n-1))$ and explicit basis vectors carrying any given unitary irrep $[m_{1,n-1}m_{2,n-1}\cdots m_{n-1,n-1}]$ of $\mathcal{U}_q(\mathfrak{u}(n-1))$. We augment this $\mathcal{U}_q(\mathfrak{u}(n-1))$ space by forming a tensor product with the one-dimensional vector space $\mathfrak{V}$ spanned by the vector $|\psi\rangle$ carrying an irrep $[m_{nn}]$ of $U(1)$, generated by $\mathcal{E}_{nn}$. The fibers then carry an irrep of $\mathcal{U}_q(\mathfrak{u}(n-1))\times U(1)$ and belong to the vector space spanned by $|(\mu)\rangle\otimes|\psi\rangle$, where $|(\mu)\rangle\in[m_{1,n-1}\cdots m_{n-1,n-1}]$, with an action on this vector space by the generators $\mathcal{E}_{ij}\otimes\mathbb{I}$ and $\mathbb{I}\otimes\mathcal{E}_{nn}$ belonging to $\mathcal{U}_q(\mathfrak{u}(n-1))\times U(1)$.

Just as in our defining example of $\mathcal{U}_q(\mathfrak{u}(2))$ in §7.4, where it proved convenient in constructing the function $\mathcal{P}(a)$ in (7.4) to formally embed the one-dimensional fiber vector carrying an irrep of $U(1)\times U(1)$ in the larger space of an irrep of $\mathcal{U}_q(\mathfrak{u}(2))$, so too is it useful to embed the vectors carrying irreps of $\mathcal{U}_q(\mathfrak{u}(n-1))\times U(1)$ in the larger space of an irrep of $\mathcal{U}_q(\mathfrak{u}(n))$. Accordingly, we assume that the vector space of the fibers carries an irrep $[m_{1n}m_{2n}\cdots m_{nn}]$ of $\mathcal{U}_q(\mathfrak{u}(n))$ transforming irreducibly under $\mathcal{U}_q(\mathfrak{u}(n))$, whose n^2 Cartan-Weyl generators are denoted $\{E_{ij}\}, 1\leqslant i,j\leqslant n$. The fiber vectors carrying irreps of $\mathcal{U}_q(\mathfrak{u}(n-1))\times U(1)$ are a subset of the space of this $\mathcal{U}_q(\mathfrak{u}(n))$ irrep, transforming under the $\mathcal{U}_q(\mathfrak{u}(n-1))\times U(1)$ subset of the generators $\{E_{ij}\}$.

It is essential to point out that the associated bundle achieved in this construction will actually involve only that subset of $\mathcal{U}_q(\mathfrak{u}(n))$ irrep vectors which belong to a single $\mathcal{U}_q(\mathfrak{u}(n-1))\times U(1)$ irrep, in exactly the same way as the vectors in (7.4), in the construction for $\mathcal{U}_q(\mathfrak{u}(2))$, are explicitly restricted to a single one-dimensional irrep of $U(1)\times U(1)$ for the vector space of the fiber.

The base manifold of the bundle has complex dimension $n-1$ and correspondingly has $n-1$ commuting coordinates, which we may equivalently regard as $n-1$ q-boson operators $\{a_1^q, a_2^q,\ldots,a_{n-1}^q\}$ acting on the vacuum. We define a $\mathcal{U}_q(\mathfrak{u}(n-1))$ action on this base manifold by the following realization, comprising the Jordan map as explained in §2.5.1, p. 49:

$$E_{ij} = a_i^q \overline{a}_j^q,\ (i \neq j), \qquad E_{ii} = N_i, \tag{7.27}$$

for $i, j = 1, 2, \ldots, n-1$. Here $\{a_i^q, \overline{a}_i^q\}$ denotes the $n-1$ commuting q-bosons obeying (2.33) for each $i = 1, \ldots, n-1$.

The vector space $\mathfrak{F}$ of the base manifold carrying representations of $\mathcal{U}_q(\mathfrak{u}(n-1))$ is the Fock space of all polynomials $\mathcal{P}$ in the $n-1$ q-bosons $\{a_i^q\}$ acting on the vacuum ket $|0\rangle$. These two realizations of $\mathcal{U}_q(\mathfrak{u}(n-1))$ — one generated by $\{\mathcal{E}_{ij}, i, j = 1, \ldots, n-1\}$ acting on the fiber, the other by $\{a_i^q \overline{a}_j^q, N_i\}$ acting on the base manifold — admit a co-product carrying tensor product representations initially of $\mathcal{U}_q(\mathfrak{u}(n-1))$, which can then be extended to $\mathcal{U}_q(\mathfrak{u}(n-1)) \times U(1)$ by adjoining the generator $\mathcal{E}_{nn}$ of $U(1)$.

Explicitly, we have the following realization of the generators of this direct product quantum group $\mathcal{U}_q(\mathfrak{u}(n-1)) \times U(1)$:

$$\begin{aligned}
\Gamma(E_{ij}) &= \mathcal{E}_{ij} \otimes q^{\frac{1}{4}(N_i - N_j)} + q^{-\frac{1}{4}(\mathcal{E}_{ii} - \mathcal{E}_{jj})} \otimes a_i^q \overline{a}_j^q, \quad i < j, \\
\Gamma(E_{ij}) &= \mathcal{E}_{ij} \otimes q^{\frac{1}{4}(N_j - N_i)} + q^{-\frac{1}{4}(\mathcal{E}_{jj} - \mathcal{E}_{ii})} \otimes a_i^q \overline{a}_j^q, \quad i > j, \\
\Gamma(E_{ii}) &= \mathcal{E}_{ii} \otimes \mathbb{I} + \mathbb{I} \otimes N_i,
\end{aligned} \tag{7.28}$$

where i, j range over $1, 2, \ldots, n-1$. To be precise, let us note that the tensor product $\otimes$ used here implies a co-product $\Delta(E_{ij})$ expanded on the two $\mathcal{U}_q(\mathfrak{u}(n-1))$ irreps carried by $\mathcal{E}_{ij}$ and by the q-boson operator realization. The space $\mathfrak{V}$ containing the vectors $|\psi\rangle$ is ignored effectively. However for the diagonal n^{th} generator, $\Gamma(E_{nn})$, $\mathfrak{V}$ cannot be ignored and the tensor product $\otimes$ has a different significance. One finds:

$$\Gamma(E_{nn}) = \mathbb{I} \otimes \mathcal{E}_{nn} - \sum_{i=1}^{n-1} N_i \otimes \mathbb{I}. \tag{7.29}$$

We have denoted these generators of $\mathcal{U}_q(\mathfrak{u}(n-1)) \times U(1)$ by $\Gamma(E_{ij})$ to distinguish the action of these particular generators as the left action of a subset $i, j = 1 \ldots n-1; i = j = n$ of the generic generators $E_{ij} (i, j = 1, \ldots, n)$ on the bundle.

The problem posed now for completing this construction is to extend this left action on the bundle by generators of the subgroup $\mathcal{U}_q(\mathfrak{u}(n-1)) \times U(1)$ of $\mathcal{U}_q(\mathfrak{u}(n))$ to an action realizing the group $\mathcal{U}_q(\mathfrak{u}(n))$ itself. This is the construction involved in validating the use of the vector coherent state to create the polynomial $\mathcal{P}(a)$. Phrased differently, in the language of fiber bundles, the problem now to be resolved is the imposition of constraints analogous to (7.4) and (7.8) such that one obtains an associated bundle from the principal bundle. This step requires the introduction of an invariant operator $\mathcal{O}$:

DEFINITION **7.30** *Define the operator* $\mathcal{O}$ *by*

$$\mathcal{O} \stackrel{\text{def}}{=} \sum_{i=1}^{n-1} \left(q^{-\frac{1}{4}\sum_{j=1}^{i-1} \mathcal{E}_{jj} + \frac{1}{4}\sum_{j=i+1}^{n-1} \mathcal{E}_{jj}}\, E_{ni} \otimes q^{\frac{1}{4}\sum_{j=1}^{i-1} N_j - \frac{1}{4}\sum_{j=i+1}^{n-1} N_j} a_i^q \right).$$

Lemma 7.31 *The operator $\mathcal{O}$ is invariant under the commutator action of the generators $\{\Gamma(E_{ij})\}$ of $\mathcal{U}_q(\mathfrak{u}(n-1)) \times U(1)$ defined in (7.28,7.29).*

Proof: The invariance under the generators of the Chevalley basis $\Gamma(E_{ii})$ ($i = 1, \ldots, n$) and $\Gamma(E_{i,i-1})$, $\Gamma(E_{i-1,i})$ with $i = 1, \ldots, n-1$, is verified directly. By defining the elements of $\Gamma(E_{ij})$ not in this set as in §2.5 (see Eqns. (2.70a,2.70b), p. 45), we extend the invariance to all $\Gamma(E_{ij})$ in (7.28,7.29). □

The construction which we have carried out so far in this section is a tensor product of vectors belonging to irrep spaces of $\mathcal{U}_q(\mathfrak{u}(n))$ tensored with vectors in the base manifold $\mathfrak{F}$. Consider now the subset $\mathfrak{W}$ of these tensor product vectors defined by

$$\mathfrak{W} \stackrel{\text{def}}{=} \sum_{(m)} \otimes(|(m)\rangle \otimes |0\rangle),$$

where (m) is a Gel'fand-Weyl pattern belonging to an irrep of $\mathcal{U}_q(\mathfrak{u}(n))$. Under the action of the generators $\{E_{ij} \otimes \mathbb{I}\}$, with $i,j = 1,2,\ldots,n$, the space $\mathfrak{W}$ splits into a direct sum of $\mathcal{U}_q(\mathfrak{u}(n))$ irreps labelled by Young frames $[m_{1n} m_{2n} \cdots m_{nn}]$, with individual vectors in these irrep spaces labelled by the Gel'fand-Weyl patterns (m).

Consider next the vector space $\mathfrak{W}(w)$ defined by

$$\mathfrak{W}(w) \stackrel{\text{def}}{=} (\mathcal{O})^w \mathfrak{W}$$

where $(\mathcal{O})^w$ denotes $\mathcal{O}$ multiplied by itself w times, with w a non-negative integer. It is easily shown that:

Lemma 7.32 *If a vector $\boldsymbol{v}$ in $\mathfrak{W}$ is labelled under the action of $E_{ij} \otimes \mathbb{I}$ by the $\mathcal{U}_q(\mathfrak{u}(n))$ Gel'fand-Weyl labels $(m) = (m_{ij})$ with $1 \leqslant i,j \leqslant n$, then $(\mathcal{O})^w \boldsymbol{v}$ carries the same $\mathcal{U}_q(\mathfrak{u}(n-1)) \times U(1)$ labels (m_{ij}) with $1 \leqslant i,j \leqslant n-1$ under the action of the generators $\Gamma(E_{ij})$ defined in (7.28,7.29).*

Proof: This is clear from Lemma 7.31 since the action of $\Gamma(E_{ij})$ commutes with the invariant $\mathcal{O}$ and $\Gamma(E_{ij}) \to E_{ij}$ on $\mathfrak{W}$. □

In fact one can state much more about the vectors in $\mathfrak{W}(w)$:

Lemma 7.33 *Let a subspace of $\mathfrak{W}(w)$ be defined as the linear span of vectors $(\mathcal{O})^w \boldsymbol{v}$, where $\boldsymbol{v}$ belongs to the $\mathcal{U}_q(\mathfrak{u}(n))$ irrep subspace of $\mathfrak{W}$ labelled by $[m_{1n} \cdots m_{nn}]$. Under the action of $E_{ij} \otimes \mathbb{I}$, this subspace splits into a direct sum of $\mathcal{U}_q(\mathfrak{u}(n-1)) \times U(1)$ irreps with labels $[m'_{1,n-1} \cdots m'_{n-1,n-1}] \times [m'_{1n}]$, where*

$$\sum_{i=1}^{n-1} (m_{i,n-1} - m'_{i,n-1}) = w, \qquad m'_{1n} = \sum_{i=1}^{n} m_{in} - \sum_{i=1}^{n-1} m_{i,n-1} + w,$$

and the $\mathcal{U}_q(\mathfrak{u}(n-1))$ irrep labels $[m'_{1,n-1} \cdots m'_{n-1,n-1}]$ are compatible, as a sub-Gel'fand-Weyl pattern, with the $\mathcal{U}_q(\mathfrak{u}(n))$ labels $[m_{1n}, m_{2n}, \ldots, m_{nn}]$.

PROOF: This follows from the fact that the expansion of $(\mathcal{O})^w$ as a sum of monomial products in the q-boson operators $\{a_1^q, \ldots, a_{n-1}^q\}$ is homogeneous of degree w. This implies that the action of $\Gamma(E_{nn})$, on a given term in the expansion of $(\mathcal{O})^w \boldsymbol{v}$, yields:

$$\Gamma(E_{nn}) = \mathcal{E}_{nn} \otimes \mathbb{I} - \sum_{i=1}^{n-1} N_i \longrightarrow \left(\sum_{i=1}^{n} m_{in} - \sum_{i=1}^{n-1} m'_{i,n-1}\right) - w.$$

However, from Lemma 7.32 we also know that $\Gamma(E_{nn})$ has eigenvalues given by:

$$\Gamma(E_{nn}) \longrightarrow \sum_{i=1}^{n} m_{in} - \sum_{i=1}^{n-1} m_{i,n-1}. \tag{7.34}$$

Hence, taking the difference, we find:

$$\sum_{i=1}^{n-1} (m_{i,n-1} - m'_{i,n-1}) = w,$$

as asserted. This also implies that the label for the $U(1)$ irrep is:

$$m'_{1n} = \sum_{i=1}^{n} m_{in} - \sum_{i=1}^{n-1} m_{i,n-1} + w,$$

since this is the eigenvalue of $\mathcal{E}_{nn} \otimes \mathbb{I}$. Since the vectors in $\mathcal{U}_q(\mathfrak{u}(n-1)) \times U(1)$ are generated from the irrep $[m_{1n} \cdots m_{nn}]$ of $\mathcal{U}_q(\mathfrak{u}(n))$ by the *lowering operators* $E_{ni} \otimes \mathbb{I}$ in $\mathcal{O}$, it follows that the irrep labels $[m'_{1,n-1} \cdots m'_{n-1,n-1}]$ of $\mathcal{U}_q(\mathfrak{u}(n-1))$ are compatible as sub-Gel'fand-Weyl pattern labels. □

The restriction to a specified value of w in $\mathfrak{W}(w)$ can be removed. Using the q-exponential function we can now construct the vector space $\mathfrak{W}(e)$:

$$\mathfrak{W}(e) \stackrel{\text{def}}{=} \exp_q(\mathcal{O})\mathfrak{W}$$

Just as in our defining example of $\mathcal{U}_q(\mathfrak{u}(2))$ in §7.4 the space $\mathfrak{W}(e)$ is too large and we must project this space onto the $\mathcal{U}_q(\mathfrak{u}(n-1)) \times U(1)$ irrep subspace that is of lowest weight in $\mathcal{U}_q(\mathfrak{u}(n))$. For this purpose we use the projection operator $\mathbb{P}$ defined by

$$\mathbb{P} \stackrel{\text{def}}{=} \left(\sum_{m_{in},(\mu)} \left| \begin{matrix} m_{1n} m_{2n} \cdots m_{nn} \\ m_{2n} \cdots m_{nn} \\ (\mu) \end{matrix} \right\rangle \left\langle \begin{matrix} m_{1n} m_{2n} \cdots m_{nn} \\ m_{2n} \cdots m_{nn} \\ (\mu) \end{matrix} \right| \right) \otimes \mathbb{I}. \tag{7.35}$$

We assert, and will prove below, that the vector space (call it $\mathfrak{W}_q$) carrying all unitary irreps of $\mathcal{U}_q(\mathfrak{u}(n))$ in this construction is the vector space given by

$$\mathfrak{W}_q \stackrel{\text{def}}{=} \mathbb{P}\,\mathfrak{W}(e).$$

Specializing to a generic vector in the irrep $[m_{1n}m_{2n}\cdots m_{nn}]$ having the Gel'fand-Weyl labels $(m_{ij}), 1 \leqslant i,j \leqslant n$, we find:

$$|(m)\rangle_{\text{ind}} = \sum_{(\mu)\hookrightarrow[m]} \langle(\mu)|\exp_q(\mathcal{O})\,|(m)\rangle\,|(\mu)\rangle \otimes |0\rangle, \tag{7.36}$$

where $(\mu) \in \mathcal{U}_q(\mathfrak{u}(n-1)) \times U(1)$, and the meaning of $(\mu) \hookrightarrow [m]$ is that the pattern (μ) appears as a sub-pattern of ("fits into") the Gel'fand-Weyl patterns of the $\mathcal{U}_q(\mathfrak{u}(n))$ irrep $[m_{1n}\cdots m_{nn}]$ as:

$$\begin{pmatrix} m_{1n} & m_{2n} & m_{3n} & \cdots & m_{n-1n} & m_{nn} \\ & m_{2n} & m_{3n} & \cdots & m_{n-1n} & m_{nn} \\ & & \mu_{1,n-1}\ \mu_{2,n-2} & \cdots & \mu_{n-2,n-2} & \\ & & & \vdots & & \\ & & & \mu_{11} & & \end{pmatrix},$$

so that the vectors $|(\mu)\rangle$ carry the $\mathcal{U}_q(\mathfrak{u}(n-1)) \times U(1)$ irrep labels:

$$|(\mu)\rangle = \left| \begin{matrix} m_{2n} & m_{3n} & \cdots & m_{nn} \\ & \mu_{1,n-2} & \cdots & \mu_{n-2,n-2} \\ & & \vdots & \\ & & \mu_{11} & \end{matrix} \right\rangle \otimes \left|(m_{1n})\right\rangle.$$

The constraint imposed by $\mathbb{P}$ determines the "vector coherent state" function and is, in bundle language, exactly the geometric constraint imposed by the construction of an associated bundle from the principal bundle.

To be fully explicit as to the meaning of (7.36), let us note that the sum in (7.36) is a finite sum, since $\mathcal{O}$ is constructed from lowering operators E_{ni} only, and that the matrix element $\langle(\mu)|\exp_q(\mathcal{O})|(m)\rangle$ is therefore a polynomial, with numerical coefficients in $\mathbb{C}[q,q^{-1}]$, over powers of the q-boson operators $\{a_1^q,\ldots,a_{n-1}^q\}$. This operator-valued polynomial acts on $|0\rangle$ to produce an orthonormalized vector in the carrier space of $\mathcal{U}_q(\mathfrak{u}(n-1))$ with an action by the generators E_{ij} as shown in (7.27), (where $1 \leqslant i,j \leqslant n-1$). The vectors $|(\mu)\rangle$, as discussed above, are orthonormal vectors in the irrep $[\mu_{1n-1},\mu_{2n-1},\ldots,\mu_{n-1,n-1}] \times [\mu_{1n}]$ of $\mathcal{U}_q(\mathfrak{u}(n-1)) \times U(1)$, with an action by the generators E_{ij}. As also remarked we inject this (μ) pattern into $\mathcal{U}_q(\mathfrak{u}(n))$ by identifying, for $\mathcal{U}_q(\mathfrak{u}(n-1))$, $\mu_{i,n-1} = m_{i+1,n}$ (with $i = 1,2,\ldots,n-1$), and for $U(1)$, $\mu = m_{1n}$.

In order to clarify the bundle nature of (7.36), let us note that since the fiber is the vector space $\{|\mu\rangle\}$ and the base manifold is the Fock space $\mathfrak{F}$, we can identify the left action of the $\mathcal{U}_q(\mathfrak{u}(n-1)) \times U(1)$ generators as an action by the generators $\Gamma(E_{ij})$ given in Eqns. (7.28,7.29). (We will shortly extend this action to all of $\mathcal{U}_q(\mathfrak{u}(n))$.) The right action on the bundle is defined by the generators $\{E_{ij}\}$ of $\mathcal{U}_q(\mathfrak{u}(n))$ now acting directly on the vectors $|(m)\rangle$ *inside the matrix element* — yielding $\sum_{(m')}\langle(m')|E_{ij}|(m)\rangle|(m')\rangle$, where the terms $\langle\cdots\rangle$ are matrix elements of the given generator E_{ij} — and hence the right action on (7.36) realizes by definition the unitary irrep $[m_{1n}\cdots m_{nn}]$ of $\mathcal{U}_q(\mathfrak{u}(n))$. We remark again that these two actions do not commute.

What remains to be proved is that the left-action $\Gamma(E_{ij})$ extends from the quantum group $\mathcal{U}_q(\mathfrak{u}(n-1)) \times U(1)$ to $\mathcal{U}_q(\mathfrak{u}(n))$, that the map Γ is an isomorphism of the quantum group algebras, and that (7.36) defines orthonormal vectors carrying the irrep $[m]$ with Gel'fand-Weyl labels (m) under the action of the $\mathcal{U}_q(\mathfrak{u}(n))$ generators $\Gamma(E_{ij})$.

7.6.1 The Isomorphism of Quantum Group Algebras

The invariant operator $\mathcal{O}$ has several properties which will be of importance for our construction. As written, $\mathcal{O}$ is the sum of $n-1$ monomials x_i and may be expressed in the form:

$$\mathcal{O} = \sum_{i=1}^{n-1} x_i$$

with

$$x_i \stackrel{\text{def}}{=} q^{-\frac{1}{4}\sum_{j=1}^{i-1}\varepsilon_{jj}+\frac{1}{4}\sum_{j=i+1}^{n-1}\varepsilon_{jj}} E_{ni} \otimes q^{\frac{1}{4}\sum_{j=1}^{i-1}N_j-\frac{1}{4}\sum_{j=i+1}^{n-1}N_j} a_i^q.$$

We state the most significant fact about the operators $\{x_i\}$ in the form of a lemma, which can be proved by direct calculation:

LEMMA 7.37 *The operators $x_1, x_2, \ldots, x_{n-1}$ obey the q-commutation rule:*

$$x_i x_j = q^{-1} x_j x_i, \qquad i < j.$$

Accordingly, the set $\{x_i\}$ forms an operator realization of the quantum coordinates of an $n-1$ hyperplane[3].

It is useful to introduce a second set of operator coordinates, the set $\{X_i\}$ defined by:

$$X_i \stackrel{\text{def}}{=} q^{-\frac{1}{4}\sum_{j=1}^{i-1}\varepsilon_{jj}+\frac{1}{4}\sum_{j=i+1}^{n-1}\varepsilon_{jj}} E_{ni} \otimes q^{-\frac{1}{4}\sum_{j=1}^{i-1}N_j+\frac{1}{4}\sum_{j=i+1}^{n-1}N_j} a_i^q.$$

By direct verification we have:

LEMMA 7.38 *The operators $\{X_i\}$ satisfy $X_i X_j = X_j X_i$ for all $i, j = 1, \ldots, n-1$.*

We are now in a position to state a very useful identity for the q-exponential function:

LEMMA 7.39 *The q-exponential function satisfies the addition law:*

$$\exp_q(\mathcal{O})|0\rangle = \exp_q\left(\sum_{i=1}^{n-1} x_i\right)|0\rangle = \prod_{i=1}^{n-1}(\exp_q(X_i))|0\rangle.$$

Since the $\{X_i\}$ commute, the product on the right hand side may be taken in any order.

[3]In Chapter 4 non-commuting coordinates satisfy these relations with q replaced by $q^{\frac{1}{2}}$. However, in the application of the q-binomial theorem it is q which appears.

Remark 7.40 Note that this identity is an identity, not of operators, but of vectors in $\mathfrak{F}$, and so the terminating vacuum ket $|0\rangle$ must be present. This lemma generalizes the addition formula for $\exp_q$ stated and proved in Lemma 2.116 (p. 65) using the q-binomial theorem.

These ancillary results have an immediate application to the determination of the left-action $\mathcal{U}_q(\mathfrak{u}(n))$ generators denoted by $\Gamma(E_{ni})$:

Lemma 7.41 *The left-action generators corresponding to the abstract $\mathcal{U}_q(\mathfrak{u}(n))$ generators E_{ni} are given by:*

$$\Gamma(E_{ni}) = q^{\frac{1}{4}\left(\sum_{j=1}^{i-1}\varepsilon_{jj} - \sum_{j=i+1}^{n-1}\varepsilon_{jj}\right)} \otimes \overline{a}_i^q,$$

for $i = 1, 2, \ldots, n-1$.

Proof: Consider the action of $\Gamma(E_{ni})$ on the vector $|(m)\rangle_{\text{ind}}$ given in (7.36). The operators E_{jj} in the q-factor of $\Gamma(E_{ni})$ have an action on the ket vectors $|(\mu)\rangle$ in the tensor product $|(\mu)\rangle \otimes |0\rangle$, showing that the eigenvalues of E_{jj} are

$$\sum_{i=1}^{j}\mu_{ij} - \sum_{i=1}^{j-1}\mu_{i,j-1}.$$

These same eigenvalues would be produced by the action of the E_{jj} acting to the left immediately *inside* the matrix element $\langle(\mu)|\exp_q(\mathcal{O})|(m)\rangle$. Since the $\overline{a}_i^q$ act on $\exp_q(\mathcal{O})$ inside the matrix element as well, we can therefore take $\Gamma(E_{ni})$ to act *inside* the matrix element. To determine the action of $\Gamma(E_{ni})$ on $\exp_q(\mathcal{O})$ we note that this operator is terminated on the right by the vacuum ket, so that we may use Lemma 7.39.

We next observe that $\Gamma(E_{ni})$ commutes with all X_j for $j \neq i$. Hence, we need only calculate that

$$\Gamma(E_{ni})\exp_q(X_i)|0\rangle = \exp_q(X_i)E_{ni}|0\rangle, \tag{7.42}$$

which can be shown directly. Thus we find that:

$$\begin{aligned}
\Gamma(E_{ni})|(m)\rangle_{\text{ind}} &= \sum_{(\mu)}\langle(\mu)|\,\Gamma(E_{ni})\exp_q(\mathcal{O})\,|(m)\rangle\,|(\mu)\rangle\otimes|0\rangle \\
&= \sum_{(\mu)}\langle(\mu)|\,\Gamma(E_{ni})\prod_{i=1}^{n-1}\exp_q(X_i)\,|(m)\rangle\,|(\mu)\rangle\otimes|0\rangle \\
&= \sum_{(\mu)}\langle(\mu)|\prod_{i=1}^{n-1}\exp_q(X_i)E_{ni}\,|(m)\rangle\,|\mu\rangle\otimes|0\rangle, \qquad (7.43)
\end{aligned}$$

where the sum in each case is over all $(\mu) \hookrightarrow [m]$ and $(\mu) \in \mathcal{U}_q(\mathfrak{u}(n-1)) \times U(1)$, and where we used Lemma 7.39 and Eqn. (7.42) in the last two lines, respectively.

By definition, however, the matrix elements $\langle\cdot|\cdot|\cdot\rangle$ of the generators E_{ni} acting on the Gel'fand-Weyl basis vectors $\{|(m)\rangle\}$ are given by:

$$E_{ni}|(m)\rangle = \sum_{(m')}\langle(m')|E_{ni}|(m)\rangle|(m')\rangle,$$

so that (7.43) becomes

$$\begin{aligned}\Gamma(E_{ni})|(m)\rangle_{\text{ind}} &= \sum_{(m')}\langle(m')|E_{ni}|(m)\rangle\sum_{(\mu)}\langle(\mu)|\prod_{i=1}^{n-1}\exp_q(X_i)\,|(m')\rangle\,|(\mu)\rangle\otimes|0\rangle \\ &= \sum_{(m')}\langle(m')|E_{ni}|(m)\rangle|(m')\rangle_{\text{ind}},\end{aligned}\tag{7.44}$$

where in the last step we used Lemma 7.39 once again. Thus, the action of $\Gamma(E_{ni})$ on the Borel-Weil basis yields precisely the same action as the (abstract) generators E_{ni} on the (abstract) standard Gel'fand-Weyl basis. □

The determination of the generators $\Gamma(E_{in})$ is, in contrast to Lemma 7.41, a *much* more difficult task. The reason for this difficulty lies in the problems associated with the existence of the Serre relations. For any calculations one needs to find an explicit form for the generators E_{ij} of $\mathcal{U}_q(\mathfrak{u}(n))$ including those which are not in the Chevalley set $\{E_{i,i+1}, E_{i,i-1}, E_{ii}\}$. The choice of any explicit form must be required to be stable under co-multiplication, and since $\Gamma(E)$ and E (where $E \in \mathcal{U}_q(\mathfrak{u}(n))$) are to have the same abstract matrix elements, this necessary requirement can easily lead to contradictions or inconsistencies for a particular (incorrect) explicit form. Of course one could always work directly with the generic matrix elements, but this is a task of daunting complexity.

The procedure we will use is based on the following observation, which follows by direct calculation:

LEMMA **7.45** *The generators $\Gamma(E_{ni})$, $i = 1, 2, \ldots, n-1$, given in Lemma 7.41 mutually commute.*

Moreover, we also find:

LEMMA **7.46** *The generators $\Gamma(E_{ni})$ in Lemma 7.41 and $\Gamma(E_{i,i\pm1})$ in Eqns. (7.28,7.29) satisfy the relations*

$$\Gamma(E_{ni})\Gamma(E_{i,i\pm1}) - q^{\mp\frac{1}{2}}\Gamma(E_{i,i\pm1})\Gamma(E_{ni}) = q^{\pm\frac{1}{2}\Gamma(E_{ii})}\Gamma(E_{n,i\pm1}).\tag{7.47}$$

These relations are essentially the maps under Γ of the relations (2.70a, 2.70b) used to define the Cartan-Weyl generators from the Chevalley set.

REMARK 7.48 The relations (7.47) are quite special to the present construction and are not necessarily stable under co-multiplication. However, since $\Gamma(E)$ and E (where $E \in \mathcal{U}_q(\mathfrak{u}(n))$) have the *same* generic matrix elements, these particular relations are, in fact, stable.

The properties of the mapping $E \to \Gamma(E)$, $E \in \mathcal{U}_q(\mathfrak{u}(n))$, imply that the matrix elements of both E and $\Gamma(E)$, evaluated on their associated irrep bases, are not only isomorphic, but identical.

Since we explicitly assume that the generators $\{E_{ij}\}$ of $\mathcal{U}_q(\mathfrak{u}(n))$ have the standard matrix elements of the Gel'fand-Weyl basis (following from the matrix elements of the Chevalley generators given explicitly in Theorem 2.74), it follows that these generators possess an involution such that $E_{ij}^{\dagger} = E_{ji}$ (with possibly $q \to q^{-1}$). It follows that the generators $\Gamma(E_{ij})$ also possess the same involution, defined via explicit matrix elements. We conclude that:

LEMMA 7.49 *1. The generators $\Gamma(E_{in}), i = 1, 2, \ldots, n-1$, mutually commute.*

2. The generators $\Gamma(E_{in})$ and $\Gamma(E_{i\pm1,i})$ satisfy the relations

$$\Gamma(E_{i\pm1,i})\Gamma(E_{in}) - q^{\pm\frac{1}{2}}\Gamma(E_{in})\Gamma(E_{i\pm1,i}) = q^{\mp\frac{1}{2}\Gamma(E_{ii})}\Gamma(E_{i\pm1,n}).$$

We are now in a position to give the explicit form of $\Gamma(E_{in})$. The procedure is to calculate, using only those quadratic commutation relations of the $\mathcal{U}_q(\mathfrak{u}(n))$ algebra which are independent of the Serre relations, to determine that component of $\Gamma(E_{in})$ involving the operator $a_i^q \left[\mathcal{E}_{nn} - \mathcal{E}_{ii} - \sum_{k=1}^{n-1} N_k\right]$. This is a straightforward calculation, the result of which is given below. One then uses Lemma 7.49 and the known operators $\Gamma(E_{i\pm1,i})$ to determine all remaining components of $\Gamma(E_{in})$. The result of this calculation is:

LEMMA 7.50 *The left-action generators $\Gamma(E_{in}), i = 1, 2, \ldots, n-1$, corresponding to the abstract $\mathcal{U}_q(\mathfrak{u}(n))$ generators E_{in} by the mapping Γ are given by:*

$$\Gamma(E_{in}) = q^{\alpha_{ii}} a_i^q \left[\mathcal{E}_{nn} - \mathcal{E}_{ii} - \sum_{k=1}^{n-1} N_k\right] - \sum_{\substack{j=1\\ j\neq i}}^{n-1} q^{\alpha_{ij}} a_j^q \mathcal{E}_{ij}, \tag{7.51}$$

where $i = 1, 2, \ldots n-1$, and where

$$q^{\alpha_{ij}} \stackrel{\text{def}}{=} q^{-\frac{1}{4}\left(\sum_{k=1}^{j-1}\mathcal{E}_{kk} - \sum_{k=j+1}^{n-1}\mathcal{E}_{kk}\right)} q^{(\frac{+}{0})\left(\frac{1}{2}\left(\mathcal{E}_{nn} - \sum_{k=1}^{n-1} N_k\right) + \frac{3}{4}\right)} q^{(\frac{+}{0})\frac{1}{4}(N_i + N_j)} q^{(\frac{+}{0})\frac{1}{2}\sum N_k},$$

with

$$\left(\frac{+}{0}\right) = \begin{cases} + & \text{for } i < j \\ - & \text{for } i > j \\ 0 & \text{for } i = j, \end{cases}$$

and the sum $\sum N_k$ *in the last term is over values of* k *such that* $i < k < j$ *(if* $i < j$*) or* $j < k < i$ *(if* $j < i$*).*

PROOF: By construction, the generators $\Gamma(E_{in})$ validate the map $E \to \Gamma(E)$, for which the matrix elements of both E and $\Gamma(E)$ (on their associated irrep bases) are identical. Since the generators $E \in \mathcal{U}_q(\mathfrak{u}(n))$ obey the $\mathcal{U}_q(\mathfrak{u}(n))$ algebra, including all Serre relations, this is also true of the generators $\Gamma(E_{in})$, which establishes the lemma. □

REMARK **7.52** It is all too easy in a proof such as the one above to make subtle errors, so that it is reassuring to know that we have actually demonstrated, by direct calculation, that *all* of the defining relations of the $\mathcal{U}_q(\mathfrak{u}(n))$ algebra are valid for $\Gamma(E_{in}), \Gamma(E_{ni})$ and, of course, for the remaining elements of the $\Gamma(E_{ij})$ mapping (this is automatically true via co-multiplication). A direct verification is such an exceedingly tedious and uninstructive task that it is not useful even to sketch the procedure here. Surprisingly, the cubic Serre relations, and the relations asserted in Lemma 7.49 are not the most difficult. The hardest relations to prove by direct calculation are the quadratic Serre relations, each of which involve $(n-1)^2$ commutators. Almost all of these commutators evaluate to zero or cancel with the result of other commutators in pairs. There are, however, a special class of terms — beginning at $\mathcal{U}_q(\mathfrak{u}(5))$ — which do not vanish identically. For these terms one can show that the constraints imposed by the form of (7.36), the constraint being expressed by Lemma 7.33 and the projection (7.35), cause these terms when operating on the basis $\{|(m)\rangle_{\text{ind}}\}$ to vanish, thus validating the claimed commutation relations.

We may sum up all of the above results by stating them collectively as a theorem:

THEOREM **7.53** *The map* $\mathfrak{g} \to \Gamma(\mathfrak{g})$*, where* $\mathfrak{g} = \mathcal{U}_q(\mathfrak{u}(n))$*, as given in (7.28-7.29) and Lemmas 7.41,7.50, is an isomorphism of* $\mathcal{U}_q(\mathfrak{u}(n))$ *quantum group algebras.*

It is a straightforward task, now that we have established the isomorphism of this theorem, to complete the proof that the vectors $|(m)\rangle_{\text{ind}}$ of (7.36) span the carrier space of the irrep $[m]$ of the quantum group $\mathcal{U}_q(\mathfrak{u}(n))$ generated by $\Gamma(E_{ij})$, $1 \leqslant i, j \leqslant n$. We need only apply the lowering operators $\Gamma(E_{ni})$ to (7.36), thereby eliminating all of the q-boson operators $\{a_i^q\}$. Using the lowering operators $\Gamma(E_{ij})$, $i > j$, in (7.28) then carries this vector to a lowest weight vector for which $\Gamma(E_{ii}) \to m_{n+1-i,n}$. This establishes:

THEOREM **7.54** *The vectors constructed in (7.36) form the carrier space of a unitary irrep* $[m]$ *of the quantum group* $\mathcal{U}_q(\mathfrak{u}(n))$ *generated by the operators* $\Gamma(E_{ij})$ *of Theorem 7.53.*

7.6.2 An Alternative Form for the Induced Irrep Vectors

The explicit vectors carrying an arbitrary irrep $[m]$ of $\mathcal{U}_q(\mathfrak{u}(n))$ have been given in (7.36). Each such vector is a sum of vectors which are tensor products of a homogeneous holomorphic vector over the (quantum space) base manifold and fiber vectors carrying $\mathcal{U}_q(\mathfrak{u}(n-1)) \times U(1)$.

It is useful, however, to express these vectors in an alternative form which makes explicit the matrix elements of the Kähler potential which normalizes the irrep vectors. To obtain this alternative form we recall that the left action generators $\Gamma(E_{ij})$ contain a subset of generators that realize the co-product action for $\mathcal{U}_q(\mathfrak{u}(n-1)) \times U(1)$. Since the Gel'fand-Weyl pattern (m) of the irrep vector $|(m)\rangle_{\text{ind}}$ uniquely specifies the irrep labels of this subgroup, we conclude that we may write the vector $|(m)\rangle_{\text{ind}}$ in the following form:

LEMMA **7.55** *The basis vectors $|(m)\rangle_{\text{ind}}$ defined in (7.36) satisfy the relation:*

$$|(m)\rangle_{\text{ind}} = K_q \begin{pmatrix} [m_n] \\ [m_{n-1}] \end{pmatrix} \sum_{(\mu),(\mu')} {}_qC^{[\mu]\ [w\,\dot{0}]\ [m_{n-1}]}_{(\mu)\ (\mu')\ (m_{n-1})} \left(\left| \begin{matrix} [w\,\dot{0}] \\ (\mu') \end{matrix} \right\rangle_{\text{base}} \otimes \left| \begin{matrix} [\mu] \\ (\mu) \end{matrix} \right\rangle_{\text{fiber}} \right), \tag{7.56}$$

where:

1. *the numerical constant $K_q \begin{pmatrix} [m_n] \\ [m_{n-1}] \end{pmatrix}$ depends only on the $\mathcal{U}_q(\mathfrak{u}(n))$ irrep labels $[m_n]$ and the $\mathcal{U}_q(\mathfrak{u}(n-1))$ irrep labels $[m_{n-1}]$;*

2. *the q-Wigner-Clebsch-Gordan coefficient ${}_qC^{\cdot\cdot}_{\cdot\cdot\cdot}$ effects the tensor coupling:*

$$[w\,\dot{0}] \times [\mu] \to [m_{n-1}]\,.$$

3. *the irrep vector $\left| \begin{matrix} [w\,\dot{0}] \\ (\mu') \end{matrix} \right\rangle_{\text{base}}$ is homogeneous and holomorphic in the q-bosons $\{a^q_i\}$ acting on the ket $|0\rangle$ with the $\mathcal{U}_q(\mathfrak{u}(n-1))$ irrep labels: $[w\,0 \cdots 0]$, where $w = \sum_{i=1}^{n-1}(m_{i,n-1} - \mu_{i,n-1})$;*

4. *the $\mathcal{U}_q(\mathfrak{u}(n-1))$ irrep labels of the fiber vector $\left| \begin{matrix} [\mu] \\ (\mu) \end{matrix} \right\rangle_{\text{fiber}}$ are given by $\mu_{i,n-1} = m_{i+1,n}$ for $i = 1, 2, \ldots, n-1$. (These fiber vectors are actually tensored with a fixed $U(1)$ vector carrying irrep $[m_{1n}]$ but this is suppressed for simplicity.)*

PROOF: This may appear to be a rather complicated result, but the underlying structure is simple and actually implicit in the construction of (7.36). The vectors in the fiber, by construction with the projection $\mathbb{P}$, carry the minimal $\mathcal{U}_q(\mathfrak{u}(n-1))$ weight contained in the irrep $[m]$. (These are the labels $[\mu]$.) The homogeneous holomorphic vectors have a unique degree w which labels the symmetric $\mathcal{U}_q(\mathfrak{u}(n-1))$ irrep that these vectors carry, $[w\ \dot{0}]$. The label w is at the same time the total degree of the polynomial in the q-bosons $\{a_i^q\}$, and this is the number of q-boson quanta in the irrep, namely $w = \sum_i (m_{i,n-1} - \mu_{i,n-1}) \geqslant 0$. The application of the known action by the generators $\Gamma(E_{ij})$ then completes the proof. □

To determine the explicit value of the function $K_q \begin{pmatrix} [m_n] \\ [m_{n-1}] \end{pmatrix}$, we remark that the $\Gamma(E_{ij})$ realize, by construction, the standard matrix elements of the generators acting on the basis $\{|(m)\rangle_{\text{ind}}\}$. To determine the function K_q then, one need only evaluate the left-action on the q-Wigner-Clebsch-Gordan (q-WCG) coupled vectors and compare. This calculation shows that:

LEMMA 7.57 *The normalizing factor $K_q \begin{pmatrix} [m_n] \\ [m_{n-1}] \end{pmatrix}$ in Lemma 7.55 for $|(m)\rangle_{\text{ind}}$ has the value:*

$$K_q \begin{pmatrix} [m_n] \\ [m_{n-1}] \end{pmatrix} = \left(\prod_{i=1}^{n-1} \frac{[p_{in} - p_{nn} - 1]!}{[p_{i,n-1} - p_{nn}]!} \right)^{\frac{1}{2}}, \tag{7.58}$$

where $p_{ij} \stackrel{\text{def}}{=} m_{ij} + j - i$, and $[n]!$ denotes the q-factorial.

The parameters p_{ij} are known as "partial hooks" in the literature since differences of the p_{ij} are precisely Nakayama's hook parameters [191]; they also appear in matrix elements of generators and tensor operators (see Theorem 2.74 and §3.7).

REMARK 7.59 1. The q-WCG coefficients that appear in Lemma 7.55 are known, in principle, via the recursion process which assumes $\mathcal{U}_q(\mathfrak{u}(n-1))$ information to obtain $\mathcal{U}_q(\mathfrak{u}(n))$ results. It will be observed, however, that these coefficients are actually determined explicitly by evaluating the q-exponentials in (7.36) and putting the results in the form of Lemma 7.55. We will not give the explicit result here, although it is useful to remark that for $w = 1$ these results confirm the q-pattern calculus results described in §3.7.

2. For the $q = 1$ limit these results go over, as they must, into the well-known results for the induced vectors (7.15) discussed in §7.5. In particular, they reduce to vector coherent states. These explicit ($q = 1$) bases have proved to be very convenient for the optimal construction of tensor operators and their matrix elements. The recursion aspect has been of considerable significance in that the procedure relates $3j$ coefficients of, for example, the $U(n)$ group, to $6j$ coefficients in the $U(n-1)$ subgroup as discussed further in §7.8. It is reasonable to expect that structural results of this type will carry over to the generic q case, implying the existence of significant new q-analog identities.

7.7 Algebraic Induction for $U(3)$ and its Quantum Extension

We now investigate the specific case $n = 3$ of the general results described in §7.5 for the classical case and §7.6 for the quantum extension. This investigation will provide another way of understanding the rather complex structure for the basis vectors as summarized in Lemmas 7.23 and 7.55 and provides a nontrivial example of how tensor couplings may be used to construct basis vectors from the appropriate subgroup.

We firstly present explicit calculations for $U(3)$, using the notations and conventions of §7.5, and then carry out the quantum extension, following Lohe and Biedenharn [192]. It will be seen that hypergeometric functions play an essential role, and that Lemmas 7.23 and 7.55 imply significant special function identities.

7.7.1 Explicit Induced Vectors for U(3)

For $n = 3$ the induced vectors defined by (7.15) for $U(n)$ take the form

$$\left| \begin{matrix} m_{13} & & m_{23} & & m_{33} \\ & m_{12} & & m_{22} & \\ & & m_{11} & & \end{matrix} \right\rangle_{\text{ind}} = \sum_m E_m((m);a)\ |0\rangle \otimes \left| \begin{matrix} m_{13} & & m_{23} & & m_{33} \\ & m_{13} & & m_{23} & \\ & & m & & \end{matrix} \right\rangle, \tag{7.60}$$

where the coefficients are the matrix elements

$$E_m((m);a) = \left\langle \begin{matrix} m_{13} & & m_{23} & & m_{33} \\ & m_{13} & & m_{23} & \\ & & m & & \end{matrix} \right| \exp(a_1 E_{13} + a_2 E_{23}) \left| \begin{matrix} m_{13} & & m_{23} & & m_{33} \\ & m_{12} & & m_{22} & \\ & & m_{11} & & \end{matrix} \right\rangle, \tag{7.61}$$

where (m) denotes the array (m_{ij}). We now verify by explicit calculation the form of these states as given by Lemma 7.23, proceeding by determining first the matrix elements in (7.61). We do this by writing the exponential in (7.61) as a product of two exponential factors, each of which is expanded as an infinite series; however, only one term of each series contributes due to orthonormality of the Gel'fand-Weyl basis. We must therefore calculate the matrix elements of E_{13} and E_{23} to arbitrary powers in the Gel'fand-Weyl basis, and then express (7.61) as a sum. We find that this sum can be written in terms of a hypergeometric ${}_6F_5$ function. Next, we use an identity due to Whipple that expresses the ${}_6F_5$ function in terms of a ${}_3F_2$ function, which in turn can be expressed as a Wigner-Clebsch-Gordan (WCG) coefficient. This calculation leads us directly to the final form (7.24) that we seek; the fact that the final answer *must* be of this form in fact implies the existence of Whipple's identity, as well as symmetries of the hypergeometric functions.

For $n = 3$, (7.24) reads

$$\left| \begin{matrix} m_{13} & & m_{23} & & m_{33} \\ & m_{12} & & m_{22} & \\ & & m_{11} & & \end{matrix} \right\rangle_{\text{ind}} = (-1)^{m_{22}-m_{23}} \left[\frac{(m_{13} - m_{33} + 1)!(m_{23} - m_{33})!}{(m_{12} - m_{33} + 1)!(m_{22} - m_{33})!} \right]^{\frac{1}{2}}$$

$$\times \sum_m \frac{a_1^{m-m_{11}}(-a_2)^{w+m_{11}-m}}{\sqrt{(m-m_{11})!(w+m_{11}-m)!}} C^{j_1\ j_2\ \ j}_{m_1\ m_2\ m_1+m_2} |0\rangle \otimes \left| \begin{matrix} m_{13} & & m_{23} & & m_{33} \\ & m_{13} & & m_{23} & \\ & & m & & \end{matrix} \right\rangle, \tag{7.62}$$

where

$$\begin{aligned} &w = m_{13} + m_{23} - m_{12} - m_{22}, \\ &j_1 = \tfrac{1}{2}(m_{13} - m_{23}), \quad j_2 = \tfrac{1}{2}w, \quad j = \tfrac{1}{2}(m_{12} - m_{22}), \\ &m_1 = m - \tfrac{1}{2}(m_{13} + m_{23}), \quad m_2 = m_{11} - m + \tfrac{1}{2}w. \end{aligned}$$

Here we have also substituted from (7.25) for the explicit K-factor, namely,

$$K\begin{pmatrix} [\mathbf{m}_3] \\ [\mathbf{m}_2] \end{pmatrix}^2 = \prod_{i=1}^{2} \frac{(p_{i3} - p_{33} - 1)!}{(p_{i2} - p_{33})!},$$

and have included the phase factor $(-1)^\varphi$ (given explicitly in [131, Eqn. 2.16]), where

$$\begin{aligned} \varphi &= \varphi([\mu]) - \varphi([0,-w]) - \varphi([\mathbf{m}_2]) \\ &= \varphi([m_{13}, m_{23}]) - \varphi([0,-w]) - \varphi([m_{12}, m_{22}]) \\ &= m_{22} - m_{23}. \end{aligned}$$

We now provide details of the calculations, together with properties of the hypergeometric functions used, in order to obtain precisely the form (7.62). Apart from its intrinsic interest, this calculation serves as a guide for establishing similar results for the quantum group case.

In order to evaluate the matrix elements (7.61) we first require the matrix elements of E_{13} and E_{23} in the Gel'fand-Weyl basis. Explicit formulas for matrix elements in $U(n)$, first found by Gel'fand and Zetlin [107], are given by Baird and Biedenharn [110], and for the particular case $n = 3$ and for E_{23} these reduce to

$$\begin{aligned} &\left\langle \begin{matrix} m_{13} & & m_{23} & & m_{33} \\ & m_{12} & & m_{22} & \\ & & m_{11} & & \end{matrix} \right| E_{23} \\ &= \left[\frac{(m_{12} - m_{11})(m_{13} - m_{12} + 1)(m_{12} - m_{23})(m_{12} - m_{33} + 1)}{(m_{12} - m_{22} + 1)(m_{12} - m_{22})} \right]^{\frac{1}{2}} \\ &\qquad \times \left\langle \begin{matrix} m_{13} & & m_{23} & & m_{33} \\ & m_{12}-1 & & m_{22} & \\ & & m_{11} & & \end{matrix} \right| \\ &+ \left[\frac{(m_{11} - m_{22} + 1)(m_{13} - m_{22} + 2)(m_{23} - m_{22} + 1)(m_{22} - m_{33})}{(m_{12} - m_{22} + 2)(m_{12} - m_{22} + 1)} \right]^{\frac{1}{2}} \\ &\qquad \times \left\langle \begin{matrix} m_{13} & & m_{23} & & m_{33} \\ & m_{12} & & m_{22}-1 & \\ & & m_{11} & & \end{matrix} \right|. \end{aligned} \tag{7.63}$$

By repeated application of E_{23} we obtain the general form

$$\left\langle \begin{matrix} m_{13} & & m_{23} & & m_{33} \\ & m_{12} & & m_{22} & \\ & & m_{11} & & \end{matrix} \right| (E_{23})^n = \sum_{r=0}^{n} A_r^n(m) \left\langle \begin{matrix} m_{13} & & m_{23} & & m_{33} \\ & m_{12}-r & & m_{22}+r-n & \\ & & m_{11} & & \end{matrix} \right| \tag{7.64}$$

where (m) is the array $\left(\begin{smallmatrix} m_{12} & & m_{22} \\ & m_{11} & \end{smallmatrix}\right)$, and the coefficients A_r^n are to be determined. This can be done by deriving a recurrence relation for A_r^n using (7.63). We find

$$\begin{aligned} A_r^{n+1} = A_{r-1}^n & \left[\frac{(m_{12}-m_{11}-r+1)(m_{13}-m_{12}+r)}{(m_{12}-m_{22}+n-2r+3)}\right. \\ & \left.\times\frac{(m_{12}-m_{23}-r+1)(m_{12}-m_{33}-r+2)}{(m_{12}-m_{22}+n-2r+2)}\right]^{\frac{1}{2}} \\ & +A_r^n\left[\frac{(m_{11}-m_{22}+n-r+1)(m_{13}-m_{22}+n-r+2)}{(m_{12}-m_{22}+n-2r+2)}\right. \\ & \left.\times\frac{(m_{23}-m_{22}+n-r+1)(m_{22}-m_{33}-n+r)}{(m_{12}-m_{22}+n-2r+1)}\right]^{\frac{1}{2}} \end{aligned} \tag{7.65}$$

with $A_0^0 = 1$. Let us merely state the solution of these recurrence relations, as verified by direct substitution:

$$\begin{aligned} A_r^n = & \frac{n!(m_{12}-m_{22}-r)!}{r!(n-r)!(m_{12}-m_{22}+n-r+1)!} \\ & \times\left[\frac{(m_{11}-m_{22}+n-r)!(m_{23}-m_{22}+n-r)!(m_{22}-m_{33})!}{(m_{11}-m_{22})!(m_{23}-m_{22})!(m_{22}-m_{33}-n+r)!}\right. \\ & \times\frac{(m_{12}-m_{11})!(m_{12}-m_{23})!(m_{12}-m_{33}+1)!(m_{13}-m_{12}+r)!}{(m_{12}-m_{11}-r)!(m_{12}-m_{23}-r)!(m_{12}-m_{33}-r+1)!(m_{13}-m_{12})!} \\ & \left.\times\frac{(m_{13}-m_{22}+n-r+1)!(m_{12}-m_{22}+1)(m_{12}-m_{22}+n-2r+1)}{(m_{13}-m_{22}+1)!}\right]^{\frac{1}{2}}. \end{aligned} \tag{7.66}$$

We also need to evaluate the matrix elements of $(E_{13})^\ell$, for arbitrary integers ℓ, but in this case we require the action of $(E_{13})^\ell$ only on those states for which $m_{12} = m_{13}, m_{22} = m_{23}$. Since $E_{13} = [E_{12}, E_{23}]$, the matrix elements of E_{13} are easily found from those of E_{23}, given in (7.63), and those of E_{12} (given in Example 2.76 for q=1):

$$\begin{aligned} \left\langle\begin{smallmatrix} m_{13} & & m_{23} & & m_{33} \\ & m_{13} & & m_{23} & \\ & & m_{11} & & \end{smallmatrix}\right| E_{13} = & \sqrt{\frac{(m_{13}-m_{33}+1)(m_{11}-m_{23})}{(m_{13}-m_{23}+1)}}\left\langle\begin{smallmatrix} m_{13} & & m_{23} & & m_{33} \\ & m_{13}-1 & & m_{23} & \\ & & m_{11}-1 & & \end{smallmatrix}\right| \\ & -\sqrt{\frac{(m_{23}-m_{33})(m_{13}-m_{11}+1)}{(m_{13}-m_{23}+1)}}\left\langle\begin{smallmatrix} m_{13} & & m_{23} & & m_{33} \\ & m_{13} & & m_{23}-1 & \\ & & m_{11}-1 & & \end{smallmatrix}\right|. \end{aligned}$$

From this follows the general form:

$$\left\langle\begin{smallmatrix} m_{13} & & m_{23} & & m_{33} \\ & m_{13} & & m_{23} & \\ & & m_{11} & & \end{smallmatrix}\right| (E_{13})^\ell = \sum_{s=0}^{\ell} B_s^\ell(m)\left\langle\begin{smallmatrix} m_{13} & & m_{23} & & m_{33} \\ & m_{13}-s & & m_{23}-\ell+s & \\ & & m_{11}-\ell & & \end{smallmatrix}\right|, \tag{7.67}$$

where (m) is the array $\begin{pmatrix} m_{13} & & m_{23} \\ & m_{11} & \end{pmatrix}$ for some coefficients B^ℓ_s. As before, we derive a recurrence relation for B^ℓ_s, using the matrix elements for E_{13}:

$$\begin{aligned} B^{\ell+1}_s &= B^\ell_{s-1}\left[\frac{s(m_{13}-m_{23}-s+1)(m_{13}-m_{33}-s+2)(m_{11}-m_{23}-s+1)}{(m_{13}-m_{23}+\ell-2s+3)(m_{13}-m_{23}+\ell-2s+2)}\right]^{\frac{1}{2}} \\ &\quad -B^\ell_s\left[\frac{(m_{13}-m_{23}+\ell-s+2)(\ell-s+1)}{(m_{13}-m_{23}+\ell-2s+2)}\right. \\ &\quad\quad \left.\times\frac{(m_{23}-m_{33}-\ell+s)(m_{13}-m_{11}+\ell-s+1)}{(m_{13}-m_{23}+\ell-2s+1)}\right]^{\frac{1}{2}} \end{aligned} \tag{7.68}$$

with $B^0_0=1$. The solution, which again is verified by direct substitution, is

$$\begin{aligned} B^\ell_s &= (-1)^{\ell+s}\ell! \\ &\quad\times\left[\frac{(m_{13}-m_{33}+1)!(m_{11}-m_{23})!(m_{13}-m_{23}-s)!(m_{13}-m_{11}+\ell-s)!}{s!(\ell-s)!(m_{13}-m_{23}+\ell+1-s)!(m_{13}-m_{33}+1-s)!(m_{11}-m_{23}-s)!}\right. \\ &\quad\times\left.\frac{(m_{23}-m_{33})!(m_{13}-m_{23}+\ell-2s+1)}{(m_{13}-m_{11})!(m_{23}-m_{33}-\ell+s)!}\right]^{\frac{1}{2}}. \end{aligned} \tag{7.69}$$

We can now evaluate $E_m((m);a)$ by expanding the exponentials, giving

$$\begin{aligned} E_m((m);a) &= \sum_{\ell,n}\left\langle \begin{smallmatrix} m_{13} & & m_{23} & & m_{33} \\ & m_{13} & & m_{23} & \\ & & m & & \end{smallmatrix}\right| \frac{a_1^\ell a_2^n}{\ell!n!}(E_{13})^\ell(E_{23})^n \left| \begin{smallmatrix} m_{13} & & m_{23} & & m_{33} \\ & m_{12} & & m_{22} & \\ & & m_{11} & & \end{smallmatrix}\right\rangle \\ &= \sum_{\ell,n=0}^{\infty}\sum_{s=0}^{\ell}\sum_{r=0}^{n}\left\langle \begin{smallmatrix} m_{13} & & m_{23} & & m_{33} \\ & m_{13}-s-r & & m_{23}-\ell+s+r-n & \\ & & m-\ell & & \end{smallmatrix}\right| \frac{(a_1)^\ell(a_2)^n}{\ell!n!}B^\ell_s\begin{pmatrix} m_{13} & & m_{23} \\ & m & \end{pmatrix} \\ &\quad\times A^n_r\begin{pmatrix} m_{13}-s & & m_{23}-\ell+s \\ & m-\ell & \end{pmatrix}\left| \begin{smallmatrix} m_{13} & & m_{23} & & m_{33} \\ & m_{12} & & m_{22} & \\ & & m_{11} & & \end{smallmatrix}\right\rangle . \end{aligned}$$

Since the Gel'fand-Weyl states are orthonormal we can eliminate three of the four summations, namely, the sums over ℓ, n, r. We must have

$$\begin{aligned} \ell &= m-m_{11} \\ r &= m_{13}-m_{12}-s \\ n &= w-m+m_{11}, \end{aligned}$$

where $w=m_{13}+m_{23}-m_{12}-m_{22}$. The matrix element now takes the form

$$\begin{aligned} E_m((m);a) &= \sum_s\frac{a_1^{m-m_{11}}a_2^{w-m+m_{11}}}{(m-m_{11})!(w-m+m_{11})!}B^{m-m_{11}}_s\begin{pmatrix} m_{13} & & m_{23} \\ & m_{11} & \end{pmatrix} \\ &\quad\times A^{w-m+m_{11}}_{m_{13}-m_{12}-s}\begin{pmatrix} m_{13}-s & & m_{23}+s-m+m_{11} \\ & m_{11} & \end{pmatrix}. \end{aligned} \tag{7.70}$$

Next, we collect all the factors as given in (7.66) and (7.69) in order to obtain the resulting matrix elements:

$$\begin{aligned}E_m((m);a) &= a_1^{m-m_{11}} a_2^{w-m+m_{11}} (-)^{m-m_{11}} \\ &\times \left[\frac{(m_{13}-m_{33}+1)!(m_{23}-m_{33})!(m_{11}-m_{22})!(m_{23}-m_{22})!}{(m_{22}-m_{33})!(m_{12}-m_{33}+1)!(m_{12}-m_{11})!(m_{12}-m_{23})!(m_{13}-m_{12})!}\right. \\ &\times \left.\frac{(m_{12}-m_{22}+1)}{(m_{13}-m)!(m-m_{23})!(m_{13}-m_{22}+1)!}\right]^{\frac{1}{2}} \\ &\times \frac{(m_{13}-m_{23})!(m_{12}+m-m_{23}-m_{11})!(m_{13}-m_{11})!}{(m-m_{11})!(m_{13}+m-m_{11}-m_{23})!(m_{11}+m_{23}-m_{22}-m)!} \sum_s R(s),\end{aligned} \tag{7.71}$$

where $R(s)$ is a product of all those terms containing s-dependent factors, and may be written:

$$R(s) = \frac{(-)^s (1+\frac{a}{2})_s (a)_s (b)_s (c)_s (d)_s (e)_s}{s! (\frac{a}{2})_s (a-b+1)_s (a-c+1)_s (a-d+1)_s (a-e+1)_s}, \tag{7.72}$$

where

$$\begin{aligned}&a = m_{11}+m_{23}-m_{13}-m-1, \quad b = m_{11}-m, \\ &c = m_{12}-m_{13}, \quad d = m_{23}-m, \\ &e = m_{22}-m_{13}-1,\end{aligned} \tag{7.73}$$

and where it can be seen that all square roots have combined to give a rational function of the basis state labels.

The function obtained by summing the terms $R(s)$ over s can be identified as a generalized hypergeometric function with argument -1:

$$\sum_s R(s) = {}_6F_5\left(\begin{matrix}1+\frac{a}{2} & a & b & c & d & e \\ & \frac{a}{2} & a-b+1 & a-c+1 & a-d+1 & a-e+1\end{matrix}; -1\right). \tag{7.74}$$

This function can be expressed in a simpler form by means of a special case of Whipple's transformation, which itself transforms a terminating well-poised ${}_7F_6$ into a Saalschützian ${}_4F_3$ (see Bailey [58, §4.3]). This transformation is derived by Bailey (§4.4, Eqn. (2)), and reads

$$\begin{aligned}&{}_6F_5\left(\begin{matrix}1+\frac{a}{2} & a & b & c & d & e \\ & \frac{a}{2} & a-b+1 & a-c+1 & a-d+1 & a-e+1\end{matrix}; -1\right) \\ &\qquad = (-1)^c \frac{(-a-1)_{-c}}{(a-e+1)_{-c}} {}_3F_2\left(\begin{matrix}1+a-b-d, & c, & e \\ 1+a-b & 1+a-d & \end{matrix}\right),\end{aligned}$$

where we have used the fact that $-c = m_{13} - m_{12}$ is a positive integer.

We now obtain from (7.70):

$$
\begin{aligned}
&E_m((m);a)\\
&= \frac{a_1^{m-m_{11}} a_2^{w-m+m_{11}}}{\sqrt{(m-m_{11})!(w-m+m_{11})!}}(-)^{m-m_{11}+m_{12}-m_{13}}(m_{13}-m_{23})!(m_{13}-m_{11})!\\
&\times\left[\frac{(m_{13}-m_{33}+1)!(m_{23}-m_{33})!(m_{11}-m_{22})!(m_{23}-m_{22})!(m_{12}-m_{22}+1)}{(m_{12}-m_{33}+1)!(m_{22}-m_{33})!(m_{13}-m_{22}+1)!(m_{13}-m)!(m_{12}-m_{11})!}\right.\\
&\times\left.\frac{1}{(m_{12}-m_{23})!(m_{13}-m_{12})!(m-m_{11})!(w+m_{11}-m)!(m-m_{23})!}\right]^{\frac{1}{2}}\\
&\times{}_3F_2\left(\begin{matrix} m-m_{13} & m_{22}-m_{13}-1 & m_{12}-m_{13} \\ & m_{23}-m_{13} & m_{11}-m_{13} \end{matrix}\right). \qquad (7.75)
\end{aligned}
$$

We expect this expression, from consideration of the general form (7.62), to be proportional to the WCG coefficient that couples the following states:

$$
\left|\begin{matrix} 0 & & -w \\ & m_{11}-m & \end{matrix}\right\rangle, \quad \left|\begin{matrix} m_{13} & & m_{23} \\ & m & \end{matrix}\right\rangle \longrightarrow \left|\begin{matrix} m_{12} & & m_{22} \\ & m_{11} & \end{matrix}\right\rangle.
$$

The WCG coefficient that performs this coupling is $C^{j_1\ j_2\ \ j}_{m_1\ m_2\ m_1+m_2}$, where

$$
\begin{aligned}
&j_1 = \tfrac{1}{2}(m_{13}-m_{23}), \quad j_2 = \tfrac{1}{2}w, \quad j = \tfrac{1}{2}(m_{12}-m_{22}),\\
&m_1 = m - \tfrac{1}{2}(m_{13}+m_{23}), \quad m_2 = m_{11} - m + \tfrac{1}{2}w. \qquad (7.76)
\end{aligned}
$$

A standard expression for the WCG coefficient is the van der Waerden form, which can be expressed as an ${}_3F_2$ function as follows (see for example [103, p. 30], or the $q = 1$ case of (3.59)):

$$
\begin{aligned}
C^{j_1\ j_2\ \ j}_{m_1\ m_2\ m_1+m_2} &= \left[\frac{(m_{12}-m_{22}+1)(m_{12}-m_{33})!(m_{23}-m_{22})!(m-m_{23})!}{(m_{13}-m_{22}+1)!(m_{13}-m)!(w+m_{11}-m)!}\right.\\
&\times\left.\frac{(m-m_{11})!(m_{11}-m_{22})!(m_{12}-m_{11})!}{(m_{13}-m_{12})!}\right]^{\frac{1}{2}}\\
&\times\frac{1}{(m_{12}+m-m_{13}-m_{23})!(m_{12}+m-m_{13}-m_{11})!}\\
&\times{}_3F_2\left(\begin{matrix} m_{12}-m_{13} & m-m_{13} & -w-m_{11}+m \\ m_{12}-m_{13}-m_{23}+m+1, & m_{12}-m_{11}+m-m_{13}+1 & \end{matrix}\right), \qquad (7.77)
\end{aligned}
$$

where we used the definitions given in (7.76). The ${}_3F_2$ function in (7.75) is not in the form (7.77), however, there is a direct transformation between the two forms, which appears in Bailey [58, p. 85] and may be written

$$
\begin{aligned}
{}_3F_2\left(\begin{matrix} a & b & c \\ & e & f \end{matrix}\right) &= \frac{(a-f)!(c-f)!(c-e)!(a-e)!}{(-e)!(-f)!(a+c-e)!(a+c-f)!}\\
&\times{}_3F_2\left(\begin{matrix} a & c & a+b+c-e-f+1 \\ & a+c-f+1 & a+c-e+1 \end{matrix}\right), \qquad (7.78)
\end{aligned}
$$

where c is a negative integer (the parameters a, b, e, f here differ from those defined in (7.73)). Now, if we substitute into (7.78) for the parameters

$$\begin{aligned} &a = m - m_{13}, \quad b = m_{22} - m_{13} - 1, \\ &c = m_{12} - m_{13}, \quad e = m_{23} - m_{13}, \quad f = m_{11} - m_{13}, \end{aligned} \tag{7.79}$$

then we transform the ${}_3F_2$ function in (7.75) into the form (7.77) for the WCG coefficient, and consequently express the induced state (7.60) in precisely the required form (7.62).

7.7.2 Algebraic Induction for $\mathcal{U}_q(\mathrm{u}(3))$

In constructing induced vectors for $\mathcal{U}_q(\mathrm{u}(3))$, analogous to (7.60) and (7.61) for $U(3)$, we extend the exponential to $\exp_q$ as for $\mathcal{U}_q(\mathrm{u}(2))$, and need to determine only the appropriate q-extension of the argument of $\exp_q$ in (7.61). $\mathcal{U}_q(\mathrm{u}(3))$ is defined in terms of elements corresponding to the simple roots, namely, E_{12} and E_{23}, and we require an appropriate definition of E_{13} in order to generalize (7.61). We choose the definition (2.70a) (with $q \leftrightarrow q^{-1}$):

$$E_{13} = q^{-\frac{1}{2}E_{22}}(E_{12}E_{23} - q^{-\frac{1}{2}}E_{23}E_{12}), \tag{7.80}$$

then as a consequence of the defining relations of $\mathcal{U}_q(\mathrm{u}(3))$ we have, as already noted in (2.71),

$$[E_{13}, E_{23}] = 0 = [E_{13}, E_{12}].$$

Similarly, we choose the operator E_{31} according to

$$E_{31} = q^{\frac{1}{2}E_{22}}(E_{32}E_{21} - q^{\frac{1}{2}}E_{21}E_{32}), \tag{7.81}$$

which satisfies

$$[E_{31}, E_{21}] = 0 = [E_{31}, E_{32}].$$

We now define the $\mathcal{U}_q(\mathrm{u}(3))$ induced vectors by

$$\left| \begin{smallmatrix} m_{13} & & m_{23} & & m_{33} \\ & m_{12} & & m_{22} & \\ & & m_{11} & & \end{smallmatrix} \right\rangle_{\mathrm{ind}} = \sum_m E^q_m((m); a^q)\,|0\rangle \otimes \left| \begin{smallmatrix} m_{13} & & m_{23} & & m_{33} \\ & m_{13} & & m_{23} & \\ & & m & & \end{smallmatrix} \right\rangle, \tag{7.82}$$

where

$$E^q_m((m); a^q) = \left\langle \begin{smallmatrix} m_{13} & & m_{23} & & m_{33} \\ & m_{13} & & m_{23} & \\ & & m & & \end{smallmatrix} \right| \exp_q(a^q_1 E_{13}) \exp_q(a^q_2 E_{23}) \left| \begin{smallmatrix} m_{13} & & m_{23} & & m_{33} \\ & m_{12} & & m_{22} & \\ & & m_{11} & & \end{smallmatrix} \right\rangle, \tag{7.83}$$

and where we now employ q-boson operators. The product of exponentials can be combined using the q-addition formula of Lemma 2.116 (p. 65) to obtain

$$\exp_q(a^q_1 E_{13}) \exp_q(a^q_2 E_{23})|0\rangle = \exp_q\left(q^{-\frac{N_2}{2}} a^q_1 E_{13} + q^{\frac{N_1}{2}} a^q_2 E_{23}\right)|0\rangle.$$

The argument of $\exp_q$ on the right hand side was denoted $\mathcal{O}$ in Definition 7.30, however, there we chose different explicit q-factors and also chose lowering generators instead of raising generators as in these explicit calculations for $\mathcal{U}_q(\mathfrak{u}(3))$ (this corresponds to the conventions chosen for $U(n)$ in §7.5).

Having now defined q-analogs of the induced vectors for $n = 3$, we next establish the $n = 3$ version of Lemma 7.12, that is, we calculate the realization $\Gamma(E_{ij})$ of the $\mathcal{U}_q(\mathfrak{u}(3))$ generators and verify that these operators satisfy the algebraic relations of $\mathcal{U}_q(\mathfrak{u}(3))$. The realization is expressed in terms of the q-boson operators a_1^q, a_2^q and their conjugates and associated number operators, and the generators $\mathcal{E}_{ij}$ of the subgroup $\mathcal{U}_q(\mathfrak{u}(2)) \times U(1)$. The results are summarized in the following lemmas:

LEMMA **7.84** *The realization $\Gamma(E_{ij})$ of the $\mathcal{U}_q(\mathfrak{u}(3))$ generators acting on the basis $|(m)\rangle_{\text{ind}}$ is given by*

$$\begin{aligned}
\Gamma(E_{11}) &= \mathcal{E}_{11} - N_1 \\
\Gamma(E_{22}) &= \mathcal{E}_{22} - N_2 \\
\Gamma(E_{33}) &= \mathcal{E}_{33} + N_1 + N_2 \\
\Gamma(E_{13}) &= \overline{a}_1^q \\
\Gamma(E_{23}) &= \overline{a}_2^q \\
\Gamma(E_{12}) &= q^{\frac{N_2}{2}}\mathcal{E}_{12} - q^{\frac{1}{2}(\mathcal{E}_{22}+1)}a_2^q\overline{a}_1^q \\
\Gamma(E_{21}) &= q^{-\frac{N_1}{2}}\mathcal{E}_{21} - q^{\frac{1}{2}(\mathcal{E}_{11}+1)}a_1^q\overline{a}_2^q \\
\Gamma(E_{32}) &= q^{-\frac{1}{2}\mathcal{E}_{33}}a_1^q\mathcal{E}_{12} + a_2^q[\mathcal{E}_{22} - \mathcal{E}_{33} - N_1 - N_2] \\
\Gamma(E_{31}) &= q^{\frac{1}{2}\mathcal{E}_{33}}a_2^q\mathcal{E}_{21} + a_1^q[\mathcal{E}_{11} - \mathcal{E}_{33} - N_1 - N_2].
\end{aligned}$$

PROOF: The expressions for $\Gamma(E_{i3}), i = 1, 2$ are immediate since

$$\overline{a}_i^q \exp_q(a_i^q E_{i3})|0\rangle = \exp_q(a_i^q E_{i3})E_{i3}|0\rangle \tag{7.85}$$

for $i = 1, 2$, using (7.9) and $[E_{13}, E_{23}] = 0$. The expressions for $\Gamma(E_{ii})$, $i = 1, 2, 3$ follow in the same way as for $\mathcal{U}_q(\mathfrak{u}(2))$, using

$$[E_{11} - N_1,\, a_1^q E_{13}] = 0 = [E_{11} - N_1,\, a_2^q E_{23}]$$

and

$$[E_{33} + N_1 + N_2,\, a_1^q E_{13}] = 0 = [E_{33} + N_1 + N_2,\, a_2^q E_{23}].$$

Let us now state an identity which is useful when considering generators satisfying the Serre relations. We may prove by induction on n that if

$$A^2B - [2]ABA + BA^2 = 0,$$

then

$$A^nB = [n]ABA^{n-1} - [n-1]BA^n, \quad n \in \mathbb{N}. \tag{7.86}$$

The application of this identity with $A = E_{23}$ and $B = E_{12}$ leads to

$$E_{23}^n E_{12} = q^{\frac{n}{2}} E_{12} E_{23}^n - [n] q^{\frac{1}{2}(E_{22}+1)} E_{13} E_{23}^{n-1}.$$

From this follows

$$\exp_q(a_2^q E_{23}) E_{12} = E_{12} q^{\frac{N_2}{2}} \exp_q(a_2^q E_{23}) q^{-\frac{N_2}{2}} - q^{\frac{1}{2}(E_{22}+1)} E_{13} a_2^q \exp_q(a_2^q E_{23}).$$

We multiply this operator equation on the right by $\exp_q(a_1^q E_{13})$, and then use (7.85) to replace $E_{13} \exp_2(a_2^q E_{23})$ by $\overline{a}_1^q \exp_q(a_2^q E_{23})$, and let the result act on the vacuum to give the required expression for $\Gamma(E_{12})$. Similarly, beginning with the formula

$$E_{13}^n E_{21} = q^{-\frac{n}{2}} E_{21} E_{13}^n - q^{-\frac{1}{2}(E_{11}+1)} [n] E_{23} E_{13}^{n-1}, \quad n \in \mathbb{N},$$

we obtain $\Gamma(E_{21})$. In order to calculate $\Gamma(E_{32})$, we first prove

$$E_{13}^n E_{32} = q^{\frac{n}{2}} E_{32} E_{13}^n + q^{-\frac{1}{2}E_{33}} [n] E_{12} E_{13}^{n-1}$$

by induction on n. Hence

$$\exp_q(a_1^q E_{13}) E_{32} = E_{32} q^{\frac{N_1}{2}} \exp_q(a_1^q E_{13}) q^{-\frac{N_1}{2}} + q^{-\frac{1}{2}E_{33}} E_{12} a_1^q \exp_q(a_1^q E_{13}).$$

Next we use

$$\exp_q(a_2^q E_{23}) E_{32} = (E_{32} + a_2^q [E_{22} - E_{33} - N_2]) \exp_q(a_2^q E_{23}),$$

which follows from

$$E_{23}^n E_{32} = E_{32} E_{23}^n + [E_{22} - E_{33} - n + 1][n] E_{23}^{n-1}.$$

By combining these formulas, we get $\Gamma(E_{32})$ as required, and $\Gamma(E_{31})$ is obtained by using (7.81). □

REMARK **7.87** Although it is not perhaps immediately obvious, this realization of $\mathcal{U}_q(\mathfrak{u}(3))$ does indeed reduce to that given by Lemma 7.19 for $U(3)$. However, we chose different conventions in §7.6 so that formulas there for $n = 3$ do not precisely match those in this lemma; note that we may insert operator-valued q-factors in the definition of $\mathcal{O}$ in 7.30 which are convention dependent.

LEMMA **7.88** *The map Γ, given by Lemma 7.84, is an isomorphism of $\mathcal{U}_q(\mathfrak{u}(3))$.*

PROOF: We first check that the definition (7.80) of E_{13} is preserved under the mapping Γ, that is, we verify

$$\Gamma(E_{13}) = q^{-\frac{1}{2}\Gamma(E_{22})} \left(\Gamma(E_{12})\Gamma(E_{23}) - q^{-\frac{1}{2}} \Gamma(E_{23})\Gamma(E_{12}) \right),$$

and this follows directly upon using the definition of the q-boson operator a_2^q. It is also immediate that

$$[\Gamma(E_{13}), \Gamma(E_{23}) = 0 = [\Gamma(E_{13}), \Gamma(E_{12})].$$

We also find

$$\begin{aligned}[\Gamma(E_{12}), \Gamma(E_{21})] &= q^{-\frac{1}{2}(N_1-N_2)}[E_{12}, E_{21}] - q^{\frac{1}{2}(E_{22}-E_{11})}[a_1^q \overline{a}_2^q, a_2^q \overline{a}_1^q] \\ &= [\Gamma(E_{11}) - \Gamma(E_{22})],\end{aligned}$$

as required, and

$$\begin{aligned}[\Gamma(E_{23}), \Gamma(E_{32})] &= \overline{a}_2^q a_2^q [E_{22} - E_{33} - N_1 - N_2] - a_2^q \overline{a}_2^q [E_{22} - E_{33} - N_1 - N_2 + 1] \\ &= [E_{22} - E_{33} - N_1 - 2N_2],\end{aligned}$$

again as required, where we used the q-boson relations (2.33) and the identity (2.12) with $a = N_2 + 1, b = E_{22} - E_{33} - N_1 - N_2 + 1, c = 1$. □

7.7.3 Explicit Induced Vectors for $\mathcal{U}_q(\mathrm{u}(3))$

In this section we undertake the q-analog of the calculation in §7.7.1 for $U(3)$ in order to obtain the $\mathcal{U}_q(\mathrm{u}(3))$ induced vectors explicitly, in a form analogous to (7.62), and the $n = 3$ case of Lemma 7.55. In order to perform this calculation we require various identities satisfied by basic hypergeometric functions which arise naturally within the calculation. Whereas the required matrix elements can be written in terms of ${}_6F_5$ functions for $U(3)$ (see (7.74)), we find that for $\mathcal{U}_q(\mathrm{u}(3))$ they can be expressed as a certain limit of basic hypergeometric functions ${}_8\phi_7$ which then reduce to ${}_3\phi_2$ functions with the help of known identities. These ${}_3\phi_2$ functions can in turn be transformed into a standard form recognizable as q-WCG coefficients, enabling us to express the induced states as a product of a K_q factor and a sum over a polynomial in a_1^q, a_2^q with a q-WCG coupling, tensored with the Gel'fand-Weyl states.

The induced vectors are defined by (7.82) and (7.83), and our first task is to evaluate the matrix elements in (7.83). The matrix elements of the simple roots E_{12}, E_{23} of the $\mathcal{U}_q(\mathrm{u}(3))$ algebra are identical to those of $U(3)$, but with integer factors replaced by corresponding q-integer factors. (We have stated the general matrix elements in Theorem 2.74, p. 47). Hence we can calculate the matrix elements of E_{13} from those of E_{12}, E_{23} by using (7.80), and we find that

$$\begin{aligned}E_{13}\left|\begin{matrix} m_{13} & & m_{23} & & m_{33} \\ & m_{12} & & m_{22} & \\ & & m_{11} & & \end{matrix}\right\rangle & \qquad (7.89) \\ = q^{-\frac{1}{2}m_{22}} \sqrt{\frac{[m_{13} - m_{12}][m_{12} - m_{23} + 1][m_{12} - m_{33} + 2][m_{11} + 1 - m_{22}]}{[m_{12} - m_{22} + 2][m_{12} - m_{22} + 1]}} & \\ \times \left|\begin{matrix} m_{13} & & m_{23} & & m_{33} \\ & m_{12}+1 & & m_{22} & \\ & & m_{11}+1 & & \end{matrix}\right\rangle &\end{aligned}$$

$$-q^{-\frac{1}{2}(m_{12}+1)}\sqrt{\frac{[m_{13}-m_{22}+1][m_{23}-m_{22}][m_{22}+1-m_{33}][m_{12}-m_{11}]}{[m_{12}-m_{22}+1][m_{12}-m_{22}]}}$$
$$\times\left|\begin{matrix} m_{13} & & m_{23} & & m_{33} \\ & m_{12} & & m_{22}+1 & \\ & & m_{11}+1 & & \end{matrix}\right\rangle .$$

From the matrix elements for E_{23} and E_{13} we can calculate the coefficients $A_r^n(m)$ and $B_s^\ell(m)$, which are defined as in (7.64) and (7.67), respectively. We do this by obtaining recurrence relations analogous to (7.65) and (7.68), and for A_r^n the only change is to again replace integer factors $(\dots)$ by q-integers $[\dots]$ to obtain the solution analogous to (7.66), and the recurrence relations are satisfied upon using the identity (2.12). The coefficients B_s^ℓ involve explicit q-factors, as is apparent from the matrix elements (7.89), and satisfy

$$\begin{aligned} B_s^{\ell+1} = {} & B_{s-1}^\ell q^{-\frac{1}{2}(m_{23}-\ell+s-1)} \\ & \times\left(\frac{[s][m_{13}-m_{23}-s+1][m_{13}-m_{33}-s+2][m_{11}-m_{23}-s+1]}{[m_{13}-m_{23}+\ell-2s+3][m_{13}-m_{23}+\ell-2s+2]}\right)^{\frac{1}{2}} \\ & -B_s^\ell q^{-\frac{1}{2}(m_{13}-s+1)}\left(\frac{[m_{13}-m_{23}+\ell-s+2][\ell-s+1]}{[m_{13}-m_{23}+\ell-2s+2]}\right. \\ & \left.\times\frac{[m_{23}-m_{33}-\ell+s][m_{13}-m_{11}+\ell-s+1]}{[m_{13}-m_{23}+\ell-2s+1]}\right)^{\frac{1}{2}} . \end{aligned}$$

The solution is

$$\begin{aligned} B_s^\ell = {} & q^{-\frac{1}{2}(\ell-s)(m_{13}-s)}q^{-\frac{1}{2}s(m_{23}-1)}q^{-\frac{\ell}{2}}(-)^{\ell+s}[\ell]! \\ & \times\left(\frac{[m_{13}-m_{33}+1]![m_{11}-m_{23}]![m_{13}-m_{23}-s]![m_{13}-m_{11}+\ell-s]!}{[s]![\ell-s]![m_{13}-m_{23}+\ell+1-s]![m_{13}-m_{33}+1-s]![m_{11}-m_{23}-s]!}\right. \\ & \left.\times\frac{[m_{23}-m_{33}]![m_{13}-m_{23}+\ell-2s+1]}{[m_{13}-m_{11}]![m_{23}-m_{33}-\ell+s]!}\right)^{\frac{1}{2}} , \end{aligned} \tag{7.90}$$

which is verified by direct substitution, using (2.12) with $a=\ell+1, b=m_{13}-m_{23}+\ell-s+2, c=s$.

The matrix element $E_m^q((m);a^q)$ is therefore given by the q-analog of (7.71) (that is, replace $(\dots)$ by $[\dots]$), together with the multiplicative q-factor

$$q^{-\frac{1}{2}s(s+a)-\frac{1}{2}(m_{13}+1)(m-m_{11})},$$

which comes from the coefficient B_s^ℓ in (7.90), and where a is defined in (7.73). The terms involving the summation s, analogous to $R(s)$ as given in (7.72), are

$$R(s)=\frac{[a+2s]([a])_s([b])_s([c])_s([d])_s([e])_s(-1)^s q^{-\frac{1}{2}s(a+s)}}{[s]![a]([1+a-b])_s([1+a-c])_s([1+a-d])_s([1+a-e])_s}, \tag{7.91}$$

where a, b, c, d, e are given by (7.73).

7.7.4 Basic Hypergeometric Functions and Watson's Formula

Now we proceed by making contact with the classical theory of basic hypergeometric functions, by observing that the function $\sum_s R(s)$ is the q-analog of the ${}_6F_5$ function appearing in (7.74). It is most conveniently expressed as the limit of an ${}_8\phi_7$ function, specifically:

$$\begin{aligned}
&\sum_s R(s)\\
&= \lim_{N\to\infty} {}_8\phi_7\left(\begin{matrix} q^a & q^{1+\frac{a}{2}} & -q^{1+\frac{a}{2}} & q^b & q^c & q^d & q^e & q^{-N} \\ & q^{\frac{a}{2}} & -q^{\frac{a}{2}} & q^{1+a-b} & q^{1+a-c} & q^{1+a-d} & q^{1+a-e} & q^{1+a+N} \end{matrix} ; q, \frac{q^{2a+2+N}}{q^{b+c+d+e}}\right)\\
&= \lim_{N\to\infty} \sum_s \frac{[a+2s]([a])_s([b])_s([c])_s([d])_s([e])_s([-N])_s}{[s]!\,[a]([1+a-b])_s([1+a-c])_s([1+a-d])_s([1+a-e])_s([1+c+N])_s},
\end{aligned} \tag{7.92}$$

where we used the definition (2.123) of a basic hypergeometric function, and the identity

$$\frac{(q^{1+\frac{a}{2}};q)_s(-q^{1+\frac{a}{2}};q)_s}{(q^{\frac{a}{2}};q)_s(-q^{\frac{a}{2}};q)_s} = q^s\frac{[a+2s]}{[a]}.$$

The limit $N\to\infty$ in (7.91) is taken using

$$\lim_{N\to\infty}\frac{([-N])_s}{([1+a+N])_s} = (-)^s q^{-\frac{1}{2}s(s+a)} \quad (q>1).$$

Although $\sum_s R(s)$ can be expressed in a certain way as a ${}_7\phi_6$ function, the form (7.92) is appropriate for our purposes because we can now use a formula due to Watson [193] that reduces a terminating, very well-poised ${}_8\phi_7$ series to a terminating balanced ${}_4\phi_3$ series, and is the q-analog of Whipple's formula, which transforms a well-poised ${}_7F_6$ into a Saalschützian ${}_4F_3$. By using this formula we can reduce $\sum_s R(s)$ to a ${}_3\phi_2$ function, which is identifiable with a q-WCG coefficient. Watson's formula states (see Bailey [58, p. 69] or Gasper and Rahman [64, p. 35]):

$$\begin{aligned}
&{}_8\phi_7\left(\begin{matrix} q^a & q^{1+\frac{a}{2}} & -q^{1+\frac{a}{2}} & q^b & q^c & q^d & q^e & q^{-N} \\ & q^{\frac{a}{2}} & -q^{\frac{a}{2}} & q^{1+a-b} & q^{1+a-c} & q^{1+a-d} & q^{1+a-e} & q^{1+a+N} \end{matrix} ; q, \frac{q^{2a+2+N}}{q^{b+c+d+e}}\right)\\
&\quad = \frac{(q^{1+a};q)_N(q^{1+a-d-e};q)_N}{(q^{1+a-d};q)_N(q^{1+a-e};q)_N}\,{}_4\phi_3\left(\begin{matrix} q^d & q^e & q^{1+a-b-c} & q^{-N} \\ & q^{1+a-b} & q^{1+a-c} & q^{e+d-a-N}\end{matrix};q,q\right).
\end{aligned}$$

In Bailey's notation we have replaced a by q^a, c by q^b, d by q^c, e by q^d, f by q^e, and g by q^{-N}, where N is a positive integer. In our notation Watson's formula reads

$$\sum_s \frac{[a+2s]([a])_s([b])_s([c])_s([d])_s([e])_s([-N])_s}{[s]!\,[a]([1+a-b])_s([1+a-c])_s([1+a-d])_s([1+a-e])_s([1+a+N])_s}$$

$$= \frac{([1+a])_N([1+a-c-e])_N}{([1+a-c])_N([1+a-e])_N}$$
$$\times \sum_s \frac{([1+a-b-d])_s([c])_s([e])_s([-N])_s}{[s]!([1+a-b])_s([1+a-d])_s([c+e-a-N])_s}. \tag{7.93}$$

Watson used this formula to prove the Rogers-Ramanujan identities, by taking suitable limits of the parameters; it also imples a general summation formula due to Jackson (see Gasper and Rahman [64, §2.6,2.7]). A discussion of proofs of the Rogers-Ramanujan identities has been given by Andrews [180].

We now let $N \to \infty$ in (7.93) and the left hand side reduces, according to (7.92), to $\sum_s R(s)$, while the multiplicative factors on the right hand side become

$$(-)^e q^{-\frac{ec}{2}} \frac{[-a-1]![a-c]!}{[a-c-e]![-a+e-1]!} \quad (q>1)$$

(in which positivity of the factorial arguments follows from the definitions of a, c, e in (7.73) and the inequalities satisfied by the Gel'fand-Weyl labels m_{ij}). We also use the limit

$$\lim_{N\to\infty} \frac{[\alpha-N]_s}{[\beta-N]_s} = q^{\frac{1}{2}(\beta-\alpha)} \quad (q>1)$$

and hence find

$$\sum_s R(s) = \frac{(-)^e q^{-\frac{1}{2}ec}[-a-1]![a-c]!}{[a-c-e]![-a+e-1]!} \sum_s \frac{([1+a-b-d])_s([c])_s([e])_s q^{-\frac{1}{2}s(e+c-a)}}{[s]!([1+a-b])_s([1+a-d])_s}, \tag{7.94}$$

which is the desired expression relating $\sum_s R(s)$ to a ${}_3\phi_2$ function.

Upon combining all terms, the matrix element $E^q_m((m);a^q)$ is now found to be:

$$E^q_m((m);a^q) = \sum_m (a^q_1)^{m-m_{11}} (a^q_2)^{w+m_{11}-m} (-)^{m_{12}-m_{13}+m-m_{11}}$$
$$\times q^{-\frac{1}{2}(m_{12}-m_{13})(m_{22}-m_{13}-1)} q^{-\frac{1}{2}(m_{13}+1)(m-m_{11})} \frac{[m_{13}-m_{22}]![m_{13}-m_{11}]!}{[m-m_{11}]![w-m+m_{11}]!}$$
$$\times \left(\frac{[m_{13}-m_{33}+1]![m_{23}-m_{33}]![m_{11}-m_{22}]![m_{23}-m_{22}]!}{[m_{22}-m_{33}]![m_{12}-m_{33}+1]![m_{13}-m]![m_{12}-m_{11}]![m_{12}-m_{23}]!} \right.$$
$$\left. \times \frac{[m_{12}-m_{22}+1]}{[m-m_{23}]![m_{13}-m_{12}]![m_{13}-m_{22}+1]!} \right)^{\frac{1}{2}}$$
$$\times {}_3\phi_2 \left(\begin{matrix} q^{m-m_{13}} & q^{m_{22}-m_{13}-1} & q^{m_{12}-m_{13}} \\ & q^{m_{23}-m_{13}} & q^{m_{11}-m_{13}} \end{matrix} ; q^{-1}, q^{-1} \right). \tag{7.95}$$

We wish to identify the ${}_3\phi_2$ function with that appearing in the q-WCG coefficients; a standard form for these coefficients, the q-analog of the van der Waerden form (7.77), is given in (3.59), p. 99, in which we identify the parameters, j_1, j_2, j, m_1, m_2 as in

(7.76). We seek to express the coefficients in the form shown in (7.95) and in order to do this we require the following transformation (Askey and Wilson [61, Eqn. (1.30)], Gasper and Rahman [64, Eqn. (3.2.2)]):

$$ {}_3\phi_2\left(\begin{matrix} q^{-n} & \alpha & \beta \\ & \gamma & \delta \end{matrix} ;q,q\right) = \alpha^n \frac{(\frac{\delta}{\alpha};q)_n}{(\delta;q)_n} {}_3\phi_2\left(\begin{matrix} q^{-n} & \alpha & \frac{\gamma}{\beta} \\ \gamma & \frac{\alpha}{\delta}q^{1-n} & \end{matrix} ;q,\frac{\beta q}{\delta}\right), $$

where n is a positive integer. Putting $\alpha = q^a, \beta = q^b, \gamma = q^e, \delta = q^f$, we can write this identity as

$$ \begin{aligned} &\sum_s \frac{q^{-\frac{1}{2}s(a+b-n-e-f+1)}([a])_s([b])_s([-n])_s}{[s]!([e])_s([f])_s} \\ &\quad = q^{-\frac{an}{2}} \frac{([f-a])_n}{([f])_n} \sum_s \frac{q^{-\frac{1}{2}s(b-f)}([a])_s([e-b])_s([-n])_s}{[s]!([e])_s([a+1-n-f])_s} \\ &\quad = \frac{([f-a])_n([e-a])_n}{([e])_n([f])_n} \sum_s \frac{q^{-\frac{sb}{2}}([a])_s([a+b+1-n-e-f])_s([-n])_s}{[s]!([a+1-n-f])_s([a+1-n-e])_s}, \end{aligned} $$

where we used the identity a second time in the last step. Hence we have

$$ \begin{aligned} &{}_3\phi_2\left(\begin{matrix} q^a & q^b & q^{-n} \\ q^e & q^f & \end{matrix} ;q^{-1},q^{-1}\right) \\ &\quad = \frac{([f-a])_n([e-a])_n}{([e])_n([f])_n} {}_3\phi_2\left(\begin{matrix} q^a & q^{a+b+1-n-e-f} & q^{-n} \\ q^{a+1-n-e} & q^{a+1-n-f} & \end{matrix} ;q^{-1},q^{-1}\right). \end{aligned} $$

This relation is the q-analog of (7.78), with $c = -n$ and, with the parameters identified as in (7.79), enables us to express the ${}_3\phi_2$ function in (7.94) in terms of the ${}_3\phi_2$ function in (7.95), that is, we can now explicitly identify the q-WCG coefficients in the matrix elements $E^q_m((m);a^q)$. Upon collecting all factors, we find:

LEMMA 7.96 *The induced vectors for $\mathcal{U}_q(\mathfrak{u}(3))$ defined by (7.82) have the explicit form*

$$ \left|\begin{matrix} m_{13} & & m_{23} & & m_{33} \\ & m_{12} & & m_{22} & \\ & & m_{11} & & \end{matrix}\right\rangle_{\text{ind}} = (-)^{m_{22}-m_{23}} K_q\left(\begin{matrix}[\mathbf{m}_3]\\ [\mathbf{m}_2]\end{matrix}\right) \sum_m \frac{(a^q_1)^{m-m_{11}}(-a^q_2)^{w+m_{11}-m}}{\sqrt{[m-m_{11}]![w+m_{11}-m]!}} $$

$$ \times q^{\frac{\mu}{4}}\, {}_{q^{-1}}C^{j_1\ j_2\ j}_{m_1\ m_2\ m_2+m_2} |0\rangle \otimes \left|\begin{matrix} m_{13} & & m_{23} & & m_{33} \\ & m_{13} & & m_{23} & \\ & & m & & \end{matrix}\right\rangle, $$

where the K_q-factor is given by

$$ K_q\left(\begin{matrix}[\mathbf{m}_3]\\ [\mathbf{m}_2]\end{matrix}\right)^2 = \frac{[m_{13}-m_{33}+1]![m_{23}-m_{33}]!}{[m_{12}-m_{33}+1]![m_{22}-m_{33}]!}, $$

and

$$ \mu = -(m_{13}-m_{12})(m_{13}-m_{22}+1) - (3m_{13}-m_{23}+2)(m-m_{11}) + w(m_{13}-m). $$

This form for the general $\mathcal{U}_q(\mathfrak{u}(3))$ basis vector is a special case of general results we obtained in Lemma 7.55 for the form of the induced basis vectors for $\mathcal{U}_q(\mathfrak{u}(n))$. The origin of the various factors is explained in Lemma 7.55, however, our calculation here explicitly shows how the theory of special functions and their q-analogs is interwoven with properties of the representations of the classical Lie groups and their q-analogs, the quantum groups.

7.8 Appendix: The Construction of Tensor Operators in the Classical Unitary Groups $U(n)$

Although we do not develop the q-analog of the method of algebraic induction for constructing tensor operators, we point out briefly in this section the significance of the methods of this chapter when applied to tensor operators for $q = 1$, methods which generalize to quantum groups. The utility of the algebraic induction procedure is that it expresses properties of operators in $U(n)$ in terms of those for $U(n-1)$, for example it shows that matrix elements of projective operators (defined in (3.8), p. 78) become, in a suitable limit, matrix elements of $U(n-1)$ projective operators which are $6j$ and $9j$ coefficients in this subgroup. Furthermore, these limit properties determine the specific numerical assignments to the Γ patterns which distinguish inequivalent tensor operators.

As indicated in the discussion of tensor operators in §3.2, two fundamental properties of tensor operators are those of equivariance and derivation. It is easily verified that the Borel-Weil realization of $U(n)$ — with generators given by Lemma 7.19 and basis vectors by Lemma 7.23 — has the derivation property on a tensor product of induced vectors. This elementary but important observation proves that the matrix realization of unit tensor operators for $U(n)$ can indeed be implemented in the Borel-Weil framework.

Consider the generic induced vector (7.24). Since it is the transformation properties which are essential we can replace, using equivariance, intrinsic space vectors by normalized unit tensor operators in $U(n-1)\otimes U(1)$. Such tensor operators are denoted by

$$\left\langle \begin{matrix} (\Gamma)_{n-2} \\ [M_{1n}\ldots M_{n-1,n}]_{n-1} \\ (\mu)_{n-2} \end{matrix} \right\rangle \otimes \langle M_{nn}\rangle,$$

where, for clarity, we denote Young frames in $U(n-1)$ by $[\ldots]_{n-1}$ and Gel'fand-Weyl (or operator) patterns in $U(n-2)$ by $(\ldots)_{n-2}$. Since the Borel-Weil construction promotes the $U(1)\times U(n-1)$ labels to $U(n)$ labels, we must require $M_{i,n} \geqslant M_{nn}$ for $i = 1\ldots, n-1$ for consistency.

The associated $U(n)$ tensor operator in the Borel-Weil construction then has the form:

$$\left\langle \begin{matrix} (\Gamma)_{n-1} \\ [M_{1n} \dots M_{nn}]_n \\ (M)_{n-1} \end{matrix} \right\rangle = K \begin{pmatrix} [M]_n \\ [M]_{n-1} \end{pmatrix} \left[B^{[\dot{0},-W]}_{(\cdot)}(a) \times \left\langle \begin{matrix} (\Gamma)_{n-2} \\ [M_{1n} \dots M_{n-1,n}]_{n-1} \\ (\cdot)_{n-2} \end{matrix} \right\rangle \right]_{(M)_{n-1}} \otimes \langle M_{nn} \rangle, \quad (7.97)$$

where the square bracket once again denotes a vector coupling in $U(n-1)$. By construction, the operators defined in (7.97) transform equivariantly as the unitary irrep $[M]$ of $U(n)$ with Gel'fand-Weyl pattern (M). The operator pattern labels (Γ) in (7.97) can be determined by a limit process (see [132]).

The operators given in (7.97), which are built by analogy to the generic induced vectors (7.24), are not the most general possible since we have used only polynomials in $\{a_\alpha\}$ and not more general polynomials over both $\{a_\alpha\}$ and $\{\overline{a}_\alpha\}$. The realization of the $U(n)$ generators given by Lemma 7.19 shows that, as an operator, $\{\overline{a}_\alpha\}$ indeed transforms as $[1, \dot{0}]$ in $U(n-1)$, but in $U(n)$ is actually a member of the tensor operator set $[1, \dot{0}, -1]$. Technical modifications are required in order to use polynomials over both $\{a_\alpha\}$ and $\{\overline{a}_\alpha\}$ in the construction of more general tensor operators, but there are special simplifications which occur for polynomials over either $\{a_\alpha\}$ or $\{\overline{a}_\alpha\}$ separately. The realization of Lemma 7.19 shows that in this case the $U(n)$ operator irrep labels $[M]$ are determined by the intrinsic space labels belonging to $U(n-1) \times U(1)$, either maximally tied (that is, labels take extreme values allowed in the Gel'fand pattern) for $\{a_\alpha\}$, or minimally tied for $\{\overline{a}_\alpha\}$. Further details may be found in Le Blanc and Biedenharn [131].

It is an important result that the two classes of tensor operators, the maximally tied and the minimally tied classes, have fully explicit matrix elements completely defined in terms of $U(n-1)$ constructs, to within a multiplicative ratio of K factors which are $U(n-1)$ invariant quantities. This is quite remarkable since it is entirely unexpected that any such general property could possibly hold true. Hence, we are able to determine explicit matrix elements for a significant class of tensor operators in $U(n)$ and, by taking a suitable limit, obtain information about the Γ patterns which label the tensor operators in $U(n)$. We refer to [131, 132] for further discussion.

[illegible equation]

where the square bracket once again denotes a vector coupling in $U(n-1)$. By construction, the operators defined in (7.97) transform equivariantly as $[\nu]$, an irrep $[\nu]$ of $U(n)$ with Gel'fand-Weyl pattern (M). The operator pattern labels (γ) in (7.97) can be determined by a limit process (see [132]).

The operators given in (7.97), which are built by analogy to the generic state vectors (7.24), are not the most general possible since we have used only polynomials in $\{a_{ij}\}$ and not more general polynomials over both $\{a_{ij}\}$ and $\{\bar{a}_{ij}\}$. The realization of the $U(n)$ generators given by Lemma 7.11 shows that, as an operator, $\{\bar{a}_{ij}\}$ indeed transforms as $[1, \dot{0}]$ in $U(n-1)$, but in $U(n)$ is actually a member of the tensor operator set $[1, \dot{0}, -1]$. Technical modifications are required in order to use polynomials over both $\{a_{ij}\}$ and $\{\bar{a}_{ij}\}$ in the construction of more general tensor operators, but there are special simplifications which occur for polynomials over either $\{a_{ij}\}$ or $\{\bar{a}_{ij}\}$ only. The realization of Lemma 7.10 shows that in this case the $U(n)$ operator pattern labels (M) are determined by the intrinsic space labels belonging to $U(n-1) \times U(1)$, either maximally tied (that is, labels take extremal values allowed in the Gel'fand patterns) for $\{a_{ij}\}$, or minimally tied for $\{\bar{a}_{ij}\}$. Further details may be found in Le Blanc and Biedenharn [131].

It is an important result that the two classes of tensor operators, the maximally tied and the minimally tied classes, have fully explicit matrix elements completely defined in terms of $U(n-1)$ constructs, to within a multiplicative ratio of h factors which are $U(n-1)$ invariant quantities. This is quite remarkable since it is entirely unexpected that any such general property could possibly hold true! Hence, we are able to determine explicit matrix elements for a significant class of tensor operators in $U(n)$ and, by taking a suitable limit, obtain information about the Γ patterns which label the tensor operators in $U(n)$. We refer to [131, 132] for further discussion.

Chapter 8

Special Topics

We have chosen several topics for further development which follow naturally from previous considerations, and illustrate the application of the properties of tensor operators. The first topic is the q-analog of the harmonic oscillator, a model which has appeared in an astonishing variety of physical applications as a basic theoretical tool, and which continues to be studied in new applications. In our development, oscillator algebras are used mainly to construct group representations but elsewhere have found a natural application in areas such as quantum optics and in the description of coherent radiation and, of course, in quantum field theory. We turn our attention then in §8.1.1 to the q-analog of coherent states and their construction using q-boson operators. We have already encountered a generalized form of coherent states in the method of algebraic induction described in Chapter 7.

Another topic which we discuss in §8.2 is a possible physical interpretation of noncommuting coordinates, an area of investigation on which there is no consensus at present. We take the point of view that quantum groups can offer a generalization of the Lorentz and also Poincaré groups as space-time symmetry groups, in which spatial coordinates appear as noncommuting components of tensor operators. We suggest one way in which such noncommuting coordinates could nevertheless lead to real measured quantities (expectation values) which of course do commute. This interpretation extends to all Hamiltonians which are quantum group invariants and are constructable from noncommuting q-tensor components.

In §8.3 we discuss the group of linear automorphisms of the canonical commutation relations and construct tensor operators with respect to this group. It is remarkable how rich the algebraic properties of the Heisenberg-Weyl algebra are, and we discuss one aspect, the Weyl-ordered polynomials of position and momentum operators which can be viewed as angular momentum tensor operators. This interpretation can be extended to the q-analog of the Heisenberg-Weyl algebra, so that q-Weyl-ordered polynomials can be regarded as q-tensor operators with respect to a quantum group. This provides an interesting example of the q-tensor operator properties which we introduced and developed in Chapter 4.

8.1 The q-Harmonic Oscillator

By analogy with the $q = 1$ case we can define q-momentum and q-position operators P, Q from the q-boson operators $a, \overline{a}$ in the same way as for boson operators. Therefore, we put

$$P \stackrel{\text{def}}{=} i\sqrt{\frac{m\hbar\omega}{2}}(a - \overline{a}), \qquad Q \stackrel{\text{def}}{=} \sqrt{\frac{\hbar}{2m\omega}}(a + \overline{a}). \tag{8.1}$$

The commutator $[P, Q]$ is then

$$i[P, Q] = \hbar\,[\overline{a}, a] = \hbar\,([N+1] - [N]),$$

and the eigenvalues of the right hand side are

$$\hbar\,([n+1] - [n]) = \hbar\,\frac{\cosh\left(\frac{1}{4}(2n+1)\log q\right)}{\cosh\left(\frac{1}{4}\log q\right)}.$$

One sees that the uncertainty is minimal (and independent of q) only in the limit $q \to 1$; the uncertainty increases with n for $q \neq 1$.

The q-harmonic oscillator Hamiltonian is defined from P, Q according to

$$\mathcal{H} \stackrel{\text{def}}{=} \frac{P^2}{2m} + \frac{m\omega^2}{2}Q^2 = \frac{\hbar\omega}{2}(\overline{a}a + a\overline{a}).$$

From the q-boson properties $a\overline{a} = [N]$, $\overline{a}a = [N+1]$, we find

$$\mathcal{H} = \frac{\hbar\omega}{2}\,([N+1] + [N]),$$

showing that the eigenvalues of $\mathcal{H}$ are

$$E(n) = \frac{\hbar\omega}{2}\,([n+1] + [n]) = \frac{\hbar\omega}{2}\left(\frac{[n+\frac{1}{2}]}{[\frac{1}{2}]}\right). \tag{8.2}$$

The normalized eigenstates are $|n\rangle = ([n]!)^{-\frac{1}{2}}\, a^n|0\rangle$. Eq. (8.2) shows that for $q \neq 1$ the energy levels are no longer uniformly spaced.

The generalization to n degrees of freedom can be carried out using the q-boson calculus as we described in §2.4 and §2.5.1. One possible Hamiltonian is

$$\mathcal{H} = \frac{\hbar\omega}{2}\sum_{i=1}^{n}\left(\overline{a}_i a_i + a_i\overline{a}_i\right) = \frac{\hbar\omega}{2}\sum_{i=1}^{n}\left([N_i + 1] + [N_i]\right).$$

Another, which is an invariant of $\mathcal{U}_q(\mathfrak{su}(n))$ when the generators are given as in §2.5.1 (see Eqn. (2.77), p. 49), is

$$\mathcal{H}' = \hbar\omega[N_1 + N_2 + \ldots + N_n],$$

and is expressible explicitly in q-boson form. The eigenstates in this case are given by (2.78) and have $m_{1i} - m_{1,i-1}$ quanta in the ith mode.

An explicit realization of the q-harmonic oscillator has been investigated by Atakishiev and Suslov [194] and Askey et al. [2, p. 57-63] where the Hermite eigenfunctions are generalized to their quantum analogs, and the momentum operator is realized as a finite difference operator, together with a generalization of the Fourier transform. For further discussion of recent work on harmonic oscillators we refer to [195].

Apart from (8.1), other definitions of momentum and position operators have been given, for example by Gel'fand and Fairlie [196], also Zumino [197] and Wess [3, p. 105], in which the Heisenberg algebra is replaced by the deformation

$$PQ - qQP = i\lambda. \tag{8.3}$$

These relations are identical to the q-boson relations expressed in the form (2.31), p. 27, upon identifying $P = i\lambda\overline{A}$, $Q = A$, although for real q and λ the relations (8.3) do not allow P and Q to be Hermitean operators, except for $q^2 = 1$; however, if $\overline{q} = q^{-1}$ and $\lambda = q\overline{\lambda}$ we can have $P = P^\dagger$ and $Q = Q^\dagger$. By contrast, the momentum and position operators defined by (8.1) are Hermitean and their closures can be extended to self-adjoint operators on a suitable domain, and have a continuous spectrum. This extension has been discussed by Burban and Klimyk [198], who have also determined the explicit deficiency subspaces, taking several alternative definitions of q-boson operators as starting points.

REMARK **8.4** In three spatial dimensions we can define q-analog position and momentum operators as spin 1 tensor operators with respect to $\mathcal{U}_q(\mathfrak{su}(2))$, using the formulas developed in Chapter 3. By means of a q-WCG coupling we can then construct the q-analog of orbital angular momentum generators, taking into account the factors arising because the generators are not themselves spin 1 tensor operators. We do not develop this important topic here but note that the subject of q-deformations of kinematical groups has received considerable attention in the literature, and refer to articles in [2, 4].

8.1.1 q-Coherent States

Coherent states appear in many applications to physics, a typical example being that of quantum optics, where the quantum states of the photon field can be prepared in coherent states and have properties close to those of classical waves.

We may identify two key characteristics of coherent states (following [199]):

(a) *continuity* of the coherent state $|z\rangle$ as a function of z, and

(b) *resolution of unity:*

$$\mathbb{1} = \int |z\rangle \, \langle z| \, d\mu(z),$$

where the integration takes place with respect to a positive measure $d\mu(z)$.

The best known example of coherent states, which certainly satisfy these two characteristics, are the canonical coherent states generated by boson creation and annihilation operators a and $\overline{a}$. These canonical coherent states are defined by

$$|z\rangle \stackrel{\text{def}}{=} \epsilon^{-\frac{1}{2}(|z|^2)} e^{za}|0\rangle = e^{-\frac{1}{2}(|z|^2)} \sum_{n=0}^{\infty} \frac{z^n}{\sqrt{n!}} |n\rangle,$$

where $|n\rangle$ denotes the orthonormal vectors generated by the creation operator a in the Fock space.

We can immediately write down q-coherent states $|z\rangle_q$ by replacing the boson operator by its q-boson analog, and the exponential function by the q-exponential function $\exp_q$:

$$\begin{aligned} |z\rangle_q &= \left(\exp_q(|z|^2)\right)^{-\frac{1}{2}} \exp_q(za^q)|0\rangle \\ &= \left(\exp_q(|z|^2)\right)^{-\frac{1}{2}} \sum_{n=0}^{\infty} \frac{z^n}{\sqrt{[n]!}} |n\rangle_q. \end{aligned} \tag{8.5}$$

These states satisfy $\overline{a}^q \; |z\rangle_q = z \; |z\rangle_q$ showing that the q-coherent state $|z\rangle_q$ is an eigenstate of the annihilation operator $\overline{a}^q$ with eigenvalue z and, since $z = {}_q\langle z|\overline{a}^q|z\rangle_q$ (assuming the states $|z\rangle_q$ are normalized), the label z is the mean of $\overline{a}^q$ in the state $|z\rangle_q$. The energy distribution in such a q-coherent state is now a q-analog Poisson distribution.

This definition of q-coherent states is not a unique q-extension of ordinary coherent states, for we could have chosen other exponential functions, as discussed in §2.8.2; this would introduce explicit q-factors in equations such as $\overline{a}^q|z\rangle_q = z|z\rangle_q$. The q-harmonic oscillator model leads naturally to the q-coherent states (8.5), given the position and momentum operators Q, P, according to:

$$a^q = \tfrac{1}{\sqrt{2}}(Q - iP), \quad \overline{a}^q = \tfrac{1}{\sqrt{2}}(Q + iP)$$

(in suitable units). The states (8.5) were considered in [79] and subsequently also in [88, 89, 90, 200]. In fact, as has happened with many q-analog concepts, q-coherent states were proposed before the appearance of quantum groups, in this case by Arik and Coon [201] in 1976 including the definition of q-boson operators!

Let us now mention the two characteristic properties of coherent states, continuity and completeness. The continuity properties of $|z\rangle_q$ as a function of z follow immediately from continuity of $\exp_q$. The resolution of unity within the Hilbert space, in terms of the states $|z\rangle_q$, has been considered by Gray and Nelson [89] and also Bracken et al. [88]. The q-analog of Euler's formula for $\Gamma(x)$ is required, and is expressed in terms of the q-integral defined in §2.8.1:

$$\int_0^\zeta \exp_q(-x)x^n d_q x = [n]! \tag{8.6}$$

where ζ is the largest zero of $\exp_q(x)$. Note that, unlike e^x, $\exp_q(x)$ alternates in sign as $x \to -\infty$ and that $|\exp_q(x)|$ becomes arbitrarily large as $-x$ increases. The q-integration does not converge, however, a natural restriction is $|z|^2 < \zeta$ and then, with the help of (8.6), the resolution of unity can be derived [89]:

$$\mathbb{1} = \int |z\rangle_q \, {}_q\langle z| \, d\mu(z) \tag{8.7}$$

where the measure $d\mu(z)$ is given by

$$d\mu(z) = \tfrac{1}{2\pi} \exp_q(|z|^2) \exp_q(-|z|^2) \, d_q|z|^2 \, d\theta,$$

where $\theta = \arg(z)$. It follows from (8.7) that an arbitrary state can be expanded in terms of the states $|z\rangle_q$ and that q-coherent states are overcomplete, since an arbitrary q-coherent state is non-orthogonal to $|z\rangle_q$, for any z; we have

$${}_q\langle \alpha \mid \beta \rangle_q = \exp_q\left(|\alpha|^2\right)^{-\frac{1}{2}} \exp_q\left(|\beta|^2\right)^{-\frac{1}{2}} \exp_q(\alpha^* \beta),$$

and in general $\exp_q(\alpha^*\beta) \neq 0$.

For further details on coherent states we refer to Klauder and Skagerstam [199], Perelomov [202], Inomata et al. [203], and for applications to quantum optics, Vogel and Welsch [204], together with those articles already cited for the q-analog generalization and its applications; here we also include [101, 205] and [2, p. 705].

8.2 Physical Interpretation of Noncommuting Coordinates

The essential (and well understood) difference between quantum and classical physics is that for quantum physics the phase-space coordinates $\{p_i, q_i\}$ of Hamiltonian mechanics are interpreted as noncommuting operators. Observables, in Dirac's formulation of quantum mechanics [206], comprise a complete set of commuting operators. It is the miracle of quantum mechanics that any given complete set supplies only *half* of the required Cauchy data of the classical approach, yet this information nonetheless suffices in quantum mechanics to reconstruct the physics.

Consider now a situation in which the physical phase-space coordinate operators obey a quantum group symmetry structure. The fact that the momenta do not commute is true even for many $q = 1$ cases, and causes no problem; the observables are, as usual, the complete set of commuting momentum observables. But how are we to interpret a (presumably equivalent) formulation in which the subset of commuting momentum observables is replaced by the subset of commuting coordinate operators? This latter subset will almost certainly be a *smaller* number of independent commuting operators than in the "classical" $q = 1$ case, where *all* coordinate operators commute. We apparently face a loss of information in using coordinates (and not momenta) as observables for a quantum group. To be more explicit, let us consider a physical problem in which the relevant physics is fully described by a symmetry group, for example, a quantum rotator with $SO(3)$ symmetry. The momenta in this problem are the three noncommuting angular momenta $\mathbf{L}$, and the two commuting coordinate operators ϑ and ϕ of the two-sphere S^2. The complete set of momentum observables are the two commuting operators $\mathbf{L}\bullet\mathbf{L}$ and L_z. Equivalently, the complete set of coordinate observables are the two commuting coordinate operators ϑ and ϕ. (This situation persists even in the classical ($\hbar \to 0$) limit.) If we were now to extend this problem to a rotator with quantum group symmetry $SO_q(3)$, the momentum observables would still be $\mathbf{L}^q\bullet\mathbf{L}^q$ and L_z^q, but the q-coordinate observables could no longer be the *two* q-coordinates, but *one* q-coordinate (since the only available coordinate invariant is the q-determinant, which is unity).

We believe that a possible way to answer the question posed by noncommuting coordinates is to be found from the example of operator-valued representation matrices for $U(2)$, as developed in Chapter 4, and generalized to $U_q(2)$.

Consider the "classical" ($q = 1$) $U(2)$ case. The commuting coordinate operators are given — in non-Hermitian unitary form — by the commuting boson operator matrix A, which is an operator-valued $U(2)$ group element corresponding to a coordinatized point in the manifold. For quantum mechanical calculations one would need a wave-function over these commuting coordinate operators. By the Peter-Weyl theorem any such function is a linear combination of matrix elements of the irrep matrices of the boson operator-valued coordinates. To obtain the required physical complex-valued wave function we simply replace (in the irrep matrices) the boson operator matrix by a complex-valued unitary matrix corresponding to a given point of the group manifold.

This crucial step (carried out in [66]) is not quite as straightforward or trivial as it might appear at first glance. To see this, let us formalize this "evaluation mapping" by using the "umbral calculus" (Rota [207, Chap. 2]), in which a linear functional (or inner product) performs the required mapping from coordinate operators to coordinate values. This mapping can be performed explicitly by the use of coherent states to supply a suitable inner product for defining a linear functional. That is, we introduce coherent states $|z^i_j\rangle$, where $z^i_j \in \mathbb{C}$, satisfying $\overline{a}^i_j|z^i_j\rangle = z^i_j|z^i_j\rangle$, and then map $\overline{a}^i_j$ to $\langle z^i_j|\overline{a}^i_j|z^i_j\rangle$; effectively, the boson matrix A may be replaced by real-valued coordinates z^i_j. We must point out, however, that this mapping, when used to give a quantum mechanical probability (which would be a function of the boson operators), must then be followed by a classical limit, $\hbar \to 0$, on the coherent state boson calculus to eliminate dispersion.

The essential point in this complicated re-examination of $U(2)$ wave functions is that the natural form for operator-valued coordinates is that of boson operators, which are *inherently quantum mechanical*, so that the natural evaluation map by coherent states interprets the evaluation of a boson operator power (a^n) differently[1] than the nth power of the evaluation of a. Hence the necessity in the usual case to carry out, at least in principle, a classical ($\hbar \to 0$) limit after the coherent state evaluation map.

By now it should be clear that for quantum groups and the quantum plane, the proposed physical interpretation of the noncommuting coordinates is to carry out the coherent state evaluation map on the q-tensor operator coordinates *without the classical limit step* that characterized the $q = 1$ case.

We thus interpret the quantized noncommutative coordinates of the quantum group by considering functions defined on these noncommuting coordinates, that is, functions on the quantum group, which we realize in terms of elementary q-tensor operators (replacing the boson operators in the $q = 1$ case) to obtain an operator-valued expression for the wave-function Ψ. (In the limit $q \to 1$, Ψ is still operator-valued except that the coordinate operators commute.) By the q-extension of the Peter-Weyl theorem, valid over noncommuting coordinates [156, 162], we may expand Ψ over $\mathbb{C}$ in terms of the quantum group irrep matrices ${}^q\mathfrak{D}(T)$ defined in (4.53), p. 147. We then form the operator-valued expectation value $\mathcal{E} = \Psi^\dagger \mathcal{O} \Psi$, where $\mathcal{O}$ is a quantum mechanical operator expressed as a function of q-coordinates, and evaluate $\mathcal{E}$ in the Hilbert space of q-coherent states $|z^i_j\rangle$ to obtain the physical expected value $\langle z^i_j|\mathcal{E}|z^i_j\rangle$. This is the quantum group measurement process which we propose for interpreting noncommuting quantum group coordinates.

It is clear that this physical expected value will show dispersion over the z^i_j coordinate values; this dispersion will depend on q and vanishes in the "classical" limit $q = 1$. Thus, this proposed quantum group coordinate interpretation would imply a true "second quantization". This outlines the construction we propose for interpreting physically the noncommuting q-coordinates of $SU_q(2)$, were this symmetry to be considered as physically relevant.

It is rather speculative to assert at the present time that quantum groups — especially any q-version of the Poincaré group — are fundamental to physics. Quantum groups however preserve the most essential elements of symmetry (irrep dimensions, for example) *continuously* for all real values of the parameter q. As the proposed evaluation map for quantum group coordinates indicates, these noncommuting q-coordinates show the existence of dispersion phenomena characteristic of an independent quantization parameter. Thus, with quantum groups one has introduced a form of coordinate quantization while preserving, continuously, all group symmetries. It has always been easy to achieve coordinate quantization using, say, discrete lattices. But prior to quantum groups, no one could achieve quantization without at the same

[1] This fact — used in determining the WCG coefficients — was the reason for developing boson operator rotation matrices in [66].

time destroying the continuous space-time symmetries essential to quantum physics. It is this remarkable achievement that makes quantum groups such an attractive possibility for constructing, albeit heuristically, new physical models.

8.3 Group Invariance of the Canonical Commutation Relations

There is a realization of the angular momentum commutation relations which is in a sense the simplest possible realization, for it can be expressed in terms of a single boson operator and its conjugate, satisfying the Heisenberg-Weyl algebra[2]. As a consequence, polynomials in the boson operators can be classified as angular momentum tensor operators, and satisfy the usual transformation properties and symmetries of tensor operators. Equivalently, one can express this realization in terms of position and momentum operators Q, P satisfying the canonical commutation relation $[Q, P] = i$ and when the Hermiticity properties of these operators are taken into account, one finds that the Lie algebra of the symplectic group $\mathrm{Sp}(2, \mathbb{R})$ has been generated. This is related to the fact that, for general n, the group of linear automorphisms of the canonical commutation relations among n pairs of position-momentum operators is $\mathrm{Sp}(2n, \mathbb{R})$. This invariance has been discussed in detail in the literature, see for example Itzykson [208], and Moshinsky et al. [209, 210] and Simon et al. [211].

The realization in terms of boson operators was analyzed by Biedenharn and Louck [103, p. 243], who termed this construction the symplecton and in particular derived properties of the polynomials constituting the tensor operator components. It turns out that these polynomials, when expressed in terms of position and momentum operators, are the Weyl-symmetrized polynomials introduced at an early stage in the development of quantum mechanics by Weyl as a solution to the problem of ordering noncommuting operators. Tensor operator techniques provide considerable insight into the properties of these polynomials, such as their symmetries and the algebra they satisfy and, furthermore, this approach generalizes to quantum groups, as we describe in §8.3.2.

As we have pointed out in Chapter 1, Lie groups provide a means of unifying the diverse properties of the classical functions and we illustrate this for $SU(2)$ with the symplecton and its q-analog. We can write the tensor operator components, the Weyl-symmetrized polynomials, as terminating hypergeometric functions with argument -1. Known properties of hypergeometric functions such as contiguity relations and linear transformations appear as algebraic properties of tensor operators. For the quantum group, realized with q-boson operators, the q-tensor polynomials appear as basic hypergeometric functions and satisfy a variety of identities as a consequence of the quantum group structure; the appearance of q-analog functions in this way supports the view that quantum groups underly the fundamental properties of these functions.

---

[2]We thank Professor E. C. G. Sudarshan for his helpful comments on this section.

8.3.1 Weyl-Ordered Polynomials and the Symplecton

In their fundamental paper of 1925, Born and Jordan [212] considered the problem of ordering noncommuting position and momentum operators, a subject later discussed also by Weyl [112]. Of the many ways in which the operator product, containing m factors of P and n factors of Q, can be ordered Weyl chose the fully symmetrized, averaged, sum of monomials in P and Q. If we denote by $T_{m,n}$ the Weyl-symmetrized polynomial resulting from such an ordering of P^mQ^n then we have, for example:

$$\begin{aligned} T_{1,1} &= \tfrac{1}{2}(PQ+QP) \\ T_{3,1} &= \tfrac{1}{4}(P^3Q+P^2QP+PQP^2+QP^3) \\ T_{2,2} &= \tfrac{1}{6}(P^2Q^2+PQPQ+PQ^2P+QP^2Q+QPQP+Q^2P^2). \end{aligned}$$

In 1932 McCoy [213] determined the following explicit expression for the Weyl-symmetrized polynomials:

$$T_{m,n}(Q,P) = 2^{-m}\sum_{s=0}^{m}\binom{m}{s}P^sQ^nP^{m-s}. \tag{8.8}$$

This formula gives the expressions

$$\begin{aligned} T_{3,1} &= \tfrac{1}{8}(P^3Q+3P^2QP+3PQP^2+QP^3) \\ T_{2,2} &= \tfrac{1}{4}(P^2Q^2+2PQ^2P+Q^2P^2), \end{aligned}$$

which are equivalent to those above, as can be seen upon using identities such as $PQP^2+P^2QP = P^3Q+QP^3$. Weyl-ordering differs in general from the Born-Jordan ordering which replaces the binomial weight in (8.8) with a unit weight.

In a series of papers, Bender and co-workers [214] investigated several problems involving the ordering of quantum operators by means of Weyl-symmetrized products. For example, in quantum mechanical models with one degree of freedom, the quantum Heisenberg differential equations were solved by constructing a function $F(Q,P)$ satisfying $[F,H]=i$, where $H=H(Q,P)$ is the Hamiltonian and $[Q,P]=i$. F was expanded as a sum over the operator basis elements $T_{m,n}(Q,P)$ and the coefficients of the expansion were obtained in closed form by using properties of $T_{m,n}$. These operators proved to be useful also in a finite-element approach to solving operator equations [214]. The algebra generated by $T_{m,n}$ was also discussed by Gel'fand and Fairlie [196] and shown to include several infinite-dimensional algebras previously investigated.

Weyl-symmetrized polynomials are best understood as components of a tensor operator with respect to the angular momentum group (in compact or noncompact form) and may be viewed as the fully symmetrized tensor product of the fundamental spinor operators Q, P. As such, their properties were investigated in detail in the formulation of the symplecton [215, 103]. The construction of the Weyl-symmetrized polynomials $T_{m,n}$ was carried out in terms of a boson operator a and its conjugate $\overline{a}$ satisfying $[\overline{a}, a] = 1$, but can be expressed equivalently in terms of position and momentum operators Q, P (as in [214]) by identifying $Q \leftrightarrow a, P \leftrightarrow -i\overline{a}$. This is possible because these operators satisfy the same Heisenberg-Weyl algebra. (The Hermiticity conditions differ according to which realization of the algebra we use, however, we will not use any Hermiticity properties in our development which therefore is indifferent as to whether the invariance group is compact or otherwise.)

We define $SU(2)$ generators by

$$\begin{aligned} J_+ &= -\tfrac{1}{2}a^2 = -\tfrac{1}{2}Q^2 \\ J_- &= \tfrac{1}{2}\overline{a}^2 = -\tfrac{1}{2}P^2 \\ J_z &= \tfrac{1}{4}(a\overline{a} + \overline{a}a) = \tfrac{i}{4}(QP + PQ), \end{aligned} \tag{8.9}$$

which satisfy $[J_z, J_\pm] = \pm J_\pm$, $[J_+, J_-] = 2J_z$, and where we have formally identified the position and momentum operators with q-boson operators as already explained. Realizations of the type (8.9) as bilinears in creation and annihilation operators are well known, and for n degrees of freedom generate a symplectic Lie algebra (see for example Moshinsky and Quesne [209]).

REMARK **8.10** The Casimir invariant for the realization (8.9) takes the value $-\frac{3}{16}$ which accords with the fact that for unitary irreps the group generated is the non-compact $\mathrm{Sp}(2, \mathbb{R})$. It is perhaps less well known that we can generalize (8.9) to realizations such as the following, which depends on two parameters α, β:

$$\begin{aligned} J_+ &= -\tfrac{1}{2}Q^2 \\ J_- &= -\tfrac{1}{2}P^2 + i\alpha(PQ^{-1} + Q^{-1}P) + (4\beta - 2\alpha)Q^{-2} \\ J_z &= \tfrac{i}{4}(QP + PQ) + \alpha. \end{aligned}$$

In this case the Casimir invariant is identically equal to $\alpha(\alpha+1) - 2\beta - \frac{3}{16}$, and so for suitable values of α, β we can obtain all finite-dimensional representations of $SU(2)$. For the application to tensor operators and Weyl-symmetrized polynomials, however, we consider only (8.9).

From a and $\overline{a}$ we construct a tensor operator with components $\mathcal{P}_{jm}$, where $j = 0, \frac{1}{2}, \ldots$ and $m = j, j-1, \ldots, -j$, by means of the definition (3.4):

$$\begin{aligned} [J_\pm, \mathcal{P}_{jm}] &= \sqrt{(j \mp m)(j \pm m + 1)}\ \mathcal{P}_{j,m\pm 1}, \\ [J_z, \mathcal{P}_{jm}] &= m\,\mathcal{P}_{jm}. \end{aligned} \tag{8.11}$$

In order to obtain $\mathcal{P}_{jm}$ explicitly, we first solve $[J^+, \mathcal{P}_{jj}] = 0$ to obtain $\mathcal{P}_{jj} = a^{2j}$ (up to a multiplicative constant), and then calculate $\mathcal{P}_{jm}$ by commutation with J_-:

$$\mathcal{P}_{jm}(a, \overline{a}) = \left[\frac{(j+m)!}{(2j)!(j-m)!}\right]^{\frac{1}{2}} \left[J_-, \mathcal{P}_{jj}(a, \overline{a})\right]_{(j-m)},$$

where the repeated commutator is defined recursively by

$$[A, B]_{(n)} = \left[A, [A, B]_{(n-1)}\right]$$

with $[A, B]_{(0)} \stackrel{\text{def}}{=} B$. Equivalently, we could solve for the component of lowest weight and then apply J_+ to obtain $\mathcal{P}_{jm}$. This gives another expression for $\mathcal{P}_{jm}$ which is related by symmetry, as shown below in (8.16).

The explicit form of the polynomials $\mathcal{P}_{jm}(a, \overline{a})$ is (see [215, 103]):

$$\mathcal{P}_{jm}(a, \overline{a}) = 2^m \left[\frac{(j-m)!}{(j+m)!}\right]^{\frac{1}{2}} \sum_{s=0}^{j-m} \frac{\overline{a}^{j-m-s} a^{j+m} \overline{a}^s}{s!(j-m-s)!}. \tag{8.12}$$

These polynomials are essentially the Weyl-symmetrized polynomials $T_{m,n}(Q, P)$ defined in (8.8), in fact (Lohe et al. [216]):

$$T_{j-m,j+m}(Q, P) = (i)^{-j+m} 2^{-j} \sqrt{(j-m)!(j+m)!}\ \mathcal{P}_{jm}(a, \overline{a}), \tag{8.13}$$

upon identifying $Q \rightarrow a$, $P \rightarrow -i\overline{a}$.

Properties of $\mathcal{P}_{jm}$ are derived most easily from the generating function formula

$$(\xi a + \eta \overline{a})^{2j} = \frac{(2j)!}{2^j} \sum_{m=-j}^{j} \Phi_{jm}(\xi, \eta)\ \mathcal{P}_{jm}(a, \overline{a}) \tag{8.14}$$

where

$$\Phi_{jm}(\xi, \eta) \stackrel{\text{def}}{=} \frac{\xi^{j+m} \eta^{j-m}}{\sqrt{(j+m)!(j-m)!}}. \tag{8.15}$$

We can regard the states Φ_{jm} as a basis for an irreducible representation space of $SU(2)$ generated by

$$K_+ = \xi \frac{d}{d\eta}, \quad K_- = \eta \frac{d}{d\xi}, \quad K_z = \tfrac{1}{2}\left(\xi \frac{d}{d\xi} - \eta \frac{d}{d\eta}\right).$$

The action of these generators on the basis states (8.15) is equivalent to the action of the generators $J_\pm, J_z$ on the operators $\mathcal{P}_{jm}$ by commutation, for we have

$$\begin{aligned} K_\mp(\xi a + \eta \overline{a})^{2j} &= [J_\pm, (\xi a + \eta \overline{a})^{2j}] \\ K_z(\xi a + \eta \overline{a})^{2j} &= [J_z, (\xi a + \eta \overline{a})^{2j}]. \end{aligned}$$

These equations show that if we regard (8.14) as defining $\mathcal{P}_{jm}$ then the relations (8.11) are immediately satisfied. A symmetric form for $\mathcal{P}_{jm}$ also follows from (8.14), since the left hand side of (8.14) is invariant under $\xi \leftrightarrow \eta$, $a \leftrightarrow \overline{a}$, showing that

$$\mathcal{P}_{jm}(a, \overline{a}) = \mathcal{P}_{j,-m}(\overline{a}, a) \tag{8.16}$$

or, equivalently,

$$T_{m,n}(Q,P) = T_{n,m}(P,Q). \tag{8.17}$$

The polynomials $\mathcal{P}_{jm}$ satisfy a product formula [103, p. 255]:

$$\mathcal{P}_{jm}\,\mathcal{P}_{j'm'} = \sum_{k=|j-j'|}^{j+j'} (2k+1)^{-\frac{1}{2}}\nabla(jj'k)\, C^{j'\ j\ \ k}_{m'\ m\ m+m'}\,\mathcal{P}_{k,m+m'} \tag{8.18}$$

where the triangle function $\nabla(abc)$ is defined by

$$\nabla(abc) \stackrel{\text{def}}{=} \left[\frac{(a+b+c+1)!}{(a+b-c)!(a-b+c)!(-a+b+c)!}\right]^{\frac{1}{2}}.$$

(This function is the inversion of Δ defined in §3.5). This product formula is equivalent to that given by Bender et al. [214] for the operators $T_{m,n}$, however, in (8.18) the role of the Wigner-Clebsch-Gordan coefficients is now manifest and leads to further properties of the triangle function, see [103, p. 250]. The product formula also implies the existence of recurrence relations for $\mathcal{P}_{jm}$, obtained by choosing a fixed value for j' and m' in (8.18), for example $j' = \frac{1}{2}$ for which the right hand side reduces to a sum of two terms.

The generating function can be used to determine the product formula in a direct way[3] as follows. In terms of momentum and position operators, the generating function takes the form

$$(\xi Q + \eta P)^n = \sum_{m=0}^{n} \binom{n}{m} \xi^m \eta^{n-m} T_{n-m,m}(Q,P) \tag{8.19}$$

for commuting indeterminates ξ, η, and hence,

$$e^{\xi Q+\eta P} = \sum_{n,m} \frac{\xi^m \eta^n}{m!n!}\, T_{n,m}(Q,P). \tag{8.20}$$

Now we use the well known formula

$$e^A e^B = e^{\frac{1}{2}[A,B]} e^{A+B}$$

for operators A, B which commute with $[A,\, B]$. Choosing $A = \xi Q + \eta P$ and $B = \xi' Q + \eta' P$ leads to

$$e^{\xi Q+\eta P} e^{\xi' Q+\eta' P} = e^{\frac{i}{2}(\xi\eta'-\eta\xi')} e^{(\xi+\xi')Q+(\eta+\eta')P}. \tag{8.21}$$

Upon substituting for the generating function from (8.20), and expanding products with the help of the binomial theorem, we obtain an identity comprising a product of

[3] We thank Dr. Michael Hall for kindly showing us this derivation.

Weyl-ordered polynomials equated with a sum of such polynomials, which is precisely the product formula (8.18). From this formula one can determine the commutator of any two Weyl-ordered polynomials, as was carried out in [214, 196].

The form (8.20) of the generating function is also useful as a direct means of deriving the explicit formula (8.8) for $T_{m,n}$. We first write (following Mukunda [217] and Hall [218]):

$$e^{\xi Q+\eta P} = e^{\frac{1}{2}\eta P}e^{\xi Q}e^{\frac{1}{2}\eta P},$$

and then expand the exponential factors. By comparing coefficients of $\xi^n\eta^m$ on each side we obtain the expression (8.8) for $T_{m,n}(Q,P)$.

In order to display the role of hypergeometric functions we define the operator (following [219, 216])

$$N = a\overline{a} = iQP,$$

which satisfies $[N,Q] = Q$ and $[N,P] = -P$. By induction we find that for positive integers n,

$$Q^nP^n = (-i)^n(N-n+1)_n \qquad P^nQ^n = (-i)^n(N+1)_n, \tag{8.22}$$

in which form these equations are valid for negative integers n also. By grouping factors suitably, that is, by writing $P^jQ^nP^{m-j} = (P^jQ^j)(Q^{n-j}P^{n-j})P^{m-n}$ in (8.8), and substituting from (8.22), we find that $T_{m,n}$ can be written:

$$\begin{aligned} T_{m,n}(Q,P) &= 2^{-m}(-i)^n(N-n+1)_n\,{}_2F_1\left(\begin{matrix} N+1,\ -m \\ N-n+1 \end{matrix};-1\right)P^{m-n} \\ &= 2^{-m}(-i)^n(N+1)_n\,{}_2F_1\left(\begin{matrix} -N+n-m,\ -m \\ -N-m \end{matrix};-1\right)P^{m-n}, \end{aligned} \tag{8.23}$$

where we reversed the order of summation to obtain the last form. (The fact that the parameter N in the hypergeometric function is an operator is immaterial, since it commutes with the other parameters of the function.) The expression (8.23) allows $T_{m,n}$ to be generalized to negative values of m, n, as was done in [214].

Properties of $T_{m,n}$ can now be derived from properties of the hypergeometric function. The equality in (8.17) follows from Euler's linear transformation for argument -1:

$$F(a,b,c;z) = (1-z)^{c-a-b}F(c-a,\ c-b,\ c;\ z),$$

where a, b, c are easily identified from (8.23). (On exchanging P and Q we put $N \to -N-1$ in order to preserve the commutation relations of N with P and Q.) Euler's transformation allows us to write

$$T_{m,n}(Q,P) = 2^{-n}(-i)^n(N+1)_n\,{}_2F_1\left(\begin{matrix} -N,\quad -n \\ -N-m \end{matrix};-1\right)P^{m-n}.$$

The defining equations (8.11) for $\mathcal{P}_{jm}(a,\overline{a})$ are equivalent to relations between contiguous hypergeometric functions:

$$\begin{aligned}(c-1)(c-2)F(a-2,\ b,\ c-2;\ -1)-(a-1)(a-2)\ F(a,b,c;-1)\\ =(c-a)(c-1)\ F(a-2,\ b-1,\ c-1;\ -1),\\ c(c+1)F(a,b,c;-1)-(b-c)(b-c-1)\ F(a+2,\ b,\ c+2;\ -1)\\ =4b(c+1)\ F(a+2,\ b+1,\ c+1;\ -1).\end{aligned}$$

The product formula (8.18) expressed as a hypergeometric function identity does not appear to be well known, although similar formulas have been derived by Burchnall and Chaundy [220]. The identity is

$$\begin{aligned}(-a)_n\,{}_2F_1\left(\begin{matrix}-a+m,\ -n\\ -a\end{matrix};-1\right)(-a-r+n)_s\,{}_2F_1\left(\begin{matrix}-a+n,\ -s\\ -a-r+n\end{matrix};-1\right)\\ =\sum_j\frac{(-)^j}{j!}(s-j+1)_j(m-j+1)_j\,{}_3F_2\left(\begin{matrix}-j,\quad -n,\quad -r\\ m-j+1,\ s-j+1\end{matrix};1\right)\\ \times(-a-r+j)_{n+s-j}\,{}_2F_1\left(\begin{matrix}-a+m,\ -n-s+j\\ -a-r+j\end{matrix};-1\right).\end{aligned}$$

In this form, the identity is seen to be valid for parameters a,r which are not necessarily integers.

The form (8.23) for $T_{m,n}$ provides calculational advantages also, for example, by using the formula $P^k f(N)=f(N+k)P^k$ for positive and negative k we can calculate commutators such as $[T_{m,n},P^k]$, which can then be expressed in terms of $T_{m',n'}$ for certain values of m',n' by using the contiguity relations of ${}_2F_1$.

It has been noted in [214] that for $m=n$ the polynomials $T_{m,n}$ can be expressed as Hahn polynomials S_n in $T_{1,1}=PQ+QP=-i(2N+1)$, and can be identified as generalized hypergeometric functions of unit argument. Specifically,

$$T_{n,n}=\frac{1}{(2n-1)!!}S_n(T_{1,1})=\frac{i^n}{(2n-1)!!\sqrt{n!}}\,{}_3F_2\left(\begin{matrix}-n,\ n+1,\ \frac{1}{4}-\frac{i}{4}T_{1,1}\\ \frac{1}{2}\qquad 1\end{matrix};1\right).$$

This result is a special case of (8.23), for the ${}_3F_2$ function can be transformed into the ${}_2F_1$ function given in (8.23) using a transformation due to Whipple [221] and standard properties of ${}_3F_2$ functions. This has also been noted, together with other properties, by Koornwinder [222].

Further properties of the Weyl-symmetrized polynomials follow from their group-theoretic origin, such as their transformations under the Lie group implemented by the representation matrix, and the properties of the adjoint polynomial, and are discussed in [103]. This group theoretical viewpoint for Weyl-symmetrized polynomials can be extended to more degrees of freedom, with the angular momentum group replaced by the symplectic group $\mathrm{Sp}(2n, \mathbb{R})$, and so the tensor operator techniques described here will be useful in general.

8.3.2 The q-Symplecton

Now we generalize the considerations of the previous sections to the quantum case, defining the polynomials $\mathcal{P}_{jm}$ as quantum group tensor operators and then developing their resulting properties. By means of q-boson operators satisfying

$$a\overline{a} = [N], \qquad \overline{a}a = [N+1]$$

we seek to construct the generators $J_{\pm}, J_z$ of $\mathcal{U}_q(\mathfrak{su}(2))$ in analogy with (8.9). It happens that we must use $q^{\frac{1}{2}}$-boson operators to accomplish this (following [219]), or equivalently, we can use q-boson operators to construct generators $J_{\pm}, J_z$ satisfying

$$[J_z, J_{\pm}] = \pm J_{\pm}, \quad [J_+, J_-] = [2J_z]_{q^2} = \frac{q^{2J_z} - q^{-2J_z}}{q - q^{-1}}.$$

The different q-value appears simply in order to account correctly for factors of $[2]_q$ as follows from the identity $[2n]_q = [2]_q[n]_{q^2}$. The commutation relations of the quantum group[4] $\mathcal{U}_{q^2}(\mathfrak{su}(2))$ are

$$[J_z, J_{\pm}] = \pm J_{\pm}, \quad [J_+, J_-] = \frac{[4J_z]}{[2]}. \tag{8.24}$$

We can also derive the following equations in the shifted factorial notation of (2.122):

$$a^n\overline{a}^n = ([N-n+1])_n\,, \quad \overline{a}^n a^n = ([N+1])_n\,, \tag{8.25}$$

which is valid for all integers n, both positive and negative. In particular, we have $a^2\overline{a}^2 = [N][N-1]$, $\overline{a}^2a^2 = [N+1][N+2]$.

LEMMA **8.26** *Let*

$$J_+ = -\frac{a^2}{[2]}, \quad J_- = \frac{\overline{a}^2}{[2]}, \quad J_z = \frac{1}{4}(2N+1),$$

then these operators satisfy the commutation relations of $\mathcal{U}_{q^2}(\mathfrak{su}(2))$.

PROOF: The commutator $[J_z, J_{\pm}] = J_{\pm}$ is immediate from the properties of the number operator and $[J_+, J_-]$ is evaluated from

$$[\overline{a}^2, a^2] = [N+1][N+2] - [N][N-1] = [2][2N+1]$$

[4]As for the $q = 1$ case we could more generally take the quantum group to be the special linear quantum group.

where we used the identity (2.12), p. 20, with $a = N+1$, $b = -N$, $c = 2$. □

As we shall see, the pair $\left(q^{\frac{1}{4}(N+1)}a,\, q^{\frac{1}{4}N}\overline{a}\right)$ forms a q-spinor with respect to these generators. If we postulate the q-deformed Heisenberg-Weyl algebra relation in the form (following [196] with $\lambda = -q^{\frac{1}{2}}$):

$$q^{\frac{1}{2}}QP - q^{-\frac{1}{2}}PQ = i, \tag{8.27}$$

then we may identify

$$Q \longleftrightarrow q^{\frac{1}{4}(N+1)}a, \quad P \longleftrightarrow -iq^{\frac{1}{4}N}\overline{a}, \tag{8.28}$$

as we did in the $q = 1$ case, since Q and P then satisfy (8.27) by virtue of the q-boson commutation relations. We can act on the q-spinor with the matrix quantum group described in Chapter 4, and we deduce the q-analog of the invariance of the canonical commutation relations under the special linear group:

Lemma 8.29 *The relation (8.27) is invariant under transformations of the matrix quantum group $SL_{q^2}(2)$, that is, if*

$$\begin{pmatrix} Q \\ P \end{pmatrix} \longrightarrow \begin{pmatrix} a & b \\ c & d \end{pmatrix} \begin{pmatrix} Q \\ P \end{pmatrix},$$

where the elements a, b, c, d satisfy (4.2), p. 117, with $q^{\frac{1}{2}} \to q$ and $ad - q^{-1}bc = \mathbb{1} = da - qbc$, then (8.27) is preserved.

This result is a special case of Lemma 4.10, (p. 123). A direct calculation shows: $q^{\frac{1}{2}}QP - q^{-\frac{1}{2}}PQ \longrightarrow i\det_{q^2}(T)$ which equals i in $SL_{q^2}(2)$.

We define operator-valued polynomials $\mathcal{P}_{jm}(a,\overline{a})$ in a and $\overline{a}$, which we call the q-symplecton polynomials, as tensor operator components with respect to $\mathcal{U}_{q^2}(\mathfrak{su}(2))$ and which therefore satisfy, according to Theorem 3.30 (p. 88):

$$\begin{aligned} \left(J_{\pm}\mathcal{P}_{jm} - q^{-m}\mathcal{P}_{jm}J_{\pm}\right)q^{-J_z} &= \sqrt{[j\pm m+1]_{q^2}[j\mp m]_{q^2}}\;\mathcal{P}_{j,m\pm 1}, \\ \left[J_z,\, \mathcal{P}_{jm}\right] &= m\,\mathcal{P}_{jm}. \end{aligned} \tag{8.30}$$

We calculate $\mathcal{P}_{jm}$ by solving for $\mathcal{P}_{jj}$, the highest weight component, from the equation $J_+\mathcal{P}_{jj} - q^{-j}\mathcal{P}_{jj}J_+ = 0$ and then act on $\mathcal{P}_{jj}$ with J_- to generate all polynomials $\mathcal{P}_{jm}$, using (8.30).

Lemma 8.31 *The tensor operator $\mathcal{P}_{jm}(a,\overline{a})$ is a polynomial in a of degree $j+m$ and in $\overline{a}$ of degree $j-m$, and has the explicit form:*

$$\begin{aligned} \mathcal{P}_{jm}(a,\overline{a}) &= \mathcal{N}_{j,m}q^{\frac{1}{2}jN}\sum_{s=0}^{j-m} q^{-s(j+\frac{1}{2})}\frac{\overline{a}^{j-m-s}a^{j+m}\overline{a}^{s}}{[s]![j-m-s]!} \\ &= \mathcal{N}_{j,m}q^{\frac{1}{2}jN}\sum_{s=0}^{j-m} q^{-s(j+\frac{1}{2})}\frac{([N+1])_{j-m-s}\,([N-2m-s+1])_{2m+s}}{[s]![j-m-s]!}(\overline{a})^{-2m}, \end{aligned}$$

where[5]

$$\mathcal{N}_{j,m} = q^{(j-m)(j-\frac{1}{4})}\frac{[j-m]![2j]!}{[j+m]!}\sqrt{\frac{[2j+2m]!!}{[2j-2m]!![4j]!!}}.$$

PROOF: The second form for $\mathcal{P}_{jm}$ follows by use of the property (8.25) with $n = 2m+s$ and $n = j-m-s$ after writing $\overline{a}^{j-m-s}a^{j+m}\overline{a}^{s} = (\overline{a}^{j-m-s}a^{j-m-s})(a^{2m+s}\overline{a}^{2m+s})(\overline{a})^{-2m}$. Negative powers of $\overline{a}$, which appear only for positive m, can be eliminated by further use of (8.25) to obtain a manifestly polynomial expression.

The recursive property of $\mathcal{N}_{j,m}$ which determines its value (choosing $\mathcal{N}_{j,j} = 1$) is

$$\frac{\mathcal{N}_{j,m-1}}{\mathcal{N}_{j,m}} = q^{j-\frac{1}{4}}\frac{[j-m+1][j+m]}{\sqrt{[2j+2m][2j-2m+2]}}.$$

According to (8.30) we must show that

$$\overline{a}^2\,\mathcal{P}_{jm} - q^{-m}\,\mathcal{P}_{jm}\overline{a}^2 = q^{\frac{1}{4}(2N+1)-m+1}\sqrt{[2j-2m+2][2j+2m]}\;\mathcal{P}_{j,m-1}\,,$$

where we have substituted for J_-, J_z from Lemma 8.26 and also used $[2n] = [2][n]_{q^2}$. This equation is equivalent to the following relation which we now prove directly:

$$\begin{aligned}
&[n+1][2j-n]\sum_{s=0}^{n+1} q^{-s(j+\frac{1}{2})}\frac{([N+1])_{n-s+1}\,([N+2n-2j-s+3])_{s-2n+2j-2}}{[s]![n-s+1]!}\\
&= q^{\frac{1}{2}(2j-2n-N-2)}\sum_{s=0}^{n} q^{-s(j+\frac{1}{2})}\frac{([N+3])_{n-s}\,([N+2n-2j-s+3])_{s-2n+2j}}{[s]![n-s]!}\\
&-q^{-\frac{1}{2}(2j+N+2)}\sum_{s=0}^{n} q^{-s(j+\frac{1}{2})}\frac{([N+1])_{n-s}\,([N+2n-2j-s+1])_{s-2n+2j}}{[s]![n-s]!},
\end{aligned}\tag{8.32}$$

where $n = j-m$. Let us prove the $n = 0$ case of (8.32) and in doing so obtain an identity which we can use for the general proof. For $n = 0$ the relation (8.32) reduces to

$$\begin{aligned}
&[2j]\left([N+1] + q^{-\frac{1}{2}(2j+1)}[N-2j+2]\right)\\
&= q^{-\frac{1}{2}(N-2j+2)}[N+1][N+2] - q^{-\frac{1}{2}(N+2j+2)}[N-2j+1][N-2j+2]
\end{aligned}\tag{8.33}$$

which is seen to be valid upon using

$$\begin{aligned}
[N+2] &= q^{-j}[N-2j+2] + q^{\frac{1}{2}(N-2j+2)}[2j]\\
[N-2j+1] &= q^{j}[N+1] - q^{\frac{1}{2}(N+1)}[2j].
\end{aligned}$$

[5] We define $[0]!! = 1$ and $[2n]!! = [2n][2n-2]\ldots[2]$ for integers n.

Now we rewrite the left hand side of (8.32) as a sum of two terms by means of the following identity:

$$\frac{[n+1]}{[n+1-s]} = q^{-\frac{s}{2}} \left(1 + q^{\frac{1}{2}(n+1)} \frac{[s]}{[n+1-s]}\right).$$

Finally, we apply to the resulting two terms the identity obtained from (8.33) by replacing $2j \to 2j - n$ and $N \to N + n - s$, and we find that this is sufficient to establish the equation (8.32), which proves the lemma. □

REMARK **8.34** 1. The normalization $\mathcal{N}_{j,m}$, which is equal to 1 for $j = m$, can be multiplied by any j-dependent factor without altering the tensor operator properties of $\mathcal{P}_{jm}$.

2. The highest weight polynomial $\mathcal{P}_{jj}$, which satisfies $J_+ \mathcal{P}_{jj} - q^{-j} \mathcal{P}_{jj} J_+ = 0$, is equal to $q^{\frac{1}{2}jN} a^{2j}$. We can sum the series for $m = -j$ to obtain also the q-symplecton polynomial of lowest weight, which is proportional to $q^{\frac{1}{2}jN} \overline{a}^{2j}$. In order to perform the sum we use the following identity (for $n = 2j$) which follows from the q-binomial theorem (by putting $x = q^{-\frac{1}{2}N} z, y = q^{\frac{1}{2}N} z$ in the q-binomial theorem and acting on 1), or can be proved directly by induction:

$$\sum_{s=0}^{n} \frac{q^{-\frac{s}{2}(n+1)}}{[s]![n-s]!} = q^{-\frac{n}{4}(n+1)} \frac{[2n]!!}{[n]!^2},$$

where we used $q^{\frac{n}{2}} + q^{-\frac{n}{2}} = [2n]/[n]$. We deduce

$$\mathcal{P}_{j,-j} = q^{j(j-1)} q^{\frac{1}{2}jN} \overline{a}^{2j}.$$

It is easy to verify that $J_- \mathcal{P}_{j,-j} - q^j \mathcal{P}_{j,-j} J_- = 0$, and so the action of J_- terminates the set of q-symplecton polynomials correctly.

EXAMPLE **8.35** 1. The spin $\frac{1}{2}$ q-symplecton polynomials are:

$$\mathcal{P}_{\frac{1}{2},\frac{1}{2}} = q^{\frac{1}{4}N} a, \quad \mathcal{P}_{\frac{1}{2},-\frac{1}{2}} = q^{\frac{1}{4}(N-1)} \overline{a}.$$

According to (3.38), p. 90, the combination

$$q^{\frac{1}{2}} \mathcal{P}_{\frac{1}{2},\frac{1}{2}} \mathcal{P}_{\frac{1}{2},-\frac{1}{2}} - q^{-\frac{1}{2}} \mathcal{P}_{\frac{1}{2},-\frac{1}{2}} \mathcal{P}_{\frac{1}{2},\frac{1}{2}}$$

is an invariant which by direct calculation we find is equal to $-q^{-\frac{1}{2}}$. Hence, the defining q-boson relations $\overline{a}a - q^{\frac{1}{2}} a\overline{a} = q^{-\frac{N}{2}}$ can be viewed as expressing the equality of a quantum group invariant with a constant.

2. The spin 1 q-symplecton polynomials are related to the generators given in Lemma 8.26 by the formula (3.47) (allowing for overall constants). In particular, $\mathcal{P}_{1,1} = q^{\frac{1}{2}N} a^2$ and $\mathcal{P}_{1,-1} = q^{\frac{1}{2}N} \overline{a}^2$.

If we perform transformations on the spinor components $\mathcal{P}_{\frac{1}{2},\frac{1}{2}}, \mathcal{P}_{\frac{1}{2},-\frac{1}{2}}$ by the co-action of the quantum matrix, in essentially the same way as on the components (Q, P) in Lemma 8.29, according to

$$\left(\mathcal{P}_{\frac{1}{2},\frac{1}{2}}, \mathcal{P}_{\frac{1}{2},-\frac{1}{2}}\right) \longrightarrow \left(\mathcal{P}_{\frac{1}{2},\frac{1}{2}}, \mathcal{P}_{\frac{1}{2},-\frac{1}{2}}\right) T,$$

where $T \in SL_{q^2}(2)$, then the q-symplecton polynomials are transformed by the $\mathfrak{D}$ matrix as follows:

$$\mathcal{P}_{jm}(a, \overline{a}) \longrightarrow \sum_{m'} {}^{q^2}\mathfrak{D}^j_{m'm}(T)\ \mathcal{P}_{jm'}(a, \overline{a}).$$

This can be viewed as another expression of the tensor operator properties of the q-symplecton polynomials.

Whereas the expression in Lemma 8.31 for $\mathcal{P}_{jm}$ was calculated from the highest weight polynomial, we could equally well start with $\mathcal{P}_{j,-j}$ and use J_+ to find $\mathcal{P}_{jm}$, using again (8.30). This gives a form for $\mathcal{P}_{jm}$ which by comparison with that in Lemma 8.31 displays a symmetry between a and $\overline{a}$, in which $q \to q^{-1}$, as for $q = 1$.

The q-symplecton polynomials can be expressed as basic hypergeometric functions by using the relations (8.25). We find

$$\mathcal{P}_{jm}(a, \overline{a}) = \mathcal{N}_{j,m} q^{\frac{1}{2}jN} \frac{([N+1])_{j-m}}{[j-m]!}\, {}_2\phi_1 \left(\begin{matrix} q^{-j+m},\ q^{2m-N} \\ q^{-j+m-N} \end{matrix} ; q, -q^{-j-m} \right) a^{2m},$$

which can be written in various forms using known identities (such as those in Gasper and Rahman [64, §1.4,1.5]). Properties of $\mathcal{P}_{jm}$ can be expressed therefore as properties of basic hypergeometric functions, and include special cases of the contiguous relations and linear transformation formulas. There is also an identity of basic hypergeometric functions which generalizes the product formula for symplecton polynomials.

We can determine a generating function for the q-symplecton polynomials which is the q-analog of (8.14), or (8.19) in the momentum-position operator formalism. In order to do this we introduce noncommuting coordinates denoted ξ, η and satisfying

$$\xi\eta = q^{-1}\eta\xi,$$

which commute with the position and momentum operators Q, P, which satisfy $q^{\frac{1}{2}}QP - q^{-\frac{1}{2}}PQ = i$. We can now adopt the same generating function as before (following Gel'fand and Fairlie [196]), and so the q-Weyl-ordered polynomials ${}_qT_{m,n}$ are generated by the following formula which we may take as defining ${}_qT_{m,n}$:

$$(\xi Q + \eta P)^n = \sum_{m=0}^{n} \begin{bmatrix} n \\ m \end{bmatrix} q^{\frac{1}{2}m(n-m)}\, \xi^m \eta^{n-m}\, {}_qT_{n-m,m}(Q, P). \tag{8.36}$$

In terms of the q-analog exponential function $E_q(z)$ defined in (2.117), p. 66, we have:

$$E_q(\xi Q + \eta P) = \sum_{n,m} q^{-\frac{1}{4}n(n-1)-\frac{1}{4}m(m-1)} \frac{\xi^m \eta^n}{[n]![m]!} \, {}_qT_{n,m}(Q,P).$$

However, this q-exponential does not appear to satisfy simple q-analog formulas which generalize (8.21), as Cigler [114] for example has discussed, and so another formulation of the generating function may be more appropriate.

We deduce from (8.36) that the q-Weyl-ordered polynomials satisfy

$$ {}_qT_{m,n}(Q,P) = {}_{q^{-1}}T_{n,m}(P,Q). \tag{8.37}$$

This follows by exchanging $\xi \leftrightarrow \eta$ and $Q \leftrightarrow P$ (and at the same time replacing $q \leftrightarrow q^{-1}$ in order to preserve the relation $\xi\eta = q^{-1}\eta\xi$), which leaves the left hand side of (8.36) invariant. On the right hand side we use $\xi^r \eta^s = q^{-rs}\eta^s\xi^r$ to re-order the noncommuting coordinates.

EXAMPLE **8.38** We have ${}_qT_{m,0} = P^m$ and ${}_qT_{0,n} = Q^n$. We determine the following expressions from the formula (8.36):

$$\begin{aligned}
{}_qT_{1,1} &= \tfrac{1}{[2]}\left(q^{-\frac{1}{2}}QP + q^{\frac{1}{2}}PQ\right) \\
{}_qT_{1,2} &= \tfrac{1}{[3]}\left(q^{-1}Q^2P + QPQ + qPQ^2\right) \\
{}_qT_{1,3} &= \tfrac{1}{[4]}\left(q^{-\frac{3}{2}}Q^3P + q^{-\frac{1}{2}}Q^2PQ + q^{\frac{1}{2}}QPQ^2 + q^{\frac{3}{2}}PQ^3\right) \\
{}_qT_{2,2} &= \tfrac{[2]}{[3][4]}\left(q^{-2}Q^2P^2 + q^{-1}QPQP + QP^2Q + PQ^2P + qPQPQ + q^2P^2Q^2\right).
\end{aligned}$$

The symmetry (8.37) is evident in the first and last expressions.

There is a formula for ${}_qT_{m,n}$ analogous to that given in (8.8):

$$ {}_qT_{m,n}(Q,P) = \frac{[m]!}{[2m]!!}\sum_{s=0}^{m} q^{-\frac{1}{4}(m+1)(m-2s)} \begin{bmatrix} m \\ s \end{bmatrix} P^s Q^n P^{m-s},$$

which reduces to (8.8) at $q = 1$, and which can be proved by induction on n from (8.36). Upon replacing Q and P by their q-boson equivalents as in (8.28), this expression becomes equal (up to constant invariant factors) to $\mathcal{P}_{\frac{1}{2}(n+m),\frac{1}{2}(n-m)}(a,\overline{a})$ as we noted for the $q = 1$ case in (8.13). Hence, the q-Weyl-ordered polynomials may be regarded as tensor operators with respect to the quantum group.

We may interpret the generating function formula (8.36) as an expansion in terms of basis states for an irrep of $\mathcal{U}_{q^2}(\mathfrak{su}(2))$, as for $q = 1$. Recall that in §4.7 (p. 153) we constructed irreps of $\mathcal{U}_q(\mathfrak{su}(2))$ over polynomials in noncommuting variables z_1, z_2, which we can identify with ξ, η after replacing $q^{\frac{1}{2}} \rightarrow q$. With the normalization given in (4.62) the basis vectors are:

$$\Phi_{jm}(\xi,\eta) = q^{\frac{1}{2}(j-m)(j+m)} \frac{\xi^{j+m}\eta^{j-m}}{\sqrt{[j-m]_{q^2}![j+m]_{q^2}!}},$$

and by realizing ξ and η in terms of q-boson operators, as shown in (4.60), we can define an action of the quantum group on this basis in which the generators $K_{\pm}, K_z$ take the usual bilinear form shown in (2.47), p. 33. This is the q-generalization of the action of the generators $K_{\pm}, K_z$ described for $q = 1$ in §8.3.1 and is equivalent to the induced action of the generators $J_{\pm}, J_z$ in (8.30). One can also perform transformations on the generating function by elements of the matrix quantum group by a right or left action, as described in §4.7, which is equivalent to the transformation of the q-symplecton polynomials by the $\mathfrak{D}$-matrix.

The quantum group approach to q-Weyl-ordered polynomials is also helpful in determining the product formula, the q-analog of (8.18). However, the reasoning outlined in §8.3.1, in particular the method using the addition formula (8.21) does not appear to generalize readily; for a suitable generalization one must allow the indeterminates ξ, η and ξ', η' to be elements of a quantum matrix $T \in \mathfrak{M}_{q^2}(2)$:

$$T = \begin{pmatrix} \xi & \eta \\ \xi' & \eta' \end{pmatrix},$$

and the determinant $\xi\eta' - \eta\xi'$ is then replaced by the quantum determinant.

Let us proceed for the q-symplecton polynomials, however, by assuming a form for the product law which is analogous to the $q = 1$ case:

$$\mathcal{P}_{jm}\,\mathcal{P}_{j'm'} = \sum_{k=|j-j'|}^{j+j'} N(jj'k)\; {}_qC^{j'\ j\ \ k}_{m'\ m\ m+m'}\; \mathcal{P}_{k,m+m'} \tag{8.39}$$

where $N(jj'k)$ is to be determined. From orthogonality of the q-WCG coefficients we deduce

$$N(jj'k)\,\mathcal{P}_{k,m''} = \sum_{m,m'} {}_qC^{j'\ j\ k}_{m'\ m\ m''}\; \mathcal{P}_{jm}\,\mathcal{P}_{j'm'}$$

which expresses the coupling of two tensor operators to form a third tensor operator. In particular, this equation also comprises an identity in q-boson operators. We can evaluate $N(jj'k)$ therefore by taking the vacuum expectation value of this equation (even though we do not impose the q-boson inner product otherwise) and in order to obtain a nontrivial result we choose $m'' = 0$. We require the following matrix elements, using the explicit expression for $\mathcal{P}_{j,m}$ from Lemma 8.31:

$$\begin{aligned} \mathcal{P}_{j',m}\,|0\rangle &= \mathcal{N}_{j',m}\,q^{j'm}\frac{[j'+m]!}{[j'-m]![2m]!}\,a^{2m}\,|0\rangle \\ \langle 0|\,\mathcal{P}_{j,-m} &= \mathcal{N}_{j,-m}\,q^{-\frac{1}{2}m(2j+1)}\frac{[2j+2m]!![j-m]!}{[2j-2m]!![j+m]![2m]!}\,\langle 0|\,\bar{a}^{2m}, \quad m \geqslant 0, \end{aligned}$$

where the normalization $\mathcal{N}_{j,m}$ is also given in Lemma 8.31. Each of these matrix elements is zero for $m < 0$. Hence,

$$
\begin{aligned}
N(jj'k)\mathcal{N}_{k,0} &= \sum_{m\geqslant 0} {}_qC^{j'\ j\ k}_{m\ -m\ 0} \langle 0|\ \mathcal{P}_{j,-m}\ \mathcal{P}_{j',m}\ |0\rangle \\
&= \sum_{m\geqslant 0} {}_qC^{j'\ j\ k}_{m\ -m\ 0}\ \mathcal{N}_{j,-m}\ \mathcal{N}_{j',m} q^{m(j'-j-\frac{1}{2})} \\
&\quad \times \frac{[2j+2m]!![j-m]![j'+m]!}{[2j-2m]!![j+m]![j'-m]![2m]!},
\end{aligned}
$$

which expresses the unknown function $N(jj'k)$ as a sum over q-WCG coefficients.

We do not develop this approach further by summing the expression for $N(jj'k)$ but note that, as for the $q = 1$ case, a transformation law involving the coupling of the $N(jj'k)$ functions with a q-Racah coefficient follows from the associativity of the q-symplecton product formula. Beginning with

$$
\left(\mathcal{P}_{j_1m_1}\ \mathcal{P}_{j_2m_2}\right)\mathcal{P}_{j_3m_3} = \mathcal{P}_{j_1m_1}\left(\mathcal{P}_{j_2m_2}\ \mathcal{P}_{j_3m_3}\right)
$$

we substitute (8.39) twice into each side to obtain a linear combination of the q-symplecton polynomials on each side, the coefficients of which must be equal. By means of the orthogonality properties of the q-WCG coefficients we derive an expression in which a product of two $N(\ )$ functions is equal to a sum over the product of two $N(\ )$ functions and four q-WCG coefficients, which in turn is expressible as a q-Racah coefficient using the formula (3.68), p. 103. Hence, we obtain:

$$
N(dbf)N(fac) = \sum_e N(bae)N(dec)\sqrt{[2e+1][2f+1]}\ {}_qW(abcd;ef),
$$

where $j_1 = d, j_2 = b, j_3 = a$, which generalizes the transformation law of the triangle function for $q = 1$ [103, p. 250].

Commutators of q-Weyl-ordered polynomials, which follow from the product formula, have been given in [196], where the case of q at a root of unity is also discussed. As already mentioned, the q-analog position and momentum operators are not Hermitean unless $\overline{q} = q^{-1}$, and so the root of unity case could be of significance for applications to physics and deserves further consideration.

As is evident from these remarks, the development of the q-analog of Weyl-ordered polynomials and q-symplecton polynomials is far from complete, an observation which can be applied to many aspects of quantum groups in general, and on this note we conclude.

❋ ❋ ❋ ❋

Bibliography

[1] H.-D. Doebner and J.-D. Hennig (eds.), *Quantum Groups*, Proceedings of 8th International Workshop on Mathematical Physics, *Lecture Notes in Physics* **370**, Springer-Verlag, Berlin, Germany (1990).

[2] B. Gruber (ed.), *Symmetries in Science VI, From the Rotation Group to Quantum Algebras*, Plenum Press, New York (1993).

[3] J. Keller and Z. Oziewicz (eds.), Proceedings of the XXIIth International Conference on *Differential Geometric Methods in Theoretical Physics*, Advances in Applied Clifford Algebras (Proc. Suppl.) 4(S1), FESC (1994).

[4] H.-D. Doebner and V. K. Dobrev (eds.), *Quantum Symmetries*, Proceedings of the International Workshop on Mathematical Physics (Clausthal, 1991), World Scientific, Singapore (1993).

[5] M. Gerstenhaber and J. Stasheff (eds.), *Deformation Theory and Quantum Groups with Applications to Mathematical Physics*, *Contemp. Math.* **134** (1990).

[6] T. Curtright, D. Fairlie and C. Zachos (eds.), *Proceedings of the Argonne Workshop: Quantum Groups*, ANL, Illinois, World Scientific, Singapore (1991).

[7] C. Kassel, *Quantum Groups*, Springer-Verlag, New York (1995).

[8] V. Chari and A. Pressley, *A Guide to Quantum Groups*, Cambridge University Press, Cambridge (1994).

[9] S. Shnider and S. Sternberg, *Quantum Groups (from coalgebras to Drinfeld algebras)*, International, Boston (1993).

[10] P. Kulish and N. Reshetikhin, "Quantum Linear Problem for the Sine-Gordon Equation and Higher Representations", *Zap. Nauch. Seminarov LOMI,* **101** (1981) 101-110, (*J. Soviet Math.* **23** (1983) 2435-2441).

[11] L. D. Faddeev, "Integrable Models in (1+1)-Dimensional Quantum Field Theory", in *Les Houches Lectures XXXIX*, 1982, Elsevier Science Publishers B.V., North-Holland, Amsterdam (1984) 563-608; E. Sklyanin, L. Takhatajan, and L. Faddeev, "Quantum Inverse Problem Method I.", *Theor. Math. Phys.* **40** (1979) 688-706.

[12] R. J. Baxter, *Exactly Solved Models in Statistical Mechanics*, Academic Press, London, England (1982).

[13] J. B. McGuire, "Study of Exactly Soluble One Dimensional N-Body Problems *J. Math. Phys.* **5** (1964) 622-636; C. N. Yang, "Some Exact Results for the Many-Body Problem in One Dimension with Repulsive Delta Function Interaction", *Phys. Rev. Lett.* **19** (1967) 1312-1314.

[14] S. Majid, "Quasi-Triangular Hopf Algebras and Yang-Baxter Equations", *Int. J. Mod. Phys. A***5** (1990) 1-91.

[15] E. Witten, "Gauge Theories, Vertex Models and Quantum Groups", *Nucl. Phys.* **B330** (1990) 285-346; "Gauge Theories and Integrable Lattice Models", *Nucl. Phys.* **B322** (1989) 629-697.

[16] M. J. Ablowitz and P. A. Clarkson, *Solitons, Nonlinear Evolution Equations and Inverse Scattering*, London Mathematical Society (Lecture Note Series **149**), Cambridge University Press, Cambridge (1991).

[17] A. C. Newell, *Solitons in Mathematics and Physics*, (Regional Conference Series in Applied Mathematics **48**), Society for Industrial and Applied Mathematics, Philadelphia, Pennsylvania (1985).

[18] L. D. Faddeev and L. A. Takhtajan, *Hamiltonian Methods in the Theory of Solitons*, Springer-Verlag, Berlin (1987).

[19] V. E. Korepin, N. M. Bogoliubov and A. G. Izergin, *Quantum Inverse Scattering Method and Correlation Functions*, Cambridge University Press, Cambridge (1993).

[20] L. A. Takhtajan, "Elementary Introduction to Quantum Groups", in *Important Developments in Soliton Theory*, eds. A. S. Fokas and V. E. Zakharov, Springer-Verlag, Berlin, Germany (1993) 441-467; *ibid.* "Quantum Groups and Integrable Models", *Adv. Studies in Pure Math.* **19** (1989) 435-457.

[21] C. N. Yang and M. L. Ge (eds.), *Braid Group, Knot Theory and Statistical Mechanics, Advanced Series in Mathematical Physics* **9**, World Scientific, Singapore (1989).

[22] V. G. Drinfeld, "Quantum Groups", *Proc. Int. Congr. of Math.*, MSRI Berkeley California, (1986) Vol. 1, p. 798-820; "Hopf Algebras and the Quantum Yang-Baxter Equation", *Sov. Math. Dokl.* **32** (1985) 254-258; "A New Realization of Yangians and Quantized Affine Algebras", *Sov. Math. Dokl.* **36** (1988) 212-219.

[23] M. Jimbo, "A q-Difference Analogue of U(g) and the Yang-Baxter Equation", *Lett. Math. Phys.* **10** (1985) 63-69.

[24] L. D. Faddeev, N. Yu. Reshetikhin and L. A. Takhtajan, "Quantization of Lie Groups and Lie Algebras", in *Algebraic Analysis*, eds. M. Kashiwara and T. Kawai, Academic, Boston (1989) 129-139; reprinted in "*Yang-Baxter Equation in Integrable Systems*", ed. M. Jimbo, World Scientific (1990); L. D. Faddeev and L. A. Takhtajan, "Liouville Model on the Lattice" in *Field Theory, Quan-*

tum Gravity, and Strings: Proceedings of a Seminar Series held at DAPHE, eds. H. J. DeVega and N. Sanchez *Lecture Notes in Physics* **246** (1986) 166-179.

[25] V. Pasquier, "Etiology of IRF Models", *Commun. Math. Phys.* **118** (1988) 255-364.

[26] E. Date, M. Jimbo, K. Miki and T. Miwa, "Generalized Chiral Potts Models and Minimal Cyclic Representations of $\mathcal{U}_q(\mathfrak{gl}(n,\mathbb{C}))$", *Commun. Math. Phys.* **137** (1991) 133-147.

[27] V. V. Bazhanov, R. M. Kashaev, V. V. Mangazeev and Yu. G. Stroganov, "$(Z_N\times)^{n-1}$ Generalizations of the Chiral Potts Model," *Commun. Math. Phys.* **138** (1991) 393-408.

[28] G. Moore and N. Seiberg, "Classical and Quantum Conformal Field Theory", *Commun. Math. Phys.* **123** (1989) 177-254.

[29] J. Fuchs, *Affine Lie Algebras and Quantum Groups*, (Cambridge Monographs on Mathematical Physics), Cambridge University Press, Cambridge (1992).

[30] L. Alvarez-Gaumé, C. Gomez and G. Sierra, "Topics in Conformal Field Theory" (Knizhnik Memorial Volume, World Scientific, 1989).

[31] V. Pasquier and H. Saleur, "Common Structures Between Finite Systems and Conformal Field Theories Through Quantum Groups", *Nucl. Phys.* **B330** (1990) 523-556.

[32] J. F. Fröhlich and T. Kerler, *Quantum Groups, Quantum Categories and Quantum Field Theory, Lecture Notes in Mathematics* **1542**, Springer-Verlag, Berlin, Germany (1993).

[33] E. Celeghini, R. Giachetti, E. Sorace and M. Tarlini, "Three-Dimensional Quantum Groups from Contractions of $SU_q(2)$", *J. Math. Phys.* **31** (1990) 2548-2551; *ibid.* "The three-dimensional Euclidean quantum group $E(3)_q$ and its R-matrix", *J. Math. Phys.* **32** (1991) 1159-1165.

[34] J. Lukierski, A. Nowicki and H. Ruegg, "New Quantum Poincaré Algebra and κ-Deformed Field Theory", *Phys. Lett.* **B293** (1992) 344-352.

[35] P. Podleś and S. L. Woronowicz, "Quantum Deformation of Lorentz Group", *Comm. Math. Phys.* **130** (1990) 381-431.

[36] A. Schmidke, J. Wess and B. Zumino, "A q-deformed Lorentz algebra", *Zeit. f. Phys.* **C52** (1991) 471-476; O. Ogievetsky, W. B. Schmidke, J. Wess and B. Zumino, "q-Deformed Poincaré Algebra", *Commun. Math. Phys.* **150** (1992) 495-518.

[37] J.-L. Gervais, "The Quantum Group Structure of 2D Gravity and Minimal Models I.", *Commun. Math. Phys.* **130** (1990) 257-283.

[38] L. Baulieu, V. Dotsenko, V. Kazakov and P. Windey (eds.), *Quantum Field Theory and String Theory*, NATO ASI Series B: Physics **328**, Plenum, New York and London, (1995).

[39] L. Castellani, "Differential Calculus on $ISO_q(N)$, Quantum Poincaré Algebra and q-Gravity", hep-th/9312179; "The Lagrangian of q-Poincaré Gravity", *Phys. Lett.* **B327** (1994) 22-28.

[40] N. Reshetikhin and F. Smirnov, "Hidden Quantum Group Symmetry and Integrable Perturbations of Conformal Field Theories", *Commun. Math. Phys.* **131** (1990) 157-177.

[41] "Quantum Groups and their Applications in Physics", proceedings of the International School of Physics "Enrico Fermi", held in Villa Monastero, Varenna, Italy, 28 June to 8 July 1994.

[42] N. Reshetikhin and V. Turaev, "Invariants of 3-Manifolds via Link Polynomials", *Inventiones Math.* **103** (1991) 547-597.

[43] V. G. Turaev, *Quantum invariants of knots and 3-manifolds*, W. de Gruyter, Berlin (1994).

[44] J. S. Carter, D. E. Flath and M. Saito, *The Classical and Quantum* $6j$*-symbols*, Princeton University Press, Princeton, New Jersey (1995).

[45] L. H. Kauffman, *Knots and Physics*, World Scientific, Singapore (1991).

[46] Yu. I. Manin, *Quantum Groups and Non-Commutative Geometry*, Université de Montréal, Centre de Recherches Mathématiques (1988).

[47] R. Askey, *Orthogonal Polynomials and Special Functions*, (Regional Conference Series in Applied Mathematics **21**), Society for Industrial and Applied Mathematics, Philadelphia, Pennsylvania (1975).

[48] E. P. Wigner, "Application of Group Theory to the Special Functions of Mathematical Physics", (Unpublished lecture notes) Princeton University, Princeton, New Jersey (1955).

[49] J. Talman, *Special Functions: A Group Theoretic Approach,* W. A. Benjamin, New York, New York (1968).

[50] N. Vilenkin, *Special Functions and the Theory of Group Representations.* (transl. from the Russian; Amer. Math. Soc. Transl., Vol. 22), American Mathematical Society, Providence, Rhode Island (1968).

[51] S. Helgason, *Differential Geometry and Symmetric Spaces,* Academic Press, New York, New York (1962).

[52] L. C. Biedenharn and P. J. Brussaard, *Coulomb Excitation*, The Oxford Press, Oxford, United Kingdom (1965).

[53] W. Miller, Jr., *Symmetry and Separation of Variables, Encyclopedia of Mathematics and Its Applications* **4**, ed. G.-C. Rota, Addison-Wesley, Reading, Massachusetts (1977).

[54] B. Kaufman, "Special Functions of Mathematical Physics from the Viewpoint of Lie Algebra", *J. Math. Phys.* **7** (1966) 447-457.

[55] C. Truesdell, *An Essay Toward a Unified Theory of Special Functions", Ann. of Math. Studies, No. 18*, Princeton University Press, Princeton, New Jersey (1948).

[56] L. Weisner, "Group-Theoretic Origin of Certain Generating Functions", *Pacific J. Math.* **5** (1955) 1033-1039.

[57] G. E. Andrews, "q-Series: Their Development and Application in Analysis, Number Theory, Combinatorics, Physics, and Computer Algebra", *Conf. Board Math. Sci.* (A.M.S.) **66** Providence, RI (1986).

[58] W. N. Bailey, *Generalized Hypergeometric Series,* Cambridge University Press, Cambridge, England (1935); reprinted Hafner, New York, New York (1972).

[59] L. J. Slater, *Generalized Hypergeometric Functions*, Cambridge University Press, Cambridge, (1966).

[60] V. A. Groza, I. I. Kachurik and A. U. Klimyk, "On Clebsch-Gordan Coefficients and Matrix Elements of Representations of the Quantum Algebra $U_q(su_2)$", *J. Math. Phys.* **31** (1990) 2769-2780.

[61] R. Askey and J. A. Wilson, "Some basic hypergeometric orthogonal polynomials that generalize Jacobi polynomials", *Memoirs Amer. Math. Soc.* **319** (1985).

[62] S. C. Milne, "A q-Analog of the Gauss Summation Theorem for Hypergeometric Series in $U(n)$", *Advances in Math.* **72** (1988) 59-131; *ibid.*, "Hypergeometric Series Well-Poised in $SU(n)$ and a Generalization of Biedenharn's G-Functions", *Advances in Math.* **36** (1980) 169-211.

[63] J. D. Louck and L. C. Biedenharn, "A Generalization of the Gauss Hypergeometric Series", *J. Math. Analysis and Appl.* **59** (1977) 423-431; *ibid.* "Some Properties of Generalized Hypergeometric Coefficients", *Adv. in Appl. Math.* **9** (1988) 447-464; *ibid.* "Identities for Generalized Hypergeometric Coefficients", *Contemp. Math.* **138** (1992) 87-97.

[64] G. Gasper and M. Rahman, *Basic Hypergeometric Series, Encyclopedia of Mathematics and its Applications* **35**, Cambridge University Press, Cambridge, England (1990).

[65] A. N. Kirillov and N. Yu. Reshetikhin, "Representations of the Algebra $U_q(s\ell(2))$, q-Orthogonal Polynomials and Invariants of Links", in *Infinite-Dimensional Lie Algebras and Groups,* Luminy-Marseille (1988); *Adv. Ser. Math. Phys.* **7** World Scientific, Teaneck, New Jersey (1989) 285-339.

[66] L. C. Biedenharn and J. D. Louck, *Angular Momentum in Quantum Physics, Encyclopedia of Mathematics and Its Applications* **8**, Addison-Wesley, Reading, Massachusetts (1981).

[67] M. E. Sweedler, *Hopf Algebras*, Benjamin, New York (1969).

[68] E. Abe, *Hopf Algebras*, Cambridge Tracts in Mathematics **74**, Cambridge University Press, Cambridge (1980).

[69] G. Lusztig, "Quantum Deformations of Certain Simple Modules Over Enveloping Algebras", *Adv. in Math.* **70** (1988) 237-249.

[70] M. Rosso, "Finite Dimensional Representations of the Quantum Analog of the Enveloping Algebra of a Complex Simple Lie Algebra", *Commun. Math. Phys.* **117** (1988) 581-593.

[71] L. C. Biedenharn and H. van Dam (eds.), *Selected Papers on the Quantum Theory of Angular Momentum*, Academic Press, New York, New York (1965).

[72] N. Yu. Reshetikhin and V. G. Turaev, "Ribbon Graphs and Their Invariants Derived from Quantum Groups", *Comm. Math. Phys.* **127** (1990) 1-26.

[73] R. B. Zhang, M. D. Gould and A. J. Bracken, "Quantum Group Invariants and Link Polynomials", *Comm. Math. Phys.* **137** (1991) 13-27.

[74] M. D. Gould and L. C. Biedenharn, "The Pattern Calculus for Tensor Operators in Quantum Groups", *J. Math. Phys.* **33** (1992) 3613-3635.

[75] L. Alvarez-Gaumé, C. Gomez and G. Sierra, "Duality and Quantum Groups", *Nucl. Phys.* **B330** (1990) 347-398.

[76] M. Nomura, "Relations for Clebsch-Gordan and Racah Coefficients in $su_q(2)$ and Yang-Baxter Equations", *J. Math. Phys.* **30** (1989) 2397-2405; "Yang-Baxter Relations in Terms of n-j Symbols of $su_q(2)$ Algebra", *J. Phys. Soc. Japan* **58** (1989) 2694-2704; "An Alternative Description of the Quantum Group $SU_q(2)$ and the q-Analog Racah-Wigner Algebra", *ibid.* **59** (1990) 439-448.

[77] P. Jordan, "Der Zusammenhang der symmetrischen und linearen Gruppen und das Mehrkörperproblem", *Z. Physik* **94** (1935) 531-535.

[78] J-U. H. Petersen, "Representations at a Root of Unity of q-Oscillators and Quantum Kac-Moody Algebras", PhD Thesis, University of London, 1994 (hep-th/9409079).

[79] L. C. Biedenharn, "The quantum group $\mathrm{SU}_q(2)$ and a q-analogue of the boson operators", *J. Phys. A: Math. Gen.* **22** (1989) L873-L878.

[80] A. J. Macfarlane, "On q-Analogues of the Quantum Harmonic Oscillator and the Quantum Group $SU_q(2)$", *J. Phys. A: Math. Gen.* **22** (1989) 4581-4588.

[81] C. P. Sun and H. C. Fu, "The q-Deformed Boson Realization of the Quantum Group $SU_q(n)$ and its Representations", *J. Phys. A: Math. Gen.* **22** (1989) L983-L986.

[82] T. Hayashi, "Q-Analogues of Clifford and Weyl Algebras — Spinor and Oscillator Representations of Quantum Enveloping Algebras", *Comm. Math. Phys.* **127** (1990) 129-144.

[83] E. Celeghini, R. Giachetti, E. Sorace and M. Tarlini, "The Quantum Heisenberg Group $H_q(1)$", *J. Math. Phys.* **32** (1991) 1155-1165.

[84] V. Bargmann, "On a Hilbert Space of Analytic Functions and an Associated Integral Transform Part I.", *Commun. Pure Appl. Math.* **14** (1961) 187-214.

[85] F. H. Jackson, "Generalization of the Differential Operative Symbol with an Extended Form of Boole's Equation", *Messenger Math.* **38** (1909) 57-61.

[86] F. H. Jackson, "q-Form of Taylor's Theorem", *Messenger Math.* **38** (1909) 62-64.

[87] F. H. Jackson, "On q-Definite Integrals", *Quart. J. Pure Appl. Math.* **41** (1910) 193-203.

[88] A. J. Bracken, D. S. McAnally, R. B. Zhang and M. D. Gould, "A q-analogue of Bargmann Space and its Scalar Product", *J. Phys. A: Math. Gen.* **24** (1991) 1379-1391.

[89] R. W. Gray and C. A. Nelson, "A Completeness Relation for the q-Analogue Coherent States by q-Integration", *J. Phys. A: Math. Gen.* **23** (1990) L945-L950.

[90] P. P. Kulish and E. V. Damaskinsky, "On the q oscillator and the quantum algebra $\mathrm{su}_q(1,1)$", *J. Phys. A: Math. Gen.* **23** (1990) L415-L419.

[91] A. T. Filippov, D. Gangopadhyay and A. P. Isaev, "Harmonic Oscillator Realization of the Canonical q-Transformation", *J. Phys. A: Math. Gen.* **24** (1991) L63-L68.

[92] J. Thomae, "Beiträge zur Theorie der durch die Heine'sche Reihe: $1 + \frac{1-q^q}{1-q} \cdot \frac{1-q^b}{1-q^c} x + \frac{1-q^a}{1-q} \cdot \frac{1-q^{a+1}}{1-q^2} \cdot \frac{1-q^b}{1-q^c} \cdot \frac{1-q^{b+1}}{1-q^{c+1}} \cdot x^2 + \ldots$ Darstellbaren Functionen", *J. reine und angew. Math.* **70** (1869) 258-281.

[93] M. Jimbo, "Quantum R matrix related to the generalized Toda system: an algebraic approach", *Proceedings of the Symposium on Field Theory, Quantum Gravity and Strings*, Meudon and Paris VI, 1984/85, *Lecture Notes in Physics* **246** (1986) 335-361.

[94] T. L. Curtright and C. K. Zachos, "Deforming Maps for Quantum Algebras", *Phys. Lett.* **B243** (1990) 237-244.

[95] D. B. Fairlie, "Quantum Deformations of SU(2)", *J. Phys. A: Math. Gen.* **23** (1990) L183-L187.

[96] T. Holstein and H. Primakoff, "Field Dependence of the Intrinsic Domain Magnetization of a Ferromagnet", *Phys. Rev.* **58** (1940) 1098-1113.

[97] R. Hermann, *Lie Groups for Physicists,* Benjamin, New York, New York (1966).

[98] D. P. Zhelobenko, "Compact Lie Groups and Their Representations", transl. from the Russian; *Amer. Math. Soc. Transl.,* Vol. 40, Amer. Math. Soc., Providence, Rhode Island.

[99] M. A. Lohe and C. A. Hurst, "Relation Between the Boson Calculus and Zhelobenko's Method", *J. Math. Phys.* **14** (1973) 1955-1958.

[100] M. Chaichian, D. Ellinas and P. Kulish, "Quantum Algebra as the Dynamical Symmetry of the Deformed Jaynes-Cumming Model", *Phys. Rev. Lett.* **65** (1990) 980-983.

[101] J. Katriel and A. I. Solomon, "Generalized q-bosons and their squeezed states", *J. Phys. A: Math. Gen.* **24** (1991) 2093-2105.

[102] L. C. Biedenharn and M. A. Lohe, "Some Remarks on Quantum Groups—A New Variation on the Theme of Symmetry", *From Symmetries to Strings: Forty Years of Rochester Conferences,* ed. A. Das, World Scientific (1990) 189-206.

[103] L. C. Biedenharn and J. D. Louck, *Racah-Wigner Algebra in Quantum Theory, Encyclopedia of Mathematics and its Applications* **9**, Addison-Wesley, Reading, Massachusetts (1981).

[104] M. Jimbo, "A q-Analogue of $U(gl(N{+}1))$, Hecke Algebra, and the Yang-Baxter Equation", *Lett. Math. Phys.* **11** (1986) 247-252.

[105] C. Quesne, "Raising and Lowering Operators for $U_q(n)$", *J. Phys. A: Math. Gen.* **26** (1993) 357-372.

[106] K. Ueno, T. Takebayashi and Y. Shibukawa, "Gelfand-Zetlin Basis for $U_q(gl(N+1))$ Modules", *Lett. Math. Phys.* **18** (1989) 215-221.

[107] I. M. Gel'fand and M. L. Zetlin, *Dokl. Akad. Nauk. SSSR* **71** No. 5 (1950) 825-828; see the appendix to: I. M. Gel'fand, R. A. Minlos, Z. Ya. Shapiro, *Representations of the Rotation and Lorentz Groups and their Applications* (Pergamon Press, Oxford, Macmillan, New York, 1963).

[108] J. D. Louck, "Recent Progress Toward a Theory of Tensor Operators in the Unitary Groups", *Am. J. Phys.* **38** (1970) 3-42.

[109] J. D. Louck, "Group Theory of Harmonic Oscillators in n-Dimensional Space", *J. Math. Phys.* **6** (1965) 1786-1804.

[110] G. E. Baird and L. C. Biedenharn, "On the Representations of the Semisimple Lie Groups. II", *J. Math. Phys.* **4** (1963) 1449-1466.

[111] M. Moshinsky, "Bases for the Irreducible Representations of the Unitary Groups and Some Applications", *J. Math. Phys.* **4** (1963) 1128-1139.

[112] H. Weyl, *The Theory of Groups and Quantum Mechanics,* Dover, New York, New York (1950).

[113] H. Exton, *q-Hypergeometric Functions and Applications,* Ellis Horwood, Chichester, England (1983).

[114] J. Cigler, "Operatormethoden für q-Identitäten", *Monatshefte f. Math.* **88** (1979) 87-105.

[115] P. Feinsilver, "Commutators, Anti-Commutators and Eulerian Calculus", *Rocky Mountain J. Math.* **12** (1982) 171-183; "Lie Algebras and Recurrence Relations III: q-Analogs and Quantized Algebras", *Acta App. Math.* **19** (1990) 207-251.

[116] M. Nomura and L. C. Biedenharn, "On the q-Symplecton Realization of the Quantum Group $SU_q(2)$", *J. Math. Phys.* **33** (1992) 3636-3648.

[117] H. Bacry, "SL(2,C), SU(2), and Chebyshev Polynomials", *J. Math. Phys.* **28** (1987) 2259-2267.

[118] G. E. Andrews, *The Theory of Partitions, Encyclopedia of Mathematics and its Applications* **2**, ed. G.-C. Rota, Addison-Wesley, Reading, Massachusetts (1976).

[119] M. Nomura, "Concepts of Tensors in $U_q(sl(2))$ and van der Waerden Method for Quantum Clebsch-Gordan Coefficients", *J. Phys. Soc. Japan* **60** (1991) 789-797.

[120] M. Nomura, "Generating Functions for Representation Functions of Quantum Group $U_q sl(2)$", *J. Phys. Soc. Japan* **60** (1991) 710-711.

[121] F. H. Jackson, "On q-Functions and a Certain Difference Operator", *Trans. Roy. Soc. Edin.* **46** (1908) 253-281.

[122] F. H. Jackson, "Transformations of q-series", *Messenger Math.* **39** (1910) 145-153.

[123] M. Gell-Mann and Y. Ne'eman, *The Eightfold Way*, Benjamin, New York (1964).

[124] I. M. Gel'fand and A. V. Zelevinsky, *Societé Math. de France, Astérique, hors serie* (1985) 117.

[125] D. Flath and L. C. Biedenharn, "Beyond the Enveloping Algebra of sl_3", *Can. J. Math.* **37** (1985) 710-729; L. C. Biedenharn and D. Flath, "On the Structure of Tensor Operators in $SU3$", *Comm. Math. Phys.* **93** (1984) 143-169.

[126] L. C. Biedenharn and J. D. Louck, "A Pattern Calculus for Tensor Operators in the Unitary Groups", *Comm. Math. Phys.* **8** (1968) 89-131.

[127] J. D. Louck and L. C. Biedenharn, "Some Properties of the Intertwining Number of the General Linear Group", in *Science and Computers, Adv. Math. Supp. Studies* **10**, ed. G.-C. Rota, Academic Press, Orlando, Florida (1986) 265-311.

[128] G. E. Baird and L. C. Biedenharn, "On the Representations of the Semisimple Lie Groups. III. The Explicit Conjugation Operators for SU_n", *J. Math. Phys.* **5** (1964) 1723-1730.

[129] L. C. Biedenharn, A. Giovannini and J. D. Louck, "Canonical Definition of Wigner Coefficients in U_n", *J. Math. Phys.* **8** (1967) 691-700.

[130] K. Baclawski, "A New Rule for Computing Clebsch-Gordan Series", *Adv. in Appl. Math.* **5** (1984) 416-432.

[131] R. LeBlanc and L. C. Biedenharn, "Implementation of the $U(n)$ Tensor Operator Calculus in a Vector Bargmann Hilbert Space", *J. Phys. A: Math. Gen.* **22** (1989) 31-48.

[132] L. C. Biedenharn, R. LeBlanc and J. D. Louck, "Recent Progress in Implementing the Tensor Operator Calculus", in *Symmetries in Science III*, eds. B. Gruber and F. Iachello, Plenum Press, New York (1989) 15-32.

[133] K. T. Hecht, *The Vector Coherent State Method and Its Application to Problems of Higher Symmetries, Lecture Notes in Physics* **290**, Springer-Verlag, Berlin (1987).

[134] J. D. Louck and L. C. Biedenharn, "Canonical Unit Adjoint Tensor Operators in $U(n)$", *J. Math. Phys.* **11** (1970) 2368-2414.

[135] V. Rittenberg and M. Scheunert, "Tensor operators for quantum groups and applications", *J. Math. Phys.* **33** (1992) 436-445.

[136] L. Vaksman, "q-Analogues of Clebsch-Gordan Coefficients, and the Algebra of Functions on the Quantum Group $SU(2)$", *Sov. Math. Dokl.* **39** (1989) 467-470.

[137] H. Ruegg, "A Simple Derivation of The Quantum Clebsch-Gordan Coefficients for $SU(2)_q$", *J. Math. Phys.* **31** (1990) 1085-1087.

[138] L. C. Biedenharn and M. Tarlini, "On q-Tensor Operators for Quantum Groups", *Lett. Math. Phys.* **20** (1990) 271-278.

[139] M. Nomura, "Recursion Relations for the Clebsch-Gordan Coefficient of Quantum Group $SU_q(2)$", *J. Phys. Soc. Japan* **59** (1990) 1954-1961.

[140] L. C. Biedenharn and M. A. Lohe, "Symmetries of Quantum Group Coupling Coefficients", in *Differential Geometry, Group Representations, and Quantization*, eds. J. D. Hennig, W. Lücke and J. Tolar, *Lecture Notes in Physics* **379**, Springer-Verlag, New York (1991) 193-206.

[141] J. Thomae, "Ueber die Funktionen, welche durch Reihen von der Form dergestellt werden ...", *J. reine angew. Math.* **87** (1879) 26-73.

[142] D. B. Sears, "On the Transformation Theory of Basic Hypergeometric Functions", *Proc. London Math. Soc. (2)* **53** (1951) 158-180.

[143] S. D. Majumdar, "The Clebsch-Gordan Coefficients", *Prog. of Theor. Phys.* **20** (1958) 798-803.

[144] V. Rajeswari and K. S. Rao, "Generalized basic hypergeometric functions and the q-analogues of 3-j and 6-j coefficients", *J. Phys. A: Math. Gen.* **24** (1991) 3761-3780.

[145] I. I. Kachurik and A. U. Klimyk, "On Racah coefficients of the quantum algebra $U_q(su_2)$", *J. Phys. A: Math. Gen.* **23** (1990) 2717-2728.

[146] M. E. Rose, *Multipole Fields,* Wiley, New York, New York (1955).

[147] A. Erdélyi, *Math. Rev.* **14** (1957) 642.

[148] H. A. Jahn and K. M. Howell, "New (Regge) Symmetry Relations for the Wigner $6j$-Symbol", *Proc. Cambridge Phil. Soc.* **55** (1959) 338-340.

[149] K. S. Rao, T. S. Santhanam, and K. Venkatesh, "On the Symmetries of the Racah Coefficient", *J. Math. Phys.* **16** (1975) 1528-1530.

[150] K. S. Rao and K. Venkatesh, "Representation of the Racah Coefficient as a Generalized Hypergeometric Function", in *Proceedings Fifth International Colloquium on Group Theoretical Methods in Physics,* eds. R. T. Sharp and B. Kolman, Academic Press, New York (1977) 649-656.

[151] K. Venkatesh, "Symmetries of the $6j$ Coefficient", *J. Math. Phys.* **19** (1978) 1973-1974; *ibid.* 2060-2063.

[152] L. C. Biedenharn, "An Identity Satisfied by Racah Coefficients", *J. Math. and Phys.* **31** (1953) 287-293.

[153] R. LeBlanc and K. T. Hecht, "New Perspective on the $U(n)$ Wigner-Racah Calculus: II. Elementary Reduced Wigner Coefficients for $U(n)$", *J. Phys. A: Math. Gen.* **20** (1987) 4613-4635.

[154] M. D. Gould, "Multiplicity-free Wigner Coefficients for Semisimple Lie Groups: I. The $U(n)$ Pattern Calculus", *J. Math. Phys.* **27** (1986) 1944-1963; *ibid.*, "Multiplicity-free Wigner Coefficients for Semisimple Lie Groups: II. Pattern Calculus for $U(n)$", *J. Math. Phys.* **27** (1986) 1964-1979.

[155] A. Schirrmacher, J. Wess and B. Zumino, "The two-parameter deformation of $GL(2)$, its differential calculus, and Lie algebra", *Zeit. f. Phys.* **C49** (1991) 317-324.

[156] S. L. Woronowicz, "Compact Matrix Pseudo-Groups", *Comm. Math. Phys.* **111** (1987) 613-665.

[157] N. Yu. Reshetikhin, L. A. Takhtajan and L. D. Faddeev, "Quantization of Lie Groups and Lie Algebras", *Leningrad J. Math.* **1** (1990) 193-225.

[158] V. K. Dobrev and P. Parashar, "Duality for Multiparametric Quantum Group GL(n)", *J. Phys. A: Math. Gen.* **26** (1993) 6991-7002.

[159] M. Moshinsky, "Wigner Coefficients for the SU_3 Group and Some Applications", *Rev. Mod. Phys.* **34** (1962) 813-828.

[160] M. Moshinsky and C. Quesne, "Noninvariance Groups in the Second-Quantization Picture and Their Applications", *J. Math. Phys.* **11** (1970) 1631-1639.

[161] C. Quesne, "Complementarity of $\mathrm{su}_q(3)$ and $\mathrm{u}_q(2)$ and q-boson realization of the $\mathrm{su}_q(3)$ irreducible representations", *J. Phys. A: Math. Gen.* **25** (1992) 5977-5998.

[162] T. Masuda, K. Mimachi, Y. Nakagami, M. Noumi and K. Ueno, "Representations of Quantum Groups and a q-Analogue of Orthogonal Polynomials", *C. R. Acad. Sci. Paris* **307**, Série I (1988) 559-564.

[163] M. Nomura, "Representation Functions d^j_{mm} of $U[sl_q(2)]$ as Wave Functions of 'Quantum Symmetric Tops' and Relationship to Braiding Matrices", *J. Phys. Soc. Japan* **59** (1990) 4260-4271.

[164] W. Hahn, "Über Orthogonalpolynome, die q-Differenzengleichungen genügen", *Math. Nach.* **2** (1949) 4-34.

[165] M. L. Ge, X. F. Liu, and C. P. Sun, "Representations of Quantum Matrix Algebra $M_q(2)$ and its q-Boson Realization", *J. Math. Phys.* **33** (1992) 2541-2545.

[166] E. V. Damaskinsky and M. A. Sokolov, "Some Remarks on the Gauss Decomposition for Quantum Group $GL_q(n)$ with Application to q-Bosonization", preprint hep-th/9407024 (1994).

[167] L. Alvarez-Gaumé, C. Gomez and G. Sierra, "Quantum Group Interpretation of Some Conformal Field Theories", *Phys. Lett.* **B220** (1989) 142-152.

[168] E. K. Sklyanin, "Some algebraic structures connected with the Yang-Baxter equation. Representations of quantum algebras", *Funct. Anal. Appl.* **17** (1983) 273-284.

[169] C. DeConcini and V. G. Kac, "Representations of Quantum Groups at Roots of Unity", in *Operator Algebras, Unitary Representations, Enveloping Algebras, and Invariant Theory,* Paris, (1989) *Progress in Mathematics* **92**, Birkhäuser, Boston, Massachusetts (1990) 471-506.

[170] D. Arnaudon, "Composition of Kinetic Momenta: The $\mathcal{U}_q(\mathfrak{sl}(2))$ Case", *Commun. Math. Phys.* **159** (1994) 175-193.

[171] D. Arnaudon, "Polynomial Relations in the Centre of $\mathcal{U}_q(\mathfrak{sl}(N))$", *Lett. Math. Phys.* **30** (1994) 251-257.

[172] H. S. Green, "A Generalized Method of Field Equations", *Phys. Rev.* **90** (1953) 270-273.

[173] O. W. Greenberg and A. M. L. Messiah, "Selection Rules for Parafields and the Absence of Para Particles in Nature", *Phys. Rev.* **138** (1965) B1155-B1167.

[174] Y. Ohnuki and S. Kamefuchi, *Quantum Field Theory and Parastatistics*, Berlin, Heidelberg, New York (1982).

[175] G. Lusztig, "Quantum Groups at Roots of 1", *Geom. Ded.* **35** (1990) 89-114.

[176] G. Keller, "Fusion Rules of $\mathcal{U}_q(\mathfrak{sl}(2,\mathbb{C}))$, $q^m = 1$", *Lett. Math. Phys.* **21** (1991) 273-286.

[177] D. Arnaudon and A. Chakrabarti, "Periodic and Partially Periodic Representations of $SU(N)_q$", *Commun. Math. Phys.* **139** (1991) 461-478; *ibid.* "Flat

Periodic Representations of $\mathcal{U}_q(\mathfrak{g})$", *Commun. Math. Phys.* **139** (1991) 605-617.

[178] J. R. Links and R. B. Zhang, "Induced Module Construction for Highest Weight Representations of $\mathcal{U}_q(\mathfrak{gl}(n))$ at Roots of Unity" (preprint, 1994).

[179] W. A. Schnizer,"Roots of Unity: Representations of Quantum Groups", *Commun. Math. Phys.* **163** (1994) 293-306.

[180] G. E. Andrews, "On the Proofs of the Rogers-Ramanujan Identities", in *q-Series and Partitions*, ed. D. Stanton, The IMA Volumes in Mathematics and Its Applications, Springer-Verlag, New York (1989) Vol. 18, 1-14.

[181] I. M. Gel'fand and M. A. Naimark, "Unitary Representations of the Classical Groups", *Trudy Mat. Inst. Steklov.* **36** (1950); German transl., Akademie-Verlag Berlin (1957).

[182] G. W. Mackey, *Induced Representations of Groups and Quantum Mechanics*, Benjamin, New York, New York (1968).

[183] A. Borel and A. Weil, *Représentations linéaires et espaces homogénes Kählerians des groupes de Lie compacts,* Séminaire Bourbaki, May 1954 (exposé by J.-P. Serre).

[184] D. J. Rowe, R. LeBlanc and K. T. Hecht, "Vector Coherent State Theory and its Application to the Orthogonal Groups", *J. Math. Phys.* **29** (1988) 287-304.

[185] R. LeBlanc and D. J. Rowe, "The Matrix Representations of g_2. II. Representations in an su(3) Basis", *J. Math. Phys.* **29** (1988) 767-776.

[186] L. C. Biedenharn and M. A. Lohe, "An Extension of the Borel-Weil Construction to the Quantum Group $U_q(n)$", *Comm. Math. Phys.* **146** (1992) 483-504.

[187] L. C. Biedenharn and M. A. Lohe, "Construction of the Unitary Irreps of the Quantum Group $U_q(n)$ by an Extension of the Borel-Weil Method", *Proceedings of the XXth International Conference on Differential Geometric Methods in Theoretical Physics*, Vol. 1, eds. S. Catto and A. Rocha (World Scientific, Singapore, 1992) 454-470.

[188] L. C. Biedenharn and M. A. Lohe, "Induced Representations and Tensor Operators for Quantum Groups", in *Quantum Groups*, ed. P. P. Kulish, *Lecture Notes in Mathematics* **1510**, Springer-Verlag (1992) 197-209.

[189] N. R. Wallach, *Harmonic Analysis on Homogeneous Spaces,* Dekker, New York, New York (1973).

[190] K. T. Hecht, R. LeBlanc and D. J. Rowe, "New perspective on the U(n) Wigner-Racah calculus: I. Vector coherent state theory and construction of Gel'fand bases", *J. Phys. A: Math. Gen.* **20** (1987) 2241-2250.

[191] T. Nakayama, "On Some Modular Properties of Irreducible Representations of S_n", *Jap. J. Math.* **17** (1940) I 89-108, II 411-423.

[192] M. A. Lohe and L. C. Biedenharn, "Basic Hypergeometric Functions and the Borel-Weil Construction for $U_q(3)$", *SIAM J. Math. Anal.* **25** (1994) 218-241.

[193] G. N. Watson, "A new proof of the Rogers-Ramanujan identities", *J. London Math. Soc.* **4** (1929) 4-9.

[194] N. M. Atakishiev and S. K. Suslov, "Difference analogs of the harmonic oscillator", *Theor. and Math. Phys.* **85** (1990) 1055-1062; *ibid.* "A realization of the q-harmonic oscillator", *Theor. and Math. Phys.* **87** (1991) 442-444.

[195] D. Han, Y. S. Kim and W. W. Zachary (eds.), *Workshop on Harmonic Oscillators*, NASA Conference Publication 3197 (1993).

[196] I. M. Gelfand and D. B. Fairlie, "The Algebra of Weyl Symmetrised Polynomials and its Quantum Extension", *Comm. Math. Phys.* **136** (1991) 487-499.

[197] B. Zumino, "Deformation of the quantum mechanical phase space with bosonic or fermionic coordinates", *Mod. Phys. Lett. A***6** (1991) 1225-1235.

[198] I. M. Burban and A. U. Klimyk, "On Spectral Properties of q-Oscillator Operators", *Lett. Math. Phys.* **29** (1993) 13-18.

[199] J. R. Klauder and B. S. Skagerstam, *Coherent States. Applications in Physics and Mathematical Physics*, World Scientific, Singapore (1985).

[200] C. Quesne, "Coherent states, K-matrix theory and q-boson realizations of the quantum algebra su$_q$(2)", *Phys. Lett.* **A153** (1991) 303-307.

[201] M. Arik and D. D. Coon, "Hilbert spaces of analytic functions and generalized coherent states", *J. Math. Phys.* **17** (1976) 524-527.

[202] A. Perelomov, *Generalized Coherent States and Their Applications*, Springer, Berlin (1986).

[203] A. Inomata, H. Kuratsuji and C. G. Gerry, "Path Integrals and Coherent States of SU(2) and SU(1,1)", World Scientific, Singapore (1992).

[204] W. Vogel and D.-G. Welsch, *Lectures on Quantum Optics*, Akademie Verlag, Berlin (1994).

[205] C. A. Nelson and M. H. Fields, "Number and phase uncertainties of the q-analog quantized field", *Phys. Rev. A***51** (1995) 2410-2429.

[206] P. A. M. Dirac, *The Principles of Quantum Mechanics,* Clarendon Press, Oxford, England (1947).

[207] G.-C. Rota, *Finite Operator Calculus*, Academic Press, New York, New York (1975).

[208] C. Itzykson, "Remarks on Boson Commutation Rules", *Commun. Math. Phys.* **4** (1967) 92-122.

[209] M. Moshinsky and C. Quesne, "Linear Canonical Transformations and Their Unitary Representations", *J. Math. Phys.* **12** (1971) 1772-1780.

[210] M. Moshinsky, T. H. Seligman and K. B. Wolf, "Canonical Transformations and the Radial Oscillator and Coulomb Problems", *J. Math. Phys.* **13** (1972) 901-907.

[211] R. Simon, E. C. G. Sudarshan and N. Mukunda, "Gaussian pure states in quantum mechanics and the symplectic group", *Phys. Rev. A***37** (1988) 3028-3038.

[212] M. Born and P. Jordan, "Zur Quantenmechanik", *Zeits. Phys.* **34** (1925) 858-888.

[213] N. H. McCoy, "On the Function in Quantum Mechanics which Corresponds to a Given Function in Classical Mechanics", *Proc. U.S. Nat. Acad. Sci.* **18** (1932) 674-676.

[214] C. M. Bender and G. V. Dunne, "Integration of Operator Differential Equations", *Phys. Rev. D***40** (1989) 3504-3511; C. M. Bender and G. V. Dunne, "Exact Solutions to Operator Differential Equations", *Phys. Rev. D***40** (1989) 2739-2742; C. M. Bender and G. V. Dunne, "Exact Operator Solutions to Euler Hamiltonians", *Phys. Lett.* **B200** (1988) 520-524; C. M. Bender and G. V. Dunne, "Polynomials and operator orderings", *J. Math. Phys.* **29** (1988) 1727-1731; C. M. Bender, L. R. Mead, and S. S. Pinsky, "Continuous Hahn Polynomials and the Heisenberg Algebra", *J. Math. Phys.* **28** (1987) 509-513; C. M. Bender, L. R. Mead, and S. S. Pinsky, "Resolution of the Operator-Ordering Problem by the Method of Finite Elements", *Phys. Rev. Lett.* **56** (1986) 2445-2448.

[215] L. C. Biedenharn and J. D. Louck, "An Intrinsically Self-Conjugate Boson Structure: The Symplecton", *Ann. Phys.* **63** (1971) 459-475.

[216] M. A. Lohe, L. C. Biedenharn and J. D. Louck, "Tensor operator formulation of Weyl-ordered polynomials", *Phys. Rev. D***43** (1991) 617-619.

[217] N. Mukunda, "Wigner distribution for angle coordinates in quantum mechanics", *Am. J. Phys.* **47** (1979) 182-187.

[218] M. Hall, "Weyl's rule and Wigner equivalents for phase space multinomials", *J. Phys. A: Math. Gen.* **18** (1985) 29-36.

[219] L. C. Biedenharn and M. A. Lohe, "Quantum Groups and Basic Hypergeometric Functions", *Proceedings of the Argonne Workshop on Quantum Groups*, T. Curtright, D. Fairlie and C. Zachos (eds.), World Scientific, Singapore (1990) 123-132.

[220] J. L. Burchnall and T. W. Chaundy, "The Hypergeometric Identities of Cayley, Orr, and Bailey", *Proc. Lond. Math. Soc. (2)* **50** (1949) 56-74.

[221] F. J. W. Whipple, "On series allied to the hypergeometric series with argument -1", *Proc. London Math. Soc.* **30** (2), (1930) 81-94.

[222] T. H. Koornwinder, "Meixner-Pollaczek polynomials and the Heisenberg algebra", *J. Math. Phys.* **30** (1989) 767-769.

Index